Emerging Contaminants in Soil and Groundwater Systems

Occurrence, Impact, Fate and Transport

Emerging Contaminants in Soil and Groundwater Systems

Occurrence, Impact, Fate and Transport

Edited by

Bin Gao

Department of Agricultural and Biological Engineering,
University of Florida, Gainesville, FL, United States

ELSEVIER

Elsevier
Radarweg 29, PO Box 211, 1000 AE Amsterdam, Netherlands
The Boulevard, Langford Lane, Kidlington, Oxford OX5 1GB, United Kingdom
50 Hampshire Street, 5th Floor, Cambridge, MA 02139, United States

British Library Cataloguing-in-Publication Data
A catalogue record for this book is available from the British Library

Library of Congress Cataloging-in-Publication Data
A catalog record for this book is available from the Library of Congress

ISBN: 978-0-12-824088-5

Publisher: Candice Janco
Acquisitions Editor: Louisa Munro
Editorial Project Manager: Alice Grant
Production Project Manager: Bharatwaj Varatharajan
Cover Designer: Victoria Pearson

Typeset by MPS Limited, Chennai, India

Contents

List of contributors

Linta Akhtar Department of Agriculture, University of Arkansas at Pine Bluff, Pine Bluff, AR, United States

Hao Chen Department of Agriculture, University of Arkansas at Pine Bluff, Pine Bluff, AR, United States

Xian Chen School of Chemistry and Environmental Engineering, Jiangsu University of Technology, Changzhou, P.R. China

Shunan Dong College of Agricultural Science and Engineering, Hohai University, Nanjing, P.R. China

Bin Gao Department of Agricultural and Biological Engineering, University of Florida, Gainesville, FL, United States

Jinsheng Huang Department of Agricultural and Biological Engineering, University of Florida, Gainesville, FL, United States

Jun Huang Hualan Design & Consulting Group Co. Ltd., Nanning, P.R. China; College of Civil Engineering and Architecture, Guangxi University, Nanning, P.R. China

Nan Jiang College of Natural Resources and Environment, Northwest A&F University, Yangling, P.R. China

Yanji Jiang College of Natural Resources and Environment, Northwest A&F University, Yangling, P.R. China

Nedgine D. Joseph Department of Chemical Engineering, University of Florida, Gainesville, FL, United States

Chengliang Li College of Resources and Environment, Shandong Agricultural University, Tai'An, P.R. China

Longfei Liu College of Resources and Environment, Shandong Agricultural University, Tai'An, P.R. China

Huimin Sun College of Natural Resources and Environment, Northwest A&F University, Yangling, P.R. China

Yuehui Tan College of Natural Resources and Environment, Northwest A&F University, Yangling, P.R. China

Mei Wang School of Water Conservancy and Environment, Zhengzhou University, Zhengzhou, P.R. China

Shengsen Wang College of Environmental Science and Engineering, Yangzhou University, Yangzhou, P.R. China

Xiaozhi Wang College of Environmental Science and Engineering, Yangzhou University, Yangzhou, P.R. China

Panfeng Wu Jiangsu Nantong Environmental Monitoring Center, Nantong, P.R. China

Xianglong Xi College of Natural Resources and Environment, Northwest A&F University, Yangling, P.R. China

Xianni Yang College of Environmental Science and Engineering, Yangzhou University, Yangzhou, P.R. China

Yicheng Yang Department of Agricultural and Biological Engineering, University of Florida, Gainesville, FL, United States

Xianqiang Yin College of Natural Resources and Environment, Northwest A&F University, Yangling, P.R. China

Zebin Yu School of Resources, Environment and Materials, Guangxi University, Nanning, P.R. China

Guangcai Zhang College of Natural Resources and Environment, Northwest A&F University, Yangling, P.R. China

Jian Zhang College of Environmental Science and Engineering, Yangzhou University, Yangzhou, P.R. China

Yue Zhang Department of Agricultural and Biological Engineering, University of Florida, Gainesville, FL, United States

Yulin Zheng Department of Agricultural and Biological Engineering, University of Florida, Gainesville, FL, United States

Shi Zhou College of Natural Resources and Environment, Northwest A&F University, Yangling, P.R. China

Ting Zhou College of Natural Resources and Environment, Northwest A&F University, Yangling, P.R. China

Chapter 1

Introduction

Bin Gao
Department of Agricultural and Biological Engineering, University of Florida, Gainesville, FL, United States

Chapter Outline

1.1 Overview of soils and groundwater

Since this book mainly focuses on soil and groundwater contamination, it is necessary to first talk about some basics of soils and groundwater. As we know, the natural earth environment can be roughly classified into three interactive zones layered from the top to the bottom as atmosphere, surface, and subsurface (Fig. 1.1). Soils and groundwater are the two main components of the subsurface environment and can be further divided into two zones, the vadose zone and groundwater (Fig. 1.1). In the soil and groundwater systems, soil serves as the matrix and water flows within the soil pores. In fact, soils

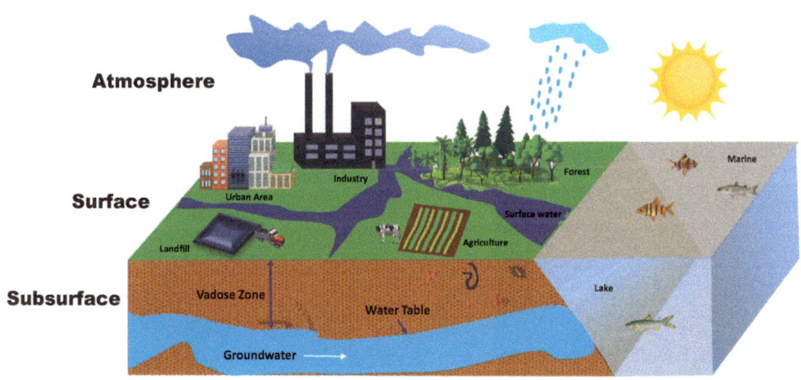

FIGURE 1.1 Overview of natural environmental system.

Emerging Contaminants in Soil and Groundwater Systems. DOI: https://doi.org/10.1016/B978-0-12-824088-5.00002-1

are the most common natural porous media that can hold and store water inside their pores through the capillary force. As a result, water exists in both the vadose zone and groundwater. Due to the combined effect of gravity and capillary forces, the vadose zone is only partially saturated with water, while groundwater is the saturated zone (aquifer) below the water table.

1.2 Typical soil and groundwater contaminants

Typical soil and groundwater contaminants can be categorized into three types: metals/metalloids, organics, and nutrients (Fig. 1.2). Typical metal and metalloid contaminants that are frequently detected in soils and groundwater including lead (Pb), chromium (Cr), arsenic (As), Zinc (Zn), cadmium (Cd), copper (Cu), mercury (Hg), and nickel (Ni), which are often called heavy metals. These heavy metals may be released from natural processes (e.g., weathering) into the soils and groundwater. However, industrial activities that involve the acquisition, production, utilization, and discharge of heavy metal–containing materials or wastes are the main contributor of this type of contaminants in the subsurface. In addition, agricultural practices such as land application of fertilizers, pesticides, manure, and biosolids that often contain trace amounts of heavy metals may also cause their accumulations in soils and eventual release into groundwater. The most common and toxic organic contaminants in soils and groundwater are persistent organic compounds including chlorinated compounds such as tetrachloroethylene

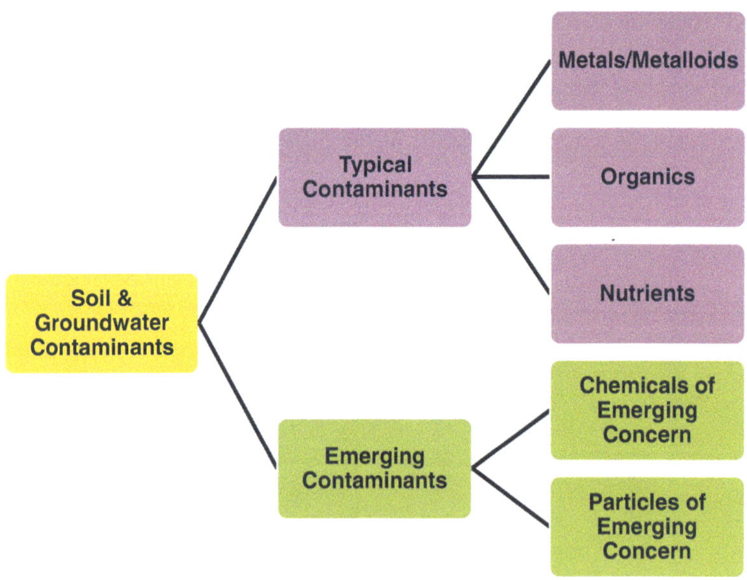

FIGURE 1.2 Categorization of soil and groundwater contaminants.

(PCE), trichloroethylene, and polychlorinated biphenyls; aromatic hydrocarbons such as benzene, toluene, xylenes, and polycyclic aromatic hydrocarbons; and pesticides such as insecticides and herbicides. They are generally released into the soil and groundwater environment due to irresponsible industrial activities such as pine line and storage tank leaking, hazardous waste and wastewater disposable, and abandoned and uncontrolled hazardous waste sites. When pesticides are sprayed on agricultural land, some of them may be carried by water flow into the subsurface to degrade soil and groundwater quality. Excess nutrients such as nitrogen (mainly nitrate) in soils and groundwater may also present serious contamination risks.

In the literature of soil and groundwater contamination, there are plenty of research papers and books on the fate and impacts of the above-mentioned typical contaminants, which are not covered in this book. Instead, this work focuses on emerging contaminants in the soil and groundwater system.

1.3 Emerging contaminants

Emerging contaminants, also called contaminants of emerging concern, refer a group of newly surfaced environmental pollutants that often have not been regulated under current environmental regulations or laws. Emerging contaminants usually are recalcitrant, toxic, bioaccumulable, and thus may have strong adverse impacts on the ecosystems and public health. They are increasingly being detected at low levels in soils and groundwater. With increasing public awareness and advancing analytical technologies, more and more emerging contaminants have been identified and added to the list.

Based on their sizes and structures, emerging contaminants can be put into two main categories, chemicals of emerging concern (CECs) and particles of emerging concern (PECs) (Fig. 1.2). CECs consist of many chemical compounds including pharmaceuticals and personal care products (PPCPs) and perfluorinated compounds (PFCs); while PECs mainly comprise two types of particulate pollutants, engineered nanoparticles (ENPs) and microplastics (MPs). The concentrations of CECs in soils and groundwater are often at the lower μg/L level, which are hard to be detected without sophisticated analytical instrument such as liquid chromatography-mass spectrometry (LC-Ms) and LC-Ms/MS. Among different types of CECs, PPCPs and PFCs are the most notorious ones that have been frequently detected in soils and groundwater. Therefore this book only covers PPCPs and PFCs as the representative CECs in soils and groundwater. For PCEs, the availability and capacity of the analytical techniques for detecting ENPs and MPs in environmental samples still need improvement. Nevertheless, good progress has been made toward the expanding of knowledge on the occurrences, fate, and impacts of ENPs and MPs in soils and groundwater, which is the other focus of this book.

1.4 Organization of this book

The aim of this book is to summarize the current state of knowledge and the most recent research advancement in emerging contaminants in soils and groundwater. The book has nine chapters center on four types of most well-known emerging contaminants, PPCPs, PFCs, ENPs, and MPs. After the introduction chapter (*Chapter 1: Introduction*), the occurrences and impacts of PPCPs in soils and groundwater are presented in *Chapter 2, Occurrences and Impacts of Pharmaceuticals and Personal Care Products in Soils and Groundwater*, which not only provides an overview of the properties of different types PPCPs, but also outlines the details on their sources, detection methods, and potential risks in soils and groundwater. *Chapter 3, Fate and Transport of Pharmaceuticals and Personal Care Products in Soils and Groundwater*, mainly focuses on the fate and transport of PPCPs in the subsurface. It summarizes findings from the most recent experimental and modeling studies on the retention, transport, and release of PPCPs in the vadose zone and groundwater. The next six chapters use the same arrangement as *Chapter 2-3* with *Chapter 4-5* on PFCs, *Chapter 6-7* on ENPs, and *Chapter 8-9* on MPs. This book project was initiated, developed, and finished during the covid-19 pandemic. It is the result of excellent teamwork with tremendous individual commitments and efforts from each of the contributors of all the chapters.

Chapter 2

Occurrences and impacts of pharmaceuticals and personal care products in soils and groundwater

Yicheng Yang, Yulin Zheng, Jinsheng Huang, Yue Zhang and Bin Gao
Department of Agricultural and Biological Engineering, University of Florida, Gainesville, FL, United States

Chapter Outline

2.1 Overview

Pharmaceutical and personal care products (PPCPs) are rapidly growing emerging environmental contaminants that have been detected in almost all environmental matrices, including soils and groundwater (Barceló & Petrovic, 2007).

Emerging Contaminants in Soil and Groundwater Systems. DOI: https://doi.org/10.1016/B978-0-12-824088-5.00006-9

PPCPs consist of two major categories based on their uses. One is pharmaceuticals, which are mainly used to diagnose, prevent, or treat the disease of humans and animals for restoring and improving functions (Patel et al., 2019). The other is personal care products (PCPs) including a group of chemicals to protect humans from potential harm for improving life quality (Brausch & Rand, 2011). Examples of some typical PPCPs include broad-spectrum antibiotics, analgesics [e.g., acetaminophen (Tylenol), ibuprofen (Advil), and aspirin], hormones, nonsteroidal anti-inflammatory drugs (NSAIDs), β-blockers, lipid regulators, mood regulators, preservatives, disinfectants, insect repellents, cosmetics, fragrances, and other chemical substances used widely in daily life for different purposes. Details of the classifications of PPCPs are summarized in Section 2.2.

The large consumption of PPCPs causes them to enter the aqueous environment directly or indirectly through various human activities including livestock breeding, aquaculture, sewage discharge, compost fertilizing, and landfill. The concentration level of PPCPs in surface water and groundwater is from ng/L to μg/L. Many PPCPs are polar, highly bioactive, and optically active. Their concentrations in the environment are usually higher than the trace concentrations. It has been recognized that continuous exposure to PPCPs even at low concentrations can result in unexpected consequences, which may become a threat to the human and ecological environment (Sui et al., 2015). Since PPCPs can cause serious environmental and health risks, they have raised continued concern in past years. There are various types of PPCPs with distinct properties, so their environmental impacts and risks could be very different. However, the potential risks of PPCPs to the environment and the ecosystem can never be underestimated as most of them share the same notorious characteristics. (1) PPCPs are pseudo-persistent because their transformation and removal rates are relatively high in the natural environment. Nevertheless, PPCPs' introduction via sewage treatment plants (STPs) and septic systems is continuous and incessant. (2) Each of the PPCPs is developed for certain biological purposes. As a result, they can make biological effects, including both the expected and the unexpected ones (side effects). (3) Some PPCPs may have harmful xenobiotics-like physicochemical behaviors, because they can persist for a long time before making a curing effect. These PPCPs are often encapsulated by certain coating materials for avoiding the substances from becoming inactive or degraded. Some PPCPs may have high lipophilicity for passing through membranes. (4) The heavy use of pharmaceuticals, especially antibiotics, is common not only in human disease treatments but also in veterinary medicine. In addition, a wide range of PCPs have been developed and used in daily life in recent years. More than 3000 PPCPs are currently available to the public and new substances enter the market continuously (Arpin-Pont et al., 2016; Barceló & Petrovic, 2007). It is thus very important to understand the fate and impacts of PPCPs in the environment. The book is on emerging contaminants in soils and groundwater, so this chapter mainly focuses on the occurrences and impacts of PPCPs in the subsurface.

An unavoidable consequence of the increased use of PPCPs is their detection in the environment, especially in the aqueous systems. Occurrences of PPCPs in surface water, STP effluent, and wastewater treatment plant (WWTP) effluent have been well documented in the literature (Esplugas et al., 2007; Kim et al., 2009; Yang et al., 2017). Because the natural earth environment is made of interactive zones (see Fig. 1.1), PPCPs in the aquatic environment can be an important source of PPCPs in soils and groundwater (Christou et al., 2017). In fact, many studies have also been conducted to explore the sources, occurrences, and pathways of PPCPs in soils and groundwater, which is one of the focuses of this chapter (Section 2.3). In general, the concentrations of PPCPs reported in most aquatic environments are low and in the ng/L to low mg/L range. PPCP concentrations in soils and groundwater can be even lower than those in surface water and wastewater (Lyu et al., 2020). Therefore appropriate analytical techniques for detecting PPCPs in soils and groundwater are essential and thus are another focus of this chapter (Section 2.4). Section 2.5 mainly covers the risks and impacts of PPCPs in soils and groundwater. Two case studies of two types of typical PPCPs in soils and groundwater are also presented in Section 2.6. At the end (Section 2.7), the perspectives and potential future research directions are summarized and discussed.

2.2 Classifications of PPCPs

2.2.1 Definition

As indicated in the previous section, PPCPs can be classified into two major groups and multiple subgroups according to their properties and applications. Pharmaceuticals generally consist of antibiotic drugs, endocrine hormones, antipyretics and analgesics, anti-inflammatory drugs, β-blockers, and blood lipid regulators. PCPs are chemicals used in cosmetics (e.g., makeups, perfumes, moisturizers, etc.), hygiene products (shampoos, toothpaste, deodorants, etc.), and protectives [insect repellents, sunscreen ultraviolet (UV) filters, etc.]. Table 2.1 lists some of the commonly used PPCPs that have been detected in the environment. To date, more than 3000 PPCPs have been used for both humans and animals to improve their living standards (Muthanna & Plosz, 2008). After being released into the soil and groundwater environment, most of the PPCPs can be degraded chemically or biologically (Reddersen et al., 2002). It is common that the degraded PPCP concentrations in soils and groundwater are significantly lower than that of the corresponding original compounds, considering that most of the PPCPs are not persistent. Some of the PPCPs, however, are very stable in soils and groundwater and thus require special attention (Hirsch et al., 1999).

TABLE 2.1 Classifications of PPCPs.

Typical classes	Representative compounds
A. Pharmaceuticals	
A1. Broad-spectrum antibiotics	
1a	Levofloxacin
1b	Penicillin
A2. Hormones	
2a	17-β-Estradiol (E2)
2b	Estriol (E3)
2c	Estrone (E1)
A3. Nonsteroidal anti-inflammatory drugs (NSAIDs)	
3a	Diclofenac
3b	Ibuprofen
3c	Naproxen
A4. β-Blockers	
4a	Metoprolol
4b	Propranolol
A5. Blood lipid regulators	
5a	Clofibric acid
5b	Gemfibrozil
B. Personal care products	
B1. Preservatives	
6a	Parabens
B2. Bactericides/disinfectants	
7a	Methyltriclosan
7b	Triclocarban (TCC)
7c	Triclosan (TCS)
B3. Insect repellents	
8a	*N,N*-diethyl-*meta*-toluamide (DEET)
B4. Fragrances	
9a	Galaxolide fragrance (HHCB)
9b	Toxalide fragrance (AHTN)

(*Continued*)

TABLE 2.1 (Continued)

Typical classes	Representative compounds
B5. Sunscreen UV filters	
10a	2-Ethyl-hexyl-4-trimethoxycinnamate (EHMC)
10b	4-Methyl-benzylidene-camphor (4-MBC)
10c	Octyl-methoxycinnamate (OMC)
10d	Octyl-triazone (OC)

Source: From Yang, Y., Ok, Y. S., Kim, K. H., Kwon, E. E., & Tsang, Y. F. (2017). Occurrences and removal of pharmaceuticals and personal care products (PPCPs) in drinking water and water/sewage treatment plants: A review. *Science of the Total Environment, 596–597*, 303–320. https://doi.org/10.1016/j.scitotenv.2017.04.102.

2.2.2 Pharmaceuticals

Pharmaceutical compounds are mainly developed for curing diseases; therefore the most commonly used pharmaceuticals are often associated with common human and animal diseases. Many compounds have been developed as drugs for human and veterinary medicine. For human pharmaceuticals alone, there are about 12,000 approved compounds and many of them have received considerable concern. Fig. 2.1 provides some examples of the most commonly detected pharmaceuticals in the environment, which can be classified into seven categories including antibiotics, NSAIDs, antihypertensives, lipid regulators, antidepressants, anticonvulsants, and hormones (Cizmas et al., 2015). Among them, antibiotics have received special attention because they are the most widely used for both humans and animals and thus are the most frequently detected in the environment including soil and groundwater systems (Chaturvedi et al., 2021; Kovalakova et al., 2020). Based on their occurrences in soils and groundwater, several subgroups of antibiotics such as sulfonamides, fluoroquinolones, tetracyclines, macrolides, and β-lactam have attracted much research interest in the literature (Dai et al., 2020; Riaz et al., 2018; Wegst-Uhrich et al., 2014). Many of them are amphoteric and ionizable compounds, granting them good solubility in water and thus relatively high mobility in soils and groundwater. In addition to antibiotics, synthetic hormones have also been frequently detected in soils and groundwater and thus received much attention (Gottschall et al., 2013). Among them, 17-estradiol is a representative that affects the ovaries and placenta (Pan et al., 2009; Xu et al., 2020).

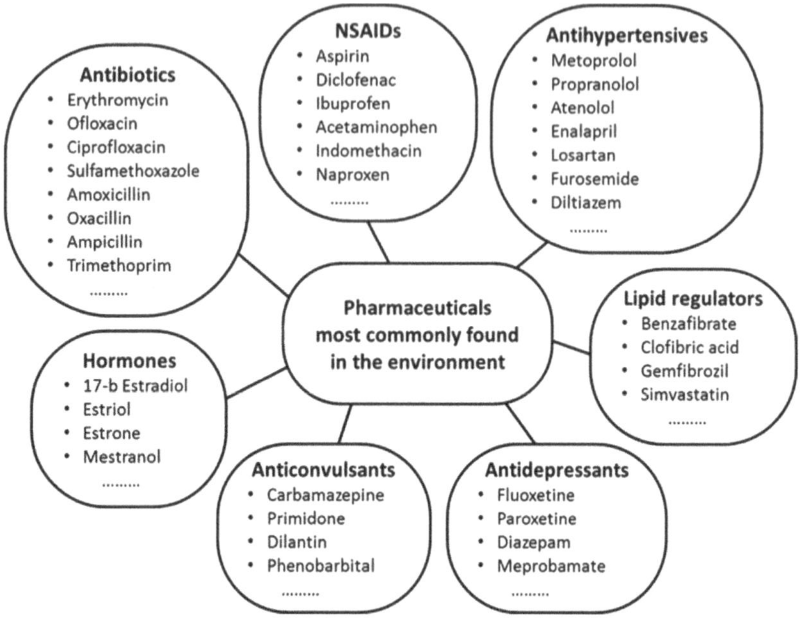

FIGURE 2.1 Most commonly detected pharmaceuticals in the environment. *From Cizmas, Leslie Sharma, Virender K Gray, Cole M McDonald, Thomas J. Pharmaceuticals and personal care products in waters: occurrence, toxicity, and risk.* Environmental chemistry letters, 13*(4), 381–394.*

2.2.3 Personal care products

PCPs often contain a variety of compounds including ones with severe, unknown, minor, or even no environmental risks, which make the classification of PCPs in the environment very difficult. In general, PCPs in the market are divided into different categories based on their uses such as oral care, skin care, hair care, body care, decorative cosmetics, etc. After the disposal, much attention has been paid to the toxic compounds in the PCPs. For PCPs in soils and groundwater, bisphenol A is one of the most commonly detected toxic compounds (Careghini et al., 2015). Bisphenol A is produced in large quantities for use widely in many consumer products in daily life. It belongs to endocrine compounds and can disrupt human hormone secretion and also affect soil ecosystem functions (Novo et al., 2018). Pesticides, fungicides, and herbicides are also commonly used in PPCs to improve life quality. For example, insect repellent (bug spray) often contains synthetic chemical compounds as the active ingredient to kill bugs. Some of the commonly used compounds include *N,N*-diethyl-*meta*-toluamide (DEET), Icaridin, *p*-menthane-3,8-diol, ethyl butylacetylaminopropionate, and 2-undecanone, which may be indirectly released into the soil and groundwater systems after the use. In addition to the indirect release, some of these active ingredients in pesticides, herbicides, and fungicides used in agricultural production can be directly released into soils to affect the quality and health of subsurface systems (Tian et al., 2016; Zhang et al., 2013). Once entering the soil and groundwater systems, it is almost impossible to determine whether these compounds are from the PCPs or agricultural practices.

2.3 Pathways and main sources of PPCPs in soils and groundwater

2.3.1 Pathways

As shown in Fig. 2.2, PPCPs can enter the soil and groundwater systems through several pathways (Boxall et al., 2012; Lambropoulou & Nollet, 2014; Price et al., 2010). The most common PPCP pollution hot spots include WWTPs/STPs, septic tanks, hospitals, animal farms, aquatic farms, and agricultural land (especially the ones with manures and biosolid applications). Furthermore, untreated household effluent and treated effluents from industries and medical services may contain some partially degraded and refractory PPCPs. This PPCPs-contaminated water may directly discharge into various receiving water bodies without proper treatment. Luo et al. (2014) reviewed the occurrences of PPCPs in aquatic environments,

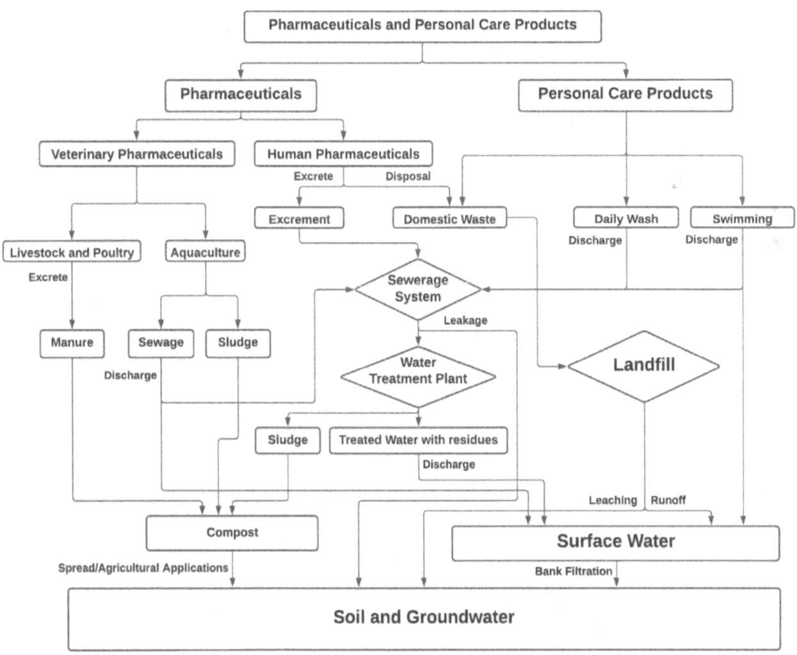

FIGURE 2.2 Summary of the pathways of PPCPs into soils and groundwater.

including sewage, surface water, and groundwater. Because soils and water are interactive, PPCP residues in can also enter the subsurface environment through the natural hydrologic cycle (Mompelat et al., 2009; Petrović et al., 2003).

Although different technologies have been applied to treat municipal wastewater, domestic sewage is still one of the major sources of PPCPs released into the soil and groundwater systems. There are several other PPCP exposure pathways including the unused medicine disposal in landfills, veterinary medicines runoff from farmyards, disposal of the animal carcasses, and reclaimed water irrigation (Awad et al., 2014; Fick et al., 2009). The management and use practices of PPCPs vary in different regions of the world. Hence, the significance of different PPCPs exposure pathways in soils and groundwater can vary geographically. For example, in certain regions around the world, the connectivity to STPs is limited. Therefore exposure models based on a specific region may not have wide applicability.

2.3.2 Domestic sewage

Domestic sewage is one of the major contributors to PPCP pollutants in soils and groundwater. Drugs for human or animal treatment can always find their way into the environment including soils and groundwater. In particular, pharmaceuticals with high stability and solubility, such as methotrexate, are excreted from the body via excretion, discharged into the sewerage systems, and finally released into soils and groundwater (Kim et al., 2011; Kimura et al., 2007). PCPs, including makeups, toothpaste, soaps, hand sanitizers, UV filters, etc., are washed off from human beings with wastewater into the STPs. In addition, sloughing during recreational activities, such as swimming, can contribute to PCP discharge (Brausch & Rand, 2011). These PPCPs and their derivatives in the sewage systems can be a major pollution source to soils and groundwater because the conventional STPs and WWTPs are not designed to and thus have limited ability to treat these emerging contaminants (Kostich et al., 2014; Yang et al., 2017).

Typical treatment processes for sewage (Fig. 2.3), including screening, degritting, primary sedimentation, aeration tanks, and final sedimentation, cannot eliminate PPCPs efficiently (Carballa et al., 2004). PPCP removal in STPs/WWTPs is complicated and closely related to the biological and physicochemical properties of contaminants, including hydrophilicity, solubility (Evgenidou et al., 2015), volatility, and biodegradability (Jones et al., 2005). Some PPCPs such as tetracycline can be effectively eliminated from sewage (Yang et al., 2017). However, in conventional STPs/WWTPs with primary and secondary treatment processes, most PPCPs can be only partially eliminated. The capability of sedimentation as the primary treatment process, in removing PPCPs, can be very limited due to the hydrophilic nature of most PPCPs (Carballa et al., 2005). Compared with that of PCPs, the removal

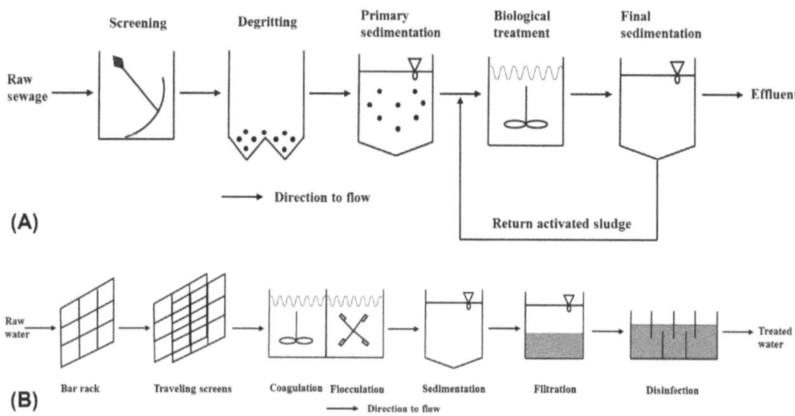

FIGURE 2.3 Typical treatment processes for sewage: (A) Flow diagram of conventional STPs; (B) Flow diagram of conventional WTPs. *From Yang, Y., Ok, Y., Kim, K., Kwon, E., & Tsang, Y. (2017). Occurrences and removal of pharmaceuticals and personal care products (PPCPs) in drinking water and water/sewage treatment plants: A review.* Science of the Total Environment, *596–597, 303–320. https://doi.org/10.1016/j.scitotenv.2017.04.102.*

efficiency of pharmaceuticals by the primary treatment is much lower. For example, tetracycline and methylparaben can be removed in sedimentation tanks but there is no considerable reduction in ibuprofen and sulfamethoxazole (Yang et al., 2017). Secondary treatment mainly refers to biological processes [e.g., activated sludge process (ASP)] and removes PPCPs through the partition, adsorption, biotransformation, and biodegradation (McClellan & Halden, 2010). The removal rate of PPCPs in ASP is closed related to the nature of PPCPs, hydraulic retention time, sludge age, adsorption capacity on sludge, and reactor design (Evgenidou et al., 2015). During the biological process, even if in the same class, the removal rates of different PPCPs can vary. Ibuprofen, ketoprofen, and diclofenac are all NSAIDs. Ibuprofen and ketoprofen biodegradation rates can be up to 75%–87%, but the diclofenac removal rate is usually lower than 25% during secondary treatment (Salgado et al., 2012; Wang et al., 2014).

Tertiary treatment processes have been used to remove the PPCPs that are poorly eliminated by the secondary treatment processes. Several advanced treatment technologies such as membrane filtration, activated carbon adsorption, and advanced oxidation processes (AOPs) have been applied in the tertiary treatment to remove PPCPs from secondary-treated wastewater. Membrane filtration processes, including nanofiltration (NF) and reverse osmosis (RO), are promising treatment methods to remove PPCPs from wastewater (Nghiem et al., 2004; Yoon et al., 2006). Ultrafiltration (UF) and microfiltration (MF) may also be used for removing PPCPs. However, their removal performances are not ideal, considering that their membrane pore sizes are considerably larger than most PPCP molecules. While NF and RO processes in wastewater treatment generally have outstanding PPCP removal efficiency, some small contaminants can still penetrate these membranes (Schäfer et al., 2011). Furthermore, the retentate of the NF/RO processes may have high concentrations of PPCPs and thus should be handled with cautions to avoid potential environmental pollutions including spilling into soils and groundwater. Granular activated carbon (GAC) and powdered activated carbon (PAC) can be used as adsorbents in tertiary treatment to remove PPCPs from sewage (Mailler et al., 2016; Yang et al., 2011). GAC is typically used in rapid-packed filters, while PAC is often applied in rapid mix processes for water treatment (Berge et al., 2018; Scheurer et al., 2010). GAC-based rapid filters are used widely in tertiary treatment in STPs (Khelladi et al., 2020). Several studies have demonstrated the effectiveness of GAC and PAC in removing various types of PPCPs from secondary effluents (Mailler et al., 2015; Stackelberg et al., 2007; Yi et al., 2020). AOPs, including ozonation, UV, photocatalysis, and Fenton reaction, have also been applied in the tertiary treatment of PPCPs in wastewater (Huber et al., 2003; Klavarioti et al., 2009; Oluwole et al., 2020). AOPs can affect the polarity and functional groups of PPCPs and further eliminate them from secondary effluents (Lin et al., 2016; Oluwole et al., 2020; Papageorgiou et al., 2014). If properly managed, these tertiary treatment technologies can dramatically decrease the occurrences of PPCPs in domestic sewage effluents, which can minimize the sources and reduce PPCP pollution in soils and groundwater.

2.3.3 Landfill

Landfill leachates are one of the representative sources of PPCPs in soils and groundwater (Clarke et al., 2015; Eggen et al., 2010). PPCPs in landfills are usually at relatively high levels but may vary by several orders of magnitude (Merel & Snyder, 2016). This is because many unused PPCPs are not recycled and are often discarded as garbage with municipal solid wastes (MSWs) (Bound & Voulvoulis, 2005; Daughton, 2003; Sasu et al., 2012; Slack et al., 2005). Most of these unused PPCPs in MSWs are sent to landfills because they are the primary disposal method for garbage and solid wastes in the world. As a result, PPCPs are abundant and frequently detected in landfills. For example, Musson and Townsend detected 22 PPCPs in a landfill in Florida with a total concentration of 8.1 mg/kg (Musson & Townsend, 2009). Several PPCPs of notably high concentrations have also been found in landfills in many other places. Song et al. (2016) reported that the average concentrations of oxytetracycline, tetracycline, and sulfamethoxazole in the landfills in China were 100.9, 63.8, and 47.9 μg/kg, respectively. In landfills, PPCPs may undergo one or multiple processes such as degradation, adsorption, or entering the leachates. Compared to other typical soil and groundwater pollutants such as heavy metals and nutrients, PPCPs often have much lower concentrations in the leachates of landfills. However, the potential risks and toxicities of PPCPs could be very high even at very low levels (Brausch & Rand, 2011; Cizmas et al., 2015). Therefore their occurrences in landfill leachates should also be concerning. Since landfill leachates directly enter the underlying soils and groundwater, landfills are another important source of PPCP pollutants in the subsurface environment (Yu et al., 2020).

2.4 Analytical techniques

2.4.1 Overview

Accurate identification and quantification of PPCPs in soils and groundwater are challenging due to low analyte concentrations and complex matrix effects (especially for soil samples). Generally, the preparation is a required step where the analyte is isolated from confounding matrix chemicals and concentrated to the point where accurate analysis can be done. A typical analytical procedure for PPCPs in soils and groundwater includes several essential steps: sample collection and preservation, sample preparation (purification and concentration), instrument analysis, detection, and finally data analysis (Fig. 2.4). There are several review articles covering these steps with more details (Ballesteros-Gómez & Rubio, 2011; Dimpe & Nomngongo, 2016; Pérez-Fernández et al., 2017; Petrović et al., 2005; Petrovic et al., 2010; Seifrtová et al., 2009; Thiele-Bruhn, 2003). It is impossible to use one single method for all different PPCPs due to physiochemical variations in analyte and matrix chemicals. Each method's accuracy should be confirmed in different aspects, including extraction efficiencies, method repeatability, limits of detection, and quantification. In addition, the method validation should be conducted to confirm the performance parameters determined during the method development (Ruiz-Angel et al., 2014).

2.4.2 Sample collection and preservation

Sampling can cause measurement uncertainty in PPCP analysis due to a contaminant spatial and temporal variability (Ballesteros-Gómez & Rubio, 2011). Therefore, to get accurate and meaningful analytical results of PPCPs in soils and groundwater, processing of a large number of samples is required. Successful sampling means that the sample amount or volume collected should be large enough to accurately represent that region of the sites under study. The lower the analyte concentrations of PPCPs are in the soils and groundwater, the larger the sample amount is required. Relatively clean soil or groundwater requires at least a kilogram or liter of sample. However, for soils and groundwater with more concentrated PPCPs, required sample amounts/volumes are usually lower.

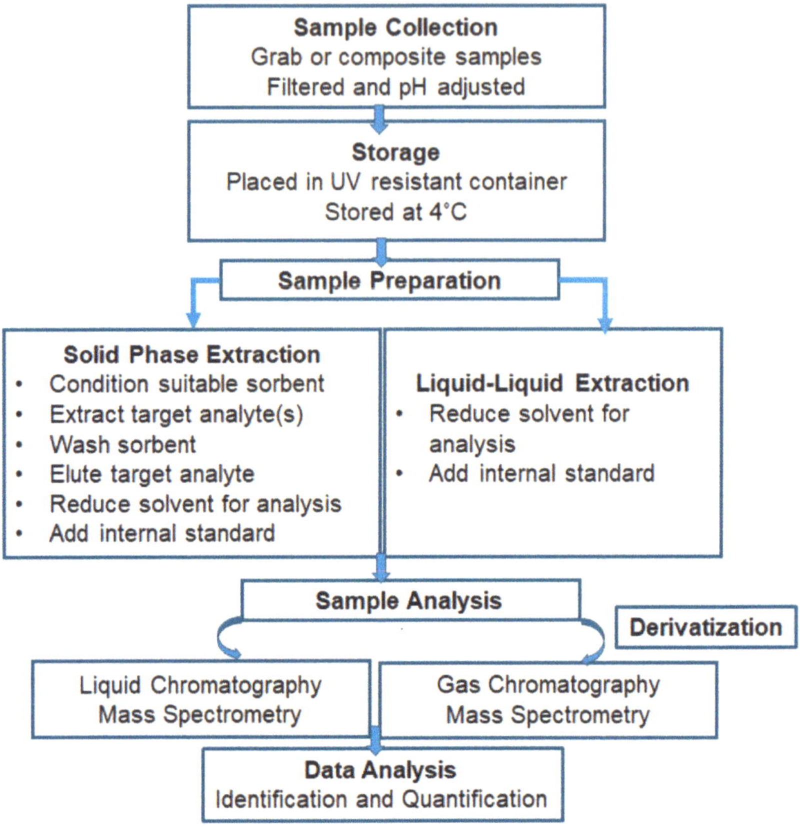

FIGURE 2.4 Typical analytical procedure of PPCPs in soils and groundwater. *From Patel, M., Kumar, R., Kishor, K., Mlsna, T., Pittman, C., & Mohan, D. (2019). Pharmaceuticals of emerging concern in aquatic systems: Chemistry, occurrence, effects, and removal methods.* Chemical Reviews, 119*(6), 3510–3673. https://doi.org/10.1021/acs.chemrev.8b00299.*

Grab sampling is the most popular method to analyze PPCPs in the environment including soils and groundwater (Evans & Kasprzyk-Hordern, 2014). However, this sampling method can introduce errors. A nonrepresentative location or improper timing and frequency of sampling can cause variation in analyte concentration. The variation may be misrepresentative, considering that PPCPs release in soils and groundwater often occur periodically, rather than continuously. There is no doubt that grab sample collection is easy and quick, reducing stereoselective degradation when correct storage procedures are employed (Evans & Kasprzyk-Hordern, 2014). Errors in sample pretreatment and storage can cause contamination or decomposition of analyte chemicals, affecting analysis results (Batley, 1999). Primary filtration of groundwater samples is necessary to remove particulates that have a negative influence on further sample preparation steps (Madrid & Zayas, 2007).

Sample preservation is also important because analyte decomposition can occur due to UV radiation exposure, microbial activities, and unwanted chemical reactions. Samples are often stored in brown amber glass bottles at 4°C before analysis. In addition, chemical preservatives can be added to samples to inhibit analyte decomposition (Śliwka-Kaszyńska et al., 2003). Acidification is another common treatment to inhibit microbial activities. Sample size should be considered because huge samples may cause difficulties in transportation and storage. Therefore there is a trend that the analysis method of PPCPs in soils and groundwater should be developed to achieve satisfactory results by using smaller sample volumes, faster preparation processes, and shorter analysis times (de Fatima Alpendurada, 2000; Mirali et al., 2014).

2.4.3 Sample preparation

The detectable concentrations of some PPCPs in soils and groundwater can be as low as 1 ng/L, and confounding matrix effects make sample preparation indispensable. Generally, sample preparations include concentrating and purifying samples. Sample preparation should be optimized due to the infinite matrix chemical variability of analyte properties. Therefore the sample preparation of PPCPs in soils and groundwater should be re-evaluated if matrix chemicals change, even though the targeted PPCP compound is the same one. Sample preparation often includes pH adjustment, extraction, and elution in a suitable solvent. After the final step of the preparation process, the soil or groundwater sample is ready for PPCP analysis.

Recently, solid phase extraction (SPE) has generally replaced liquid − liquid extraction (LLE) for PPCP analyte preparations (concentration and purification steps) (Pérez-Fernández et al., 2017; Płotka-Wasylka et al., 2016). Compared with LLE, SPE advantages include better selectivity, sensitivity, and reproducibility as well as lower cost because it requires less time and solvent (Białk-Bielińska et al., 2016). SPE is the treatment in a cartridge where the analyte dissolved in a mobile phase is bound to a solid stationary phase as pass through the cartridge. It requires four steps, including (1) conditioning the cartridge with a suitable solvent that wets the stationary phase surface; (2) loading the sample onto the cartridge, where the target analyte is retained while unwanted matrix chemicals pass through with the solvent (mobile phase), (3) washing cartridge to further remove impurities before the analyte is eluted in a solvent, which is often buffered to an appropriate pH; and (4) adjusting the solvent to a known volume for instrumental analysis.

Most SPE materials are commercially available. To choose a suitable SPE sorbent, it is necessary to consider the physical and chemical properties of both the target analyte(s) and matrix chemicals. Solution pH as a factor can affect SPE results of PPCPs. Many pharmaceuticals are acidic with pK_a values below 4.5. These drugs are found mainly in their ionized form in the natural environment and are retained poorly by lipophilic sorbents. In SPE, the attraction between the packing material and the analyte should be strong. Attraction can change significantly with pH value. Therefore precise pK_a values should be known for the adjustment of the procedure's pH accordingly.

In PPCP extractions from soil samples, there is usually an additional extraction step where the solid is mixed or sonicated with a polar organic solvent. Then, after filtration, samples can be advanced to the SPE step (Seifrtová et al., 2009). Based on SPE, there are several modified extraction methods, including solid phase microextraction (SPME) and membrane SPME (M-SPME). SPME is the treatment that takes a coated fiber to collect and preconcentrate target analytes. After samples are equilibrated in SPME, the adsorbed analytes can be desorbed thermally or by a solvent for instrumental analysis. Different from traditional SPE, SPME has lower sample volumes and more efficient sample processing and extraction. However, in the SPME step, limits of detection and matrix effects may increase (Yang & Peppard, 1994). The proposes of M-SPME are to isolate and enrich polar organic compounds from aqueous solutions (Ballesteros-Gómez & Rubio, 2011). Furthermore, recently, molecularly imprinted sorbents are used to analyze PPCPs in soils and groundwater due to improved selectivity (Ansari & Karimi, 2017).

2.4.4 Instrument analysis and detection

A chromatographic separation step and mass spectrometry (MS) detection constitute instrumental analysis. First, through a chromatographic separation step, the target PPCPs can be further isolated from matrix chemicals. Then, MS detection is followed to analyze target PPCPs. Both liquid chromatography (LC) and gas chromatography (GC) are used (Pérez-Fernández et al., 2017; Seifrtová et al., 2009; Togola & Budzinski, 2007). Among them, LC is more common because the volatility of many polar PPCPs is low. Different from LC, during GC analysis, higher column temperatures caused by high boiling points may make PPCPs decompose. Thus an additional derivatization step is often necessary before GC-MS analysis of PPCPs of high boiling points (Evans & Kasprzyk-Hordern, 2014).

There are several detectors that can be used for PPCP analysis after chromatographic separation. Among them, MS and MS/MS are dominant. The advantages of MS/MS detection over MS detection include the low requirement of PPCP separation from matrix chemicals and improved accuracy. Electrospray ionization (EI) and atmospheric pressure chemical ionization (APCI) are two main common techniques that matrix chemicals can affect. Generally, matrix effects can reduce the signal intensity of the PPCP analyte, leading to significantly reduced limits of detection. To improve accuracy and precision, using chelating agents during sample preparation is another way to further remove confounding matrix chemicals (Seifrtová et al., 2009). Several other common LC analysis detectors that can be used for PPCPs include time-of-flight mass spectrometry (TOF-MS), UV, and fluorescence (Pérez-Fernández et al., 2017). Besides, Raman spectroscopy and the flow injection analysis (FIA) technique have also been used to analyze PPCPs in soil and groundwater samples (Hadad et al., 2011). To detect or analyze target PPCP compounds, several analytical techniques including chromatographic, electrophoretic, electrochemical, spectroscopic, and titrimetric can be employed together (Siddiqui et al., 2017).

2.4.5 Data analysis

Calibration curves can be drawn by using standards spiked with the target PPCPs in a range of method-relevant concentrations. Through calibration curves, PPCP concentrations in the soil and groundwater samples can be identified and quantified. Matrix effects are one of the most critical factors that can negatively affect the analysis. Therefore calibration standards with a full array of matrix chemicals can reduce possible matrix effect errors. To reduce interference, there are several widely used calibration methods, including internal standard-based calibration, second-order multivariate calibration algorithms, standard addition calibration, and matrix matched calibration (Ballesteros-Gómez & Rubio, 2011). In order to reduce the noise from limits of detection and quantification, internal standards are usually employed. An ideal internal standard can efficiently improve the accuracy of the analysis. The internal standard should be the chemicals that cannot be found in the natural environment. The physiochemical properties of the internal standard should be similar to the target analyte. Sometimes, an internal standard can be added prior to sample preparation to know the recovery rate during the analytical procedure. An internal standard can also be input to the final extracts before instrumental analysis.

2.4.6 Quality control

Method calibrations should be performed in the situation where relevant matrix chemicals exist to simulate the matrix effect. Method extraction efficiencies of PPCPs can be summed up based on recovery studies. Acceptable values are generally under a range from 80% to 120%. Repeating extractions (typically 3–5 times) of PPCPs from soil and groundwater samples can determine method variability. Instrument repeatability and precision are dominantly affected by the injection of multiple PPCP concentrations (Yang et al., 2017).

2.5 Risks and impacts

2.5.1 Overview

The widespread occurrence of PPCPs in soils and groundwater has been a growing concern due to their negative effects on environmental and human health. PPCPs exist widely and interactively in different environmental compartments such as sewage, rivers, lakes, soil, and groundwater, imposing great risks to the ecosystems. PPCPs and their residues in soils and groundwater can eventually enter and accumulate in the food chain due to biological functions of soils and groundwater such as agricultural production and water supply (Rajapaksha et al., 2014). Despite the concentrations of PPCPs in soils and groundwater are low, ranging between ng/L and μg/L, PPCP residues can still have critical adverse health impacts. For example, chronic exposure to estrogenic in water can cause the enlargement of fish livers (Gunnarsson et al., 2009). Single and mixed PPCP residues can negatively affect reproduction and result in histopathological changes in zebrafish (Galus et al., 2013; Overturf et al., 2015). In addition, the impacts and toxicities of PPCPs to humans after a long-time exposure are still largely unknown (Boxall et al., 2012).

Many PPCPs could rapidly decompose in soils and groundwater. However, overdose and continuous release into the soil and groundwater systems cause their pseudo-persistence with serious influences on aquatic organisms (Kostich et al., 2014). PPCPs should be detected and monitored frequently in soils and groundwater due to their wide uses, improper disposal, and biochemically active structures (McClellan & Halden, 2010; Ternes et al., 2004), which would be helpful to better understand their environmental risks and impacts.

2.5.2 Translocation of PPCPs from soils and groundwater into the food chain

PPCPs and their residues in soils and groundwater can be translocated into the food chain such as in the edible tissues of plants that grew in polluted soils and/or irrigated with polluted water including groundwater (Fig. 2.5) (Dodgen et al., 2015; Keerthanan et al., 2021; Rajapaksha et al., 2014). An individual PPCP risk to humans could be too low to affect, but the additive

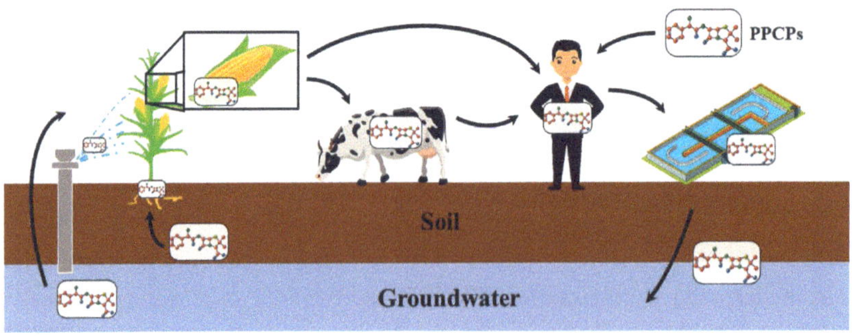

FIGURE 2.5 Translocation of PPCPs from soils and groundwater into food chain.

effect of PPCPs through the food chain accumulation can potentially be hazardous. So far, the negative impacts of ingesting PPCPs present in plants by humans are still unknown. Gu and Karthikeyan reported a few adverse effects of consuming antibiotic residues in fresh vegetables and fruits (Gu & Karthikeyan, 2008). It has been demonstrated that exposure to certain PPCPs by humans, especially children, could cause serious allergies or toxicity (Al-Farsi et al., 2017). Furthermore, some interaction effects could result from the ingestion of two different antibiotics at the same time. Consuming antibiotic-contaminated food may cause resistance to human antibacterial activity, resulting in unexpected infections and diseases that cannot be treated with antibiotic drugs. Antibiotic resistance can be developed through several ways such as an overdose of antibiotics and improper use of antibiotics frequently (Schmidt et al., 2008; Shenker et al., 2011). The development of antibiotic resistance can impose great risks to human beings because it would limit the choice and availability of the antibiotic drugs that can be used in common medicine, surgery, transplant, and other treatments. Gu and Karthikeyan found that the development of antibiotic resistance bacteria of dairy calves fed with milk contaminated with antibiotics was much higher than that with noncontaminated milk (Gu & Karthikeyan, 2008). Human exposure of PPCPs in the food chain can be estimated using the US EPA model (Kinney et al., 2006; Söffker & Tyler, 2012). For example, it was estimated that per capita annual exposure of PPCPs was 20.0, 15.1, 11.8, 8.3, 7.2, 7.0, 7.0, and 0.6 g/day for tomato, carrot, bell pepper, cabbage, cucumber, celery, lettuce, and spinach (0.6), respectively (Al-Farsi et al., 2017).

2.5.3 Risks and impacts of pharmaceuticals

While the negative effects of antibiotics such as on living organisms are well documented in the literature including several recent reviews (Anh et al., 2021; Huang et al., 2021; Limbu et al., 2021; Xu et al., 2021), this section mainly outlines the risks and impacts of hormones and NSAIDs on aquatic organisms.

As an essential component of the endocrine system, reproductive steroids are extremely important to living organisms. They play an important role in influencing behaviors, sexual differentiation, and the development of sex reproductive characteristics and organs (Kime, 1999). In addition, the influences of reproductive steroids on aquatic organisms are controlled by various environmental factors, which can alter the reproduction of living organisms. Therefore reproductive steroids in the soils and groundwater can have strong impacts on organisms and ecosystems by affecting their behaviors and reproduction.

Estrogens are the most commonly used medicines in therapy and combined oral contraceptives. They are often classified as natural [e.g., estrone and 17b-estradiol (E2)] and synthetic [e.g., 17a-ethinylestradiol (EE2)] estrogens based on the origins. In fact, estrone and E2 are the most common natural produced estrogens. Both natural and synthetic estrogens and their derivatives (metabolites) can enter soils and groundwater through WWTP/STP effluent and land application of biosolids (Ternes et al., 1999; Zhou et al., 2007). They may have a strong impact on the health of aquatic organisms. It was reported that exposure to E2 (187.2 ng/L, 21 days) caused Japanese medaka (*Oryzias latipes*) to reduce its fertility and fecundity (Sun et al., 2009). Marine Java medaka (*Oryzias javanicus*), however, showed a different response to E2 exposure (68 ng/L) with a decreased fertility but increased fecundity (Imai et al., 2005). Cripe et al. found that a low dosage of E2 (82 ng/L) decreased the fecundity of marine sheepshead minnow (*Cyprinodon variegatus*) and its first-generation offspring, but not the second-generation until the E2 level was increased to 290 ng/L (Cripe et al., 2009). Dammann et al. observed a reduction in fecundity of fathead minnow (*Pimephales promelas*) due to the expose of low levels of E2 (26 ng/L) and estrone (54 ng/L) (Dammann et al., 2011). On the other hand, high levels of estrone (1188 and 3701 ng/L) exposure caused a reduction in fecundity and fertilization (Imai et al., 2007).

NSAIDs including diclofenac, ibuprofen, naproxen, and indomethacin are commonly used to inhibit the two isoforms of the cyclooxygenase (COX) enzymes (Vane & Botting, 1998). COX enzymes are very important to vertebrates and can catalyze the production of prostaglandins, which play an important role in many physiological processes of aquatic organisms such as homeostasis, water balance, thermoregulation, and glomerular filtration (Fujimori et al., 2011; Grosser et al., 2002). Because of their wide uses, NSAIDs are found in the environment including soils and groundwater. Previous studies have shown that, after being released into the aquatic systems, ibuprofen caused decreased spawning frequency, increased number of spawning eggs, and delayed hatching of medaka even at low exposure levels (Flippin et al., 2007; Han et al., 2010). Morthorst et al. found that Zebrafish exposed to 21–506 μg/L of ibuprofen only showed a reduction in prostaglandin E2, but not any other negative signs on the reproductive parameters as reported by other researchers (Morthorst et al., 2013). For example, Ji et al. reported a significant negative effect of ibuprofen on the reproduction of Zebrafish, including lower egg production, lower hatchability, and higher embryo mortality (Ji et al., 2013). These studies suggest that the risks and impacts of ibuprofen exposure to aquatic organisms depend on the speciation type and exposure time. In addition to ibuprofen, other NSAIDs have also been shown varied risks and impacts on the health of living organisms.

2.5.4 Risks and impacts of PCPs

PCPs are a diverse group of chemicals used in daily life products, such as soaps, lotions, toothpaste, fragrances, and sunscreens. Different from pharmaceuticals for internal use, PCPs are usually used externally, such as on the human body, and are often not metabolized. Therefore large quantities of pristine PCPs can enter the environment including soils and groundwater directly through regular usage (Gottschall et al., 2012; Ternes et al., 2004). A study by the Environmental Working Group of the United States showed that 1,4-dioxane, as a known carcinogen, was found in 28% of 27,000 personal care products (Environmental Working Group, 2008). In addition, a survey of 20 teen girls (14–19 years old) showed that 16 hazardous chemicals, including 2-benzenedicarboxylic salt, synthetic musk, and Triclosan (TCS), were found in the girls' bodies because of using cosmetic products.

The use of disinfectants in PCPs has increased in the last decade. TCS is the most common disinfectant compound used in toothpaste, mouthwashes, antimicrobial soaps, sanitizers, deodorants, and other household items (Chalew & Halden, 2009). The wide use makes TCS enter the environment (Chalew & Halden, 2009); however, only a few studies have explored its environmental risks and impacts. One study found that there was no influence on Japanese medaka reproduction after the fish were exposed to 20–200 mg/L of TCS for 21 days (Ishibashi et al., 2004). Other disinfectants, including triclocarban, biphenylol, chlorophene, and bromophene, are also often found in soils and groundwater. Their environmental risks and impacts are also mostly unknown. Therefore more research should be done to investigate the potential risks of disinfectants and their residuals in the soil and groundwater environment.

Chemical UV-filters such as 3-benzylidene camphor (3BC), benzophenone-1 (BP1), benzophenone-2 (BP2), and benzophenone-3 (BP3) are employed in a variety of PCPs, especially sunscreen agents, protecting consumers from UV radiation. These chemicals can enter the soil and water systems either directly by washing-off from the skin or indirectly through wastewater effluents (Fent et al., 2010). Previous studies have shown that exposure to 3BC at 74 and 285 mg/L for 21 days or 1.2 mg/L of BP2 for 15 days reduced fecundity of fathead minnows (Kunz et al., 2006; Weisbrod et al., 2007). In Japanese medaka, exposure to BP3 can affect egg production and decrease hatching success and fecundity (Coronado et al., 2008; Kim et al., 2014). There is strong evidence showing the negative impacts of sunscreen agents; however, further investigations are still needed to assess the potential toxicities and risks of chemical UV-filters in soils and groundwater.

2.5.5 Antibiotic resistance

The development of antibiotic-resistant bacteria (ARB) and antibiotic-resistant genes (ARG) due to the release of antibiotic drugs in soils and groundwater has not been a major concern in the assessment of the potential risks of PPCPs in the environment. The World Health Organization (WHO) has called out antibiotic resistance and listed it as one of the major public health issues of the 21st century (World Health Organization, 2014). Annual prescriptions of antibiotic drugs amount to more than 250 million in the United States alone, and many other countries do not regulate antibiotic prescriptions and make them over-the-counter medicine. As a result, a large number of antibiotics are used all over the world, inevitably causing their release into the environment including soils and groundwater (Gu et al., 2019; Kaszab et al., 2019; Rosi-Marshall & Kelly, 2015). WWTPs receive effluent not only from homes and communities, but also from hospitals and animal-feeding facilities (Devarajan et al., 2015). As a result, WWTPs are an important source of ARB and ARG in soils and groundwater as well as a hot spot for horizontal gene transfer (Devarajan et al., 2015; Martínez, 2008; Pazda et al., 2019). In addition, soils with a high clay and silt content can make DNA more stable, which may be beneficial to the potential transport of the biologically active DNA through the vadosc zone into the groundwater (Devarajan et al., 2015; Poté et al., 2003). We should give priority to assess the risks of the potential development of antibiotic resistance due to antibiotic contamination in soils and groundwater (Boxall et al., 2012; Goulas et al., 2019). Research is still needed to further determine and understand how PPCPs and their residues in soils and groundwater can cause some microbes to develop ARGs, and what are the key impacts of ARB and ARG in soils and groundwater on human beings and the ecosystems (Boxall et al., 2012). Reinthaler et al. found some similarities in the antibiotic resistance patterns between Escherichia *coli* isolated from sewage sludge and human patients under antibiotic treatment in 2000 and 2009 (Reinthaler et al., 2013). However, the authors also reported that there was not a correlation between fluoroquinolone resistance between sewage sludge and humans under the same conditions. Nevertheless, it is generally accepted by the scientific communities that the development of antibiotic resistance in the environment can promote antibiotic resistance in urban and medical settings. It is thus critical to consider antibiotic resistance when assessing the risks and impacts of PPCPs in soils and groundwater.

2.6 Case studies of PPCPs in soils and groundwater

2.6.1 Sulfonamides

Although there are various types of PPCPs in soils and groundwater, previous sections mainly discuss those used by humans. This casc study gives an example of the occurrences and impacts of veterinary antibiotics used in livestock farms. Among veterinary antibiotics, sulfonamides are of particular interest for a number of reasons (De Liguoro et al., 2009). They are one of the most commonly used types of drugs for animal treatment (Du & Liu, 2012; Tasho & Cho, 2016). In 2016, sulfonamides were the third most-sold class of antibiotics for food-producing species, accounting for 12% of total sales in 30 European countries and corresponding to the amount of 14.4 mg of sulfonamides sold per kilogram of animal biomass (European Medicines Agency, 2018). Sulfonamides can affect both Gram-negative and Gram-positive bacteria as well as some protozoa. In fact, sulfonamides have been used as veterinary antibiotics for many farm animals, including pigs, cattle, poultry, goats, sheep, and fish (Papich & Riviera, 2009).

As sulfonamides cannot be completely metabolized by animals, a large part of them is being excreted in animal wastes such as manure. As a result, sulfonamides and their residuals can be discharged directly or indirectly to the soils through the disposal or land application of animal wastes (Fig. 2.6) (Białk-Bielińska et al., 2014). In general, soils have a low ability to sorb sulfonamides, which have low soil partition coefficient (K_d) values (see Table 2.2). Sulfonamides are therefore highly mobile in soils and can be transported to underlying groundwater (Boxall et al., 2002; Chen et al., 2011; Dong et al., 2016; García-Galán et al., 2010; Thiele-Bruhn et al., 2004). This behavior places them in contrast to other veterinary antibiotics such as tetracycline ($K_d = 1147-2370$ L/kg) and enrofloxacin ($K_d = 260-5612$ L/kg), which have limited mobility in soils due to strong sorption (Ostermann et al., 2013; Sarmah et al., 2006). However, sulfonamide contamination of soils and groundwater does not come only from agriculture, but also from urban and industrial sources. Table 2.3 shows examples of groundwater contamination by sulfonamides from various sources in the Europe and United States.

Some sulfonamides such as sulfadiazine (SDZ), sulfamethazine (SM2) and sulfamethoxazole (SMZ) in soils can be translated into edible plants (Ahmed et al., 2015; Bizkarguenaga et al., 2016; Jin et al., 2020). It was reported that SM2 and SMZ were accumulated more easily by pakchoi cabbage than SDZ (Li et al., 2013). The dissipation half-lives of SMZ (16.8d) and SM2 (16.7d) in soil were much longer than that of DZ (10.8d). The residual concentrations of SMZ and SM2 in the soil were thus higher,

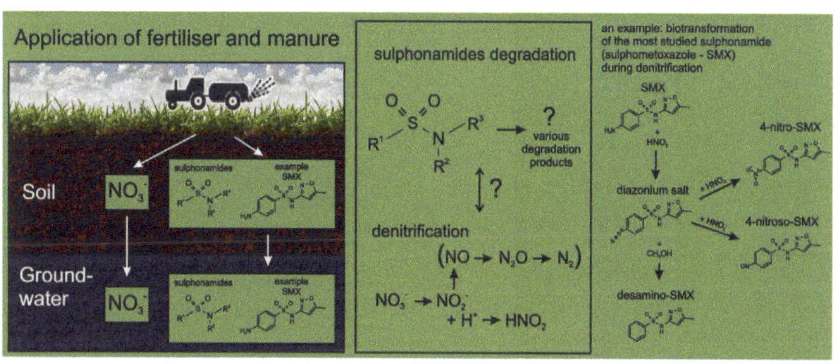

FIGURE 2.6 Sulfonamides as emerging contaminants in soils and groundwater. *From Bílková, Z., Malá, J., & Hrich, K. (2019). Fate and behaviour of veterinary sulphonamides under denitrifying conditions.* Science of The Total Environment, 695, *133824. https://doi.org/10.1016/j.scitotenv.2019.133824.*

TABLE 2.2 Soil partition coefficients of several main veterinary sulfonamides.

Compound	K_d (L/kg)	Matrices	pH	OC%	References
Sulfachloropyridazine	0.9–1.8	Sand and clay loam	6.5–6.9	NR	Boxall et al. (2002)
Sulfadiazine	1.4–2.8	Loess loam, sand and clay fraction	6.7–7.0	1.61–4.40	Thiele-Bruhn et al. (2004)
Sulfadimethoxine	2.3–4.6	Loess loam, sand and clay fraction	6.7–7.0	1.61–4.40	Thiele-Bruhn et al. (2004)
Sulfamethazine	2.1–3.0	Loess and silty loam, sand and clay fraction	6.7–7.3	1.61–4.40	Thiele-Bruhn et al. (2004), Shan et al. (2018)
Sulfamethoxazole	1.5	Clay loam	4.9	1.35	Liu et al. (2010)
Sulfanilamide	1.5–1.7	Loess loam, sand and clay fraction	6.7–7.0	1.61–4.40	Thiele-Bruhn et al. (2004)
Sulfapyridine	3.1–3.5	Loess loam, sand and clay fraction	6.7–7.0	1.61–4.40	Thiele-Bruhn et al. (2004)

K_d, Soil partition coefficient; *NR*, not reported; *OC*, organic carbon.
Source: From Bílková, Z., Malá, J., & Hrich, K. (2019). Fate and behaviour of veterinary sulphonamides under denitrifying conditions. *Science of the Total Environment, 695*, 133824. https://doi.org/10.1016/j.scitotenv.2019.133824.

TABLE 2.3 Maximum concentrations of sulfonamides detected in groundwater.

Compound	c_{max} (ng/L)	Location	References
Sulfadiazine	$0.81-1160 \times 10^3$	Denmark, the Netherlands, Spain	Holm et al. (1995), García-Galán et al. (2010, 2011), Kivits et al. (2018)
Sulfadimethoxine	0.6–130	Spain, the United States	Batt et al. (2006), García-Galán et al. (2010, 2011), Watanabe et al. (2010)
Sulfadoxine	0.6–78.3	China, Spain	García-Galán et al. (2010, 2011), Hu et al. (2010)
Sulfaguadinine	1600×10^3	Denmark	Holm et al. (1995)
Sulfamerazine	3.22–103.0	Spain, the United States	García-Galán et al. (2010, 2011), Bartelt-Hunt et al. (2011)
Sulfamethazine	$3.71-900 \times 10^3$	Europe, the United States	Holm et al. (1995), Hirsch et al. (1999), Hamscher et al. (2005), Batt et al. (2006), Barnes et al. (2008), García-Galán et al. (2010, 2011), Teijon et al. (2010), Watanabe et al. (2010), Bartelt-Hunt et al. (2011), Kivits et al. (2018)
Sulfamethizole	330×10^3	Denmark	Holm et al. (1995)
Sulfamethoxazole	8.8–1110	Asia, Europe, the United States	Hirsch et al. (1999), Sacher et al. (2001), Barnes et al. (2008), Heberer et al. (2008), Radjenović et al. (2008), García-Galán et al. (2010,

(Continued)

TABLE 2.3 (Continued)

Compound	c_{max} (ng/L)	Location	References
			2011), Hu et al. (2010), Loos et al. (2010), Teijon et al. (2010), Fram and Belitz (2011), Ma et al. (2014), Kivits et al. (2018)
Sulfamethoxypyridazine	0.2–0.77	Spain	García-Galán et al. (2010, 2011)
Sulfanilamide	300×10^3	Denmark	Holm et al. (1995)
Sulfapyridine	1.11–104	Spain	García-Galán et al. (2010, 2011), Teijon et al. (2010)
Sulfaquinoxaline	1.17–274.0	Spain	García-Galán et al. (2010, 2011)
Sulfathiazole	0.3–305	Spain, the United States	Bartelt-Hunt et al. (2011), García-Galán et al. (2011)
Sulfisoxazole	17.1	Spain	García-Galán et al. (2011)

Source: From Bílková, Z., Malá, J., & Hrich, K. (2019). Fate and behaviour of veterinary sulphonamides under denitrifying conditions. *Science of the Total Environment, 695*, 133824. https://doi.org/10.1016/j.scitotenv.2019.133824.

causing their higher accumulation in pakchoi cabbage in comparison with SDZ. The authors spiked the soil with sulfonamides and found a positive correlation between the applied concentration of SM2 and its accumulation in pakchoi cabbage. They explained that a higher residual concentration of SM2 in soils promoted the uptake of SM2 in pakchoi cabbage. In addition, they found the concentrations of different sulfonamides in the soils and the cabbage were different, probably due to the soil had different bonding abilities to these sulfonamides.

2.6.2 DEET

DEET (*N*,*N*-diethyl-*meta*-toluamide or *N*,*N*-diethyl-3-methylbenzamide) has been recognized widely as a broad spectrum insect repellent. Because of its efficacious against mosquitoes and other insects, DEET is used at least once in a season by approximately 30% of the US population (Antwi et al., 2008; Veltri et al., 1994). In the United States alone, an estimated 2–3 million kg of DEET are produced and consumed each year (Aronson et al., 2012). As a result, a large amount of DEET has been released into the environment through different ways, such as: (1) into the air during spray application onto skin or clothing; (2) into the water after washing and/or rinsing skin or clothing via effluent discharges or directly during recreational activity; (3) into the soil via overspray or irrigation with reclaimed water; and (4) into the groundwater via surface waters or leachate from landfills. In addition, DEET residues on skin and in clothes can be sloughed into sewage during swimming, bathing, and washing (Weeks et al., 2012). DEET is persistent and has good environmental stability, which means that it cannot be easily removed by typical wastewater treatment processes. As a result, DEET was found in the influents and effluents of WWTPs in Australia (French et al., 2015), Spain (Benitez et al., 2013), and China (Sui et al., 2010). Sui et al. found that DEET concentrations in the discharges of the treated wastewater in China reached 180–600 ng/L (Sui et al., 2010). In addition, a study in the United States during the summer showed high DEET concentrations (2087 ± 32 ng/L) in swimming pools (Weng et al., 2014). Because of its high stability, DEET has also been detected in human excretions. Smallwood et al. found that the DEET level in urine of the US park employees can reach as high as 5690 μg/L (Smallwood et al., 1992).

DEET is frequently detected in environmental samples collected around the world (Table 2.4). For example, a study by the United States Geological Survey listed DEET as one of the most commonly detected compounds in the US streams (Quednow & Püttmann, 2009). Similarly, it was found that DEET can be detected in 16 of the 18 river samples taken from Jakarta River in Indonesia, and concentrations were as high as 24,000 ng/L (Quednow & Püttmann, 2009). In Germany, a high level of DEET (320 μg/L) was also detected in landfill seepage water (Schwarzbauer et al., 2002). Because DEET has a medium soil partition coefficient and high stability, it can transport through the soil vadose zone to pollute the groundwater (Cordy et al., 2004). In a survey of groundwater in 23 European countries, DEET was detected in more than 80% of the samples (84% of 164 groundwater samples) (Loos et al., 2010).

In the United States, DEET was designated by the EPA as a pesticide that has certain toxicity (slight to moderate) to aquatic organisms (United States Environmental Protection Agency, 1998). Its toxicity to aquatic organisms has been measured in fish, invertebrates, amphibians, algae, and microorganisms. It was reported that DEET had strong toxicities to rainbow trout (*Onchorhynchus mykiss*), water fleas (*Daphnia magna*), and algae (Aronson et al., 2012; Weeks et al., 2012). Concerns have been raised about its potential risks to human beings such as cases of toxic encephalopathy, acute manic psychosis, seizure, and dermal toxicities (Chen-Hussey et al., 2014). Tisch et al. found that DEET may have potential carcinogenic risks to human nasal mucosal cells (Tisch et al., 2002). An occupational exposure study pointed

TABLE 2.4 Frequency of detection, median, and maximum concentrations of DEET in waters from around the world.

Water type	Country	Number of samples	Detection frequency (%)	Med. conc. (μg/L)	Max. conc. (μg/L)	References
Groundwater	The United States	47	26	–	13	Barnes et al. (2005)
	The United States	5	80	6.1	13	Barnes et al. (2004)
	Spain	–	–	–	0.03	Guardiola et al. (1989)
Open water	North Sea	14	86	0.19	1.09	Weigel et al. (2004)
Drinking water	The United States	4	25	<RL	0.07	Stackelberg et al. (2004)
Sewage (influent)	Australia	2	100	–	1.5	Costanzo et al. (2007)
	Germany	41	98	–	3.00	Knepper (2004)
Sewage (effluent)	Australia	2	100	–	0.14	Costanzo et al. (2007)
	Germany	41	100	–	1.50	Knepper (2004)
	The United States	11	82	0.18	2.1	Glassmeyer et al. (2005)
Septic waste	The United States	5	40	–	1.0	Conn et al. (2006)

Surface water	Australia	35	97	0.093	0.49	Costanzo et al. (2007)
	Germany	8	100	0.019	0.03	Schwarzbauer and Heim (2005)
	The Netherlands	8	63	–	0.04	Hendriks et al. (1994)
	The United States	5	80	0.049	0.19	Oros et al. (2003)
	The United States	55	74	0.05	1.13	Kolpin et al. (2002), Costanzo et al. (2007)
	The United States	29	66	0.097	0.64	Glassmeyer et al. (2005)
	The United States	75	4	<RL	0.13	Kolpin et al. (2004)

RL denotes Analytical Reporting Limit; – denotes information not provided in reference.
Source: From Costanzo, S. D., Watkinson, A. J., Murby, E. J., Kolpin, D. W., & Sandstrom, M. W. (2007). Is there a risk associated with the insect repellent DEET (N,N-diethyl-m-toluamide) commonly found in aquatic environments? *Science of the Total Environment, 384*(1–3), 214–220. https://doi.org/10.1016/j.scitotenv.2007.05.036.

out that DEET can increase the risk of testicular cancer (Hardell et al., 1998). These results clearly demonstrate that DEET as a repellent may impose potential risks to humans. The analyses of United States Poison Center data that reviewed more than 29,000 human exposures, however, concluded that DEET may have little to no risk when applied according to product labels (Bell et al., 2002; Veltri et al., 1994). It is worth noting that DEET is commonly used with millions of applications to humans every year, it would have already been noted if it can serious health issues to humans. Nevertheless, the long-term effects of DEET and its residuals on human beings are still unknown. Because DEET remains a very efficacious and economical insect repellent, it is thus important to further investigate their health effects and their fate and impact in the soil and groundwater systems (Swale & Bloomquist, 2019).

2.7 Conclusion and perspectives

Currently, PPCPs in soils and groundwater have been recognized as an important environmental issue. Many researchers have investigated the occurrences and sources of PPCPs in the subsurface with focuses on not only municipal water treatment systems, landfills, and agricultural practices but also the interactions among environmental water bodies, in which trace concentrations ranging from nanograms to micrograms per liter can be detected. As demonstrated in this chapter soils and groundwater polluted by PPCPs may result from sewage discharge, limited water treatment, landfill leaching, and agricultural events. In many places, PPCPs have been detected in treated water effluent from STPs/WWTPs. Generally, conventional water treatment systems may not work well with PPCPs and relatively high concentrations of PPCPs could ultimately enter into soils, groundwater and the food chain through sewage effluent, excess sludge, and reclaimed water. Recently, some advanced treatment technologies, such as membrane filtration, carbon adsorption, and AOPs, have been used for PPCP removal in wastewater, but their performance and cost vary by case. It is crucial to evaluate the effects of tertiary treatment technologies on the occurrences and fate of PPCPs in the effluent as well as the soil and groundwater systems. The results can be used for the optimization of existing treatment systems, protecting the soil and groundwater quality, and promoting the use of reclaimed water (Yang et al., 2017).

Advancement in analytical techniques makes it possible to detect and monitor trace level PPCPs in soils and groundwater. Soundable analytical procedures for PPCPs in soils and groundwater have been established in the literature. The concept of green analytical chemistry for organic contaminants, especially PPCP quality control, has attracted increasing research interest (Perez-Lemus et al., 2019). There are several goals of green analytical chemistry. They are to apply more eco-friendly and sustainable analysis, less toxic reagents, to modify current methods to increase safety, to reduce reagents consumption and waste production (Ballesteros-Gómez & Rubio, 2011). Meanwhile, analytical devices should be miniaturized, and analysis time should be shortened. In addition, solventless extraction techniques comply with principles of green analytical chemistry (Tobiszewski et al., 2010). It is clear that additional research is required to better understand the applications of green analytical chemistry in measuring PPCPs in soil and groundwater samples.

Most of PPCPs are synthetic chemical compounds with designed biological functions. As a result, a number of PPCPs in soils and groundwater may present risks and negative impacts to the ecosystems and public health. Previous research has shown that continuous releases of PPCPs and the presence of multiple toxic compounds in soils and groundwater may exacerbate their negative impacts. There is also increasing concern over the development of ARB and ARG in soils and groundwater. Because there are so many PPCP compounds, there is a lack of information regarding the environmental risk assessments of PPCP pollution in soils and groundwater. Further investigations are also needed in this direction to better understand the environmental risks and impacts of PPCPs in soils and groundwater.

References

Ahmed, M. B. M., Rajapaksha, A. U., Lim, J. E., Vu, N. T., Kim, I. S., King, H. M., Lee, S. S., & Ok, Y. S. (2015). Distribution and accumulative pattern of tetracyclines and sulfonamides in edible vegetables of cucumber, tomato, and lettuce. *Journal of Agricultural and Food Chemistry*, *63*, 398–405.

Al-Farsi, R. S., Ahmed, M., Al-Busaidi, A., & Choudri, B. (2017). Translocation of pharmaceuticals and personal care products (PPCPs) into plant tissues: A review. *Emerging Contaminants*, *3*, 132–137.

Anh, H. Q., Le, T. P. Q., Da Le, N., Lu, X. X., Duong, T. T., Garnier, J., Rochelle-Newall, E., Zhang, S. R., Oh, N. H., Oeurng, C., Ekkawatpanit, C., Nguyen, T. T., Nguyen, Q. T., Nguyen, T. D., Nguyen, T. N., Tran, T. L., Kunisue, T., Tanoue, R., Takahashi, S., Minh, T. B., Le, H. T., Pham, T. N. M., & Nguyen, T. A. H. (2021). Antibiotics in surface water of East and Southeast Asian countries: A focused review on contamination status, pollution sources, potential risks, and future perspectives. *Science of the Total Environment*, *764*.

Ansari, S., & Karimi, M. (2017). Novel developments and trends of analytical methods for drug analysis in biological and environmental samples by molecularly imprinted polymers. *Trends in Analytical Chemistry*, *89*, 146–162.

Antwi, F. B., Shama, L. M., & Peterson, R. K. (2008). Risk assessments for the insect repellents DEET and picaridin. *Regulatory Toxicology and Pharmacology*, *51*, 31–36.

Aronson, D., Weeks, J., Meylan, B., Guiney, P. D., & Howard, P. H. (2012). Environmental release, environmental concentrations, and ecological risk of N, N-diethyl-m-toluamide (DEET). *Integrated Environmental Assessment and Management*, *8*, 135–166.

Arpin-Pont, L., Bueno, M. J. M., Gomez, E., & Fenet, H. (2016). Occurrence of PPCPs in the marine environment: A review. *Environmental Science and Pollution Research*, *23*, 4978–4991.

Awad, Y. M., Kim, S.-C., Abd El-Azeem, S. A., Kim, K.-H., Kim, K.-R., Kim, K., Jeon, C., Lee, S. S., & Ok, Y. S. (2014). Veterinary antibiotics contamination in water, sediment, and soil near a swine manure composting facility. *Environmental Earth Sciences*, *71*, 1433–1440.

Ballesteros-Gómez, A., & Rubio, S. (2011). Recent advances in environmental analysis. *Analytical chemistry*, *83*(12), 4579–4613.

Barceló, D., & Petrovic, M. (2007). *Pharmaceuticals and personal care products (PPCPs) in the environment*. Springer.

Barnes, K. K., Christenson, S. C., Kolpin, D. W., Focazio, M. J., Furlong, E. T., Zaugg, S. D., Meyer, M. T., & Barber, L. B. (2004). Pharmaceuticals and other organic waste water contaminants within a leachate plume downgradient of a municipal landfill. *Groundwater Monitoring & Remediation*, *24*, 119–126.

Barnes, K. K., Kolpin, D. W., Furlong, E. T., Zaugg, S. D., Meyer, M. T., & Barber, L. B. (2008). A national reconnaissance of pharmaceuticals and other organic wastewater contaminants in the United States—I) Groundwater. *Science of the Total Environment*, *402*, 192–200.

Barnes, K. K., Kolpin, D. W., Furlong, E. T., Zaugg, S. D., Meyer, M. T., Barber, L. B., & Focazio, M. J. (2005). Studies examine contaminants: Pharmaceuticals, hormones and other organic wastewater contaminants in ground water resources. *National Driller*, *26*, 38–39.

Bartelt-Hunt, S., Snow, D. D., Damon-Powell, T., & Miesbach, D. (2011). Occurrence of steroid hormones and antibiotics in shallow groundwater impacted by livestock waste control facilities. *Journal of Contaminant Hydrology*, *123*, 94–103.

Batley, G. (1999). Quality assurance in environmental monitoring. *Marine Pollution Bulletin*, *39*, 23–31.

Batt, A. L., Snow, D. D., & Aga, D. S. (2006). Occurrence of sulfonamide antimicrobials in private water wells in Washington County, Idaho, USA. *Chemosphere*, *64*, 1963–1971.

Bell, J. W., Veltri, J. C., & Page, B. C. (2002). Human exposures to N,N-diethyl-m-toluamide insect repellents reported to the American Association of Poison Control Centers 1993–1997. *International Journal of Toxicology*, *21*, 341–352.

Benitez, F. J., Acero, J. L., Garcia-Reyes, J. F., Real, F. J., Roldan, G., Rodriguez, E., & Molina-Díaz, A. (2013). Determination of the reaction rate constants and decomposition mechanisms of ozone with two model emerging contaminants: DEET and nortriptyline. *Industrial & Engineering Chemistry Research*, *52*, 17064–17073.

Berge, A., Bulete, A., Fildier, A., Maiiler, R., Gasperi, J., Coquet, Y., Nauleau, F., Rocher, V., & Vulliet, E. (2018). Non-target strategies by HRMS to evaluate fluidized micro-grain activated carbon as a tertiary treatment of wastewater. *Chemosphere*, *213*, 587–595.

Białk-Bielińska, A., Kumirska, J., Borecka, M., Caban, M., Paszkiewicz, M., Pazdro, K., & Stepnowski, P. (2016). Selected analytical challenges in the determination of pharmaceuticals in drinking/marine waters and soil/sediment samples. *Journal of Pharmaceutical and Biomedical Analysis*, *121*, 271–296.

Białk-Bielińska, A., Maszkowska, J., Puckowski, A., & Stepnowski, P. (2014). *Exposure and hazard identification of sulphonamides in the terrestrial environment. Environmental Risk Assessment of Soil Contamination*. IntechOpen.

Bizkarguenaga, E., Zabaleta, I., Mijangos, L., Iparraguirre, A., Fernandez, L. A., Prieto, A., & Zuloaga, O. (2016). Uptake of perfluorooctanoic acid, perfluorooctane sulfonate and perfluorooctane sulfonamide by carrot and lettuce from compost amended soil. *Science of the Total Environment*, *571*, 444–451.

Bound, J. P., & Voulvoulis, N. (2005). Household disposal of pharmaceuticals as a pathway for aquatic contamination in the United Kingdom. *Environmental Health Perspectives, 113*, 1705–1711.

Boxall, A. B., Blackwell, P., Cavallo, R., Kay, P., & Tolls, J. (2002). The sorption and transport of a sulphonamide antibiotic in soil systems. *Toxicology Letters, 131*, 19–28.

Boxall, A. B., Rudd, M. A., Brooks, B. W., Caldwell, D. J., Choi, K., Hickmann, S., Innes, E., Ostapyk, K., Staveley, J. P., & Verslycke, T. (2012). Pharmaceuticals and personal care products in the environment: What are the big questions? *Environmental Health Perspectives, 120*, 1221–1229.

Brausch, J. M., & Rand, G. M. (2011). A review of personal care products in the aquatic environment: Environmental concentrations and toxicity. *Chemosphere, 82*, 1518–1532.

Bîlková, Z., Malá, J., & Hrich, K. (2019). Fate and behaviour of veterinary sulphonamides under denitrifying conditions. *Science of the Total Environment, 695*, 133824.

Carballa, M., Omil, F., & Lema, J. M. (2005). Removal of cosmetic ingredients and pharmaceuticals in sewage primary treatment. *Water Research, 39*(19), 4790–4796.

Carballa, M., Omil, F., Lema, J. M., Llompart, Ma, García-Jares, C., Rodríguez, I., Gomez, M., & Ternes, T. (2004). Behavior of pharmaceuticals, cosmetics and hormones in a sewage treatment plant. *Water Research, 38*, 2918–2926.

Careghini, A., Mastorgio, A. F., Saponaro, S., & Sezenna, E. (2015). Bisphenol A, nonylphenols, benzophenones, and benzotriazoles in soils, groundwater, surface water, sediments, and food: A review. *Environmental Science and Pollution Research, 22*, 5711–5741.

Chalew, T. E., & Halden, R. U. (2009). Environmental exposure of aquatic and terrestrial biota to triclosan and triclocarban 1. *Journal of the American Water Resources Association, 45*, 4–13.

Chaturvedi, P., Shukla, P., Giri, B. S., Chowdhary, P., Chandra, R., Gupta, P., & Pandey, A. (2021). Prevalence and hazardous impact of pharmaceutical and personal care products and antibiotics in environment: A review on emerging contaminants. *Environmental Research, 194*.

Chen, H., Gao, B., Li, H., & Ma, L. Q. (2011). Effects of pH and ionic strength on sulfamethoxazole and ciprofloxacin transport in saturated porous media. *Journal of Contaminant Hydrology, 126*, 29–36.

Chen-Hussey, V., Behrens, R., & Logan, J. G. (2014). Assessment of methods used to determine the safety of the topical insect repellent N,N-diethyl-m-toluamide (DEET). *Parasites & Vectors, 7*, 1–7.

Christou, A., Aguera, A., Bayona, J. M., Cytryn, E., Fotopoulos, V., Lambropoulou, D., Manaia, C. M., Michael, C., Revitt, M., Schroder, P., & Fatta-Kassinos, D. (2017). The potential implications of reclaimed wastewater reuse for irrigation on the agricultural environment: The knowns and unknowns of the fate of antibiotics and antibiotic resistant bacteria and resistance genes - A review. *Water Research, 123*, 448–467.

Cizmas, L., Sharma, V. K., Gray, C. M., & McDonald, T. J. (2015). Pharmaceuticals and personal care products in waters: Occurrence, toxicity, and risk. *Environmental Chemistry Letters, 13*, 381–394.

Clarke, B. O., Anumol, T., Barlaz, M., & Snyder, S. A. (2015). Investigating landfill leachate as a source of trace organic pollutants. *Chemosphere, 127*, 269–275.

Conn, K. E., Barber, L. B., Brown, G. K., & Siegrist, R. L. (2006). Occurrence and fate of organic contaminants during onsite wastewater treatment. *Environmental Science & Technology, 40*, 7358–7366.

Cordy, G. E., Duran, N. L., Bouwer, H., Rice, R. C., Furlong, E. T., Zaugg, S. D., Meyer, M. T., Barber, L. B., & Kolpin, D. W. (2004). Do pharmaceuticals, pathogens, and other organic waste water compounds persist when waste water is used for recharge? *Groundwater Monitoring & Remediation*, *24*, 58–69.

Coronado, M., De Haro, H., Deng, X., Rempel, M. A., Lavado, R., & Schlenk, D. (2008). Estrogenic activity and reproductive effects of the UV-filter oxybenzone (2-hydroxy-4-methoxyphenyl-methanone) in fish. *Aquatic Toxicology*, *90*, 182–187.

Costanzo, S., Watkinson, A., Murby, E., Kolpin, D. W., & Sandstrom, M. W. (2007). Is there a risk associated with the insect repellent DEET (N, N-diethyl-m-toluamide) commonly found in aquatic environments? *Science of the Total Environment*, *384*, 214–220.

Cripe, G. M., Hemmer, B. L., Goodman, L. R., Fournie, J. W., Raimondo, S., Vennari, J. C., Danner, R. L., Smith, K., Manfredonia, B. R., & Kulaw, D. H. (2009). Multigenerational exposure of the estuarine sheepshead minnow (*Cyprinodon variegatus*) to 17β-estradiol. I. Organism-level effects over three generations. *Environmental Toxicology and Chemistry: An International Journal*, *28*, 2397–2408.

Dai, Y. J., Liu, M., Li, J. J., Yang, S. S., Sun, Y., Sun, Q. Y., Wang, W. S., Lu, L., Zhang, K. X., Xu, J. Y., Zheng, W. L., Hu, Z. Y., Yang, Y. H., Gao, Y. W., & Liu, Z. H. (2020). A review on pollution situation and treatment methods of tetracycline in groundwater. *Separation Science and Technology*, *55*, 1005–1021.

Dammann, A., Shappell, N., Bartell, S., & Schoenfuss, H. (2011). Comparing biological effects and potencies of estrone and 17β-estradiol in mature fathead minnows, *Pimephales promelas*. *Aquatic Toxicology*, *105*, 559–568.

Daughton, C. G. (2003). Cradle-to-cradle stewardship of drugs for minimizing their environmental disposition while promoting human health. II. Drug disposal, waste reduction, and future directions. *Environmental Health Perspectives*, *111*, 775–785.

de Fatima Alpendurada, M. (2000). Solid-phase microextraction: a promising technique for sample preparation in environmental analysis. *Journal of Chromatography. A*, *889*, 3–14.

De Liguoro, M., Fioretto, B., Poltronieri, C., & Gallina, G. (2009). The toxicity of sulfamethazine to *Daphnia magna* and its additivity to other veterinary sulfonamides and trimethoprim. *Chemosphere*, *75*, 1519–1524.

Devarajan, N., Laffite, A., Graham, N. D., Meijer, M., Prabakar, K., Mubedi, J. I., Elongo, V., Mpiana, P. T., Ibelings, B. W., & Wildi, W. (2015). Accumulation of clinically relevant antibiotic-resistance genes, bacterial load, and metals in freshwater lake sediments in Central Europe. *Environmental Science & Technology*, *49*, 6528–6537.

Dimpe, K. M., & Nomngongo, P. N. (2016). Current sample preparation methodologies for analysis of emerging pollutants in different environmental matrices. *TrAC Trends in Analytical Chemistry*, *82*, 199–207.

Dodgen, L. K., Ueda, A., Wu, X. Q., Parker, D. R., & Gan, J. (2015). Effect of transpiration on plant accumulation and translocation of PPCP/EDCs. *Environmental Pollution*, *198*, 144–153.

Dong, S. N., Gao, B., Sun, Y. Y., Shi, X. Q., Xu, H. X., Wu, J. F., & Wu, J. C. (2016). Transport of sulfacetamide and levofloxacin in granular porous media under various conditions: Experimental observations and model simulations. *Science of the Total Environment*, *573*, 1630–1637.

Du, L., & Liu, W. (2012). Occurrence, fate, and ecotoxicity of antibiotics in agro-ecosystems. A review. *Agronomy for Sustainable Development*, *32*, 309–327.

Eggen, T., Moeder, M., & Arukwe, A. (2010). Municipal landfill leachates: A significant source for new and emerging pollutants. *Science of the Total Environment*, *408*, 5147–5157.

Esplugas, S., Bila, D. M., Krause, L. G. T., & Dezotti, M. (2007). Ozonation and advanced oxidation technologies to remove endocrine disrupting chemicals (EDCs) and pharmaceuticals and personal care products (PPCPs) in water effluents. *Journal of Hazardous Materials*, *149*, 631–642.

Environmental Working Group. (2008). Study: Almost Half Of All 'Natural'Personal Care Products Contain Known Carcinogen.

European Medicines Agency, E.S.O.V.A.C. (2018). Sales of veterinary antimicrobial agents in 30 European countries in 2016.

Evans, S. E., & Kasprzyk-Hordern, B. (2014). Applications of chiral chromatography coupled with mass spectrometry in the analysis of chiral pharmaceuticals in the environment. *Trends in Environmental Analytical Chemistry*, *1*, e34–e51.

Evgenidou, E. N., Konstantinou, I. K., & Lambropoulou, D. A. (2015). Occurrence and removal of transformation products of PPCPs and illicit drugs in wastewaters: A review. *Science of the Total Environment*, *505*, 905–926.

Fent, K., Zenker, A., & Rapp, M. (2010). Widespread occurrence of estrogenic UV-filters in aquatic ecosystems in Switzerland. *Environmental Pollution*, *158*, 1817–1824.

Fick, J., Söderström, H., Lindberg, R. H., Phan, C., Tysklind, M., & Larsson, D. J. (2009). Contamination of surface, ground, and drinking water from pharmaceutical production. *Environmental Toxicology and Chemistry / SETAC*, *28*, 2522–2527.

Flippin, J. L., Huggett, D., & Foran, C. M. (2007). Changes in the timing of reproduction following chronic exposure to ibuprofen in Japanese medaka, *Oryzias latipes*. *Aquatic Toxicology*, *81*, 73–78.

Fram, M. S., & Belitz, K. (2011). Occurrence and concentrations of pharmaceutical compounds in groundwater used for public drinking-water supply in California. *Science of the Total Environment*, *409*, 3409–3417.

French, V. A., King, S. C., Kumar, A., Northcott, G., McGuinness, K., & Parry, D. (2015). Characterisation of microcontaminants in Darwin Harbour, a tropical estuary of northern Australia undergoing rapid development. *Science of the Total Environment*, *536*, 639–647.

Fujimori, C., Ogiwara, K., Hagiwara, A., Rajapakse, S., Kimura, A., & Takahashi, T. (2011). Expression of cyclooxygenase-2 and prostaglandin receptor EP4b mRNA in the ovary of the medaka fish, *Oryzias latipes*: Possible involvement in ovulation. *Molecular and Cellular Endocrinology*, *332*, 67–77.

Galus, M., Kirischian, N., Higgins, S., Purdy, J., Chow, J., Rangaranjan, S., Li, H., Metcalfe, C., & Wilson, J. Y. (2013). Chronic, low concentration exposure to pharmaceuticals impacts multiple organ systems in zebrafish. *Aquatic Toxicology*, *132*, 200–211.

García-Galán, M. J., Díaz-Cruz, M. S., & Barceló, D. (2010). Determination of 19 sulfonamides in environmental water samples by automated on-line solid-phase extraction-liquid chromatography–tandem mass spectrometry (SPE-LC–MS/MS). *Talanta*, *81*, 355–366.

García-Galán, M. J., Garrido, T., Fraile, J., Ginebreda, A., Díaz-Cruz, M. S., & Barceló, D. (2010). Simultaneous occurrence of nitrates and sulfonamide antibiotics in two ground water bodies of Catalonia (Spain). *Journal of Hydrology*, *383*, 93–101.

García-Galán, M. J., Garrido, T., Fraile, J., Ginebreda, A., Díaz-Cruz, M. S., & Barceló, D. (2011). Application of fully automated online solid phase extraction-liquid chromatography-electrospray-tandem mass spectrometry for the determination of sulfonamides and their acetylated metabolites in groundwater. *Analytical and Bioanalytical Chemistry*, *399*, 795–806.

Glassmeyer, S. T., Furlong, E. T., Kolpin, D. W., Cahill, J. D., Zaugg, S. D., Werner, S. L., Meyer, M. T., & Kryak, D. D. (2005). Transport of chemical and microbial compounds from known wastewater discharges: Potential for use as indicators of human fecal contamination. *Environmental Science & Technology*, *39*, 5157–5169.

Gottschall, N., Topp, E., Edwards, M., Payne, M., Kleywegt, S., Russell, P., & Lapen, D. R. (2013). Hormones, sterols, and fecal indicator bacteria in groundwater, soil, and subsurface drainage following a high single application of municipal biosolids to a field. *Chemosphere, 91*, 275–286.

Gottschall, N., Topp, E., Metcalfe, C., Edwards, M., Payne, M., Kleywegt, S., Russell, P., & Lapen, D. R. (2012). Pharmaceutical and personal care products in groundwater, subsurface drainage, soil, and wheat grain, following a high single application of municipal biosolids to a field. *Chemosphere, 87*, 194–203.

Goulas, A., Livoreil, B., Grall, N., Benoit, P., Couderc-Obert, C., Dagot, C., Patureau, D., Petit, F., Laouenan, C., & Andremont, A. (2019). What are the effective solutions to control the dissemination of antibiotic resistance in the environment? A systematic review protocol. (vol. 7, 3, 2018) *Environmental Evidence, 8*.

Grosser, T., Yusuff, S., Cheskis, E., Pack, M. A., & FitzGerald, G. A. (2002). Developmental expression of functional cyclooxygenases in zebrafish. *Proceedings of the National Academy of Sciences, 99*, 8418–8423.

Gu, D. M., Feng, Q. Y., Guo, C. S., Hou, S., Lv, J. P., Zhang, Y., Yuan, S., & Zhao, X. (2019). Occurrence and risk assessment of antibiotics in manure, soil, wastewater, groundwater from livestock and poultry farms in Xuzhou, China. *Bulletin of Environmental Contamination and Toxicology, 103*(4), 590–596. Available from https://dx.doi.org/10.1007/s00128-019-02692-0.

Gu, C., & Karthikeyan, K. (2008). Sorption of the antibiotic tetracycline to humic-mineral complexes. *Journal of Environmental Quality, 37*, 704–711.

Guardiola, J., Ventura, F., & Rivera, J. (1989). Occurrence of industrial organic pollution in a groundwater supply: Screening, monitoring and evaluation of treatment processes. *Water Supply, 7*, 11–16.

Gunnarsson, L., Adolfsson-Erici, M., Björlenius, B., Rutgersson, C., Förlin, L., & Larsson, D. (2009). Comparison of six different sewage treatment processes—reduction of estrogenic substances and effects on gene expression in exposed male fish. *Science of the Total Environment, 407*, 5235–5242.

Hadad, G. M., Abdel-Salam, R. A., & Emara, S. (2011). Optimized and validated flow-injection spectrophotometric analysis of topiramate, piracetam and levetiracetam in pharmaceutical formulations. *Acta Pharmaceutica, 61*, 377–389.

Hamscher, G., Pawelzick, H. T., Höper, H., & Nau, H. (2005). Different behavior of tetracyclines and sulfonamides in sandy soils after repeated fertilization with liquid manure. *Environmental Toxicology and Chemistry: An International Journal, 24*, 861–868.

Han, S., Choi, K., Kim, J., Ji, K., Kim, S., Ahn, B., Yun, J., Choi, K., Khim, J. S., & Zhang, X. (2010). Endocrine disruption and consequences of chronic exposure to ibuprofen in Japanese medaka (*Oryzias latipes*) and freshwater cladocerans *Daphnia magna* and *Moina macrocopa*. *Aquatic Toxicology, 98*, 256–264.

Hardell, L., Näsman, A., Ohlson, C.-G., & Fredrikson, M. (1998). Case-control study on risk factors for testicular cancer. *International Journal of Oncology, 13*, 1299–1602.

Heberer, T., Massmann, G., Fanck, B., Taute, T., & Dünnbier, U. (2008). Behaviour and redox sensitivity of antimicrobial residues during bank filtration. *Chemosphere, 73*, 451–460.

Hendriks, A., Maas-Diepeveen, J., Noordsij, A., & Van der Gaag, M. (1994). Monitoring response of XAD-concentrated water in the Rhine delta: A major part of the toxic compounds remains unidentified. *Water Research, 28*, 581–598.

Hirsch, R., Ternes, T., Haberer, K., & Kratz, K.-L. (1999). Occurrence of antibiotics in the aquatic environment. *Science of the Total Environment, 225*, 109–118.

Holm, J. V., Ruegge, K., Bjerg, P. L., & Christensen, T. H. (1995). Occurrence and distribution of pharmaceutical organic compounds in the groundwater downgradient of a landfill (Grindsted, Denmark). *Environmental Science & Technology*, *29*, 1415–1420.

Hu, X., Zhou, Q., & Luo, Y. (2010). Occurrence and source analysis of typical veterinary antibiotics in manure, soil, vegetables and groundwater from organic vegetable bases, northern China. *Environmental Pollution*, *158*, 2992–2998.

Huang, C. H., Tang, Z. R., Xi, B. D., Tan, W. B., Guo, W., Wu, W. X., & Ma, C. Y. (2021). Environmental effects and risk control of antibiotic resistance genes in the organic solid waste aerobic composting system: A review. *Frontiers of Environmental Science & Engineering*, *15*(6), 2095–2201.

Huber, M. M., Canonica, S., Park, G.-Y., & Von Gunten, U. (2003). Oxidation of pharmaceuticals during ozonation and advanced oxidation processes. *Environmental Science & Technology*, *37*, 1016–1024.

Imai, S., Koyama, J., & Fujii, K. (2005). Effects of 17β-estradiol on the reproduction of Java-medaka (*Oryzias javanicus*), a new test fish species. *Marine Pollution Bulletin*, *51*, 708–714.

Imai, S., Koyama, J., & Fujii, K. (2007). Effects of estrone on full life cycle of Java medaka (*Oryzias javanicus*), a new marine test fish. *Environmental Toxicology and Chemistry: An International Journal*, *26*, 726–731.

Ishibashi, H., Matsumura, N., Hirano, M., Matsuoka, M., Shiratsuchi, H., Ishibashi, Y., Takao, Y., & Arizono, K. (2004). Effects of triclosan on the early life stages and reproduction of medaka *Oryzias latipes* and induction of hepatic vitellogenin. *Aquatic Toxicology*, *67*, 167–179.

Ji, K., Liu, X., Lee, S., Kang, S., Kho, Y., Giesy, J. P., & Choi, K. (2013). Effects of non-steroidal anti-inflammatory drugs on hormones and genes of the hypothalamic-pituitary-gonad axis, and reproduction of zebrafish. *Journal of Hazardous Materials*, *254*, 242–251.

Jin, C. X., Wei, S., Sun, R. L., Zou, W., Zhang, X. L., Zhou, Q. X., Liu, R., & Huang, L. Q. (2020). The forms, distribution, and risk assessment of sulfonamide antibiotics in the manure-soil-vegetable system of feedlot livestock. *The Bulletin of Environmental Contamination and Toxicology*, *105*, 790–797.

Jones, O. A. H., Voulvoulis, N., & Lester, J. N. (2005). Human pharmaceuticals in wastewater treatment processes. *Critical Reviews in Environmental Science and Technology*, *35*, 401–427.

Kaszab, E., Rado, J., Kriszt, B., Paszti, J., Lesinszki, V., Szabo, A., Toth, G., Khaledi, A., & Szoboszlay, S. (2019). Groundwater, soil and compost, as possible sources of virulent and antibiotic-resistant *Pseudomonas aeruginosa*. *International Journal of Environmental Health Research*.

Keerthanan, S., Jayasinghe, C., Biswas, J. K., & Vithanage, M. (2021). Pharmaceutical and personal care products (PPCPs) in the environment: Plant uptake, translocation, bioaccumulation, and human health risks. *Critical Reviews in Environmental Science and Technology*, *51*, 1221–1258.

Khelladi, R. M. B., Fellah, A. C., Pontie, M., Shabani, M., & Guellil, F. Z. (2020). Performance of sand and granular activated carbon filtration coupling in tertiary urban wastewater treatment in Algeria. *Desalination and Water Treatment*, *205*, 111–123.

Kim, J.-W., Jang, H.-S., Kim, J.-G., Ishibashi, H., Hirano, M., Nasu, K., Ichikawa, N., Takao, Y., Shinohara, R., & Arizono, K. (2009). Occurrence of pharmaceutical and personal care products (PPCPs) in surface water from Mankyung River, South Korea. *Journal of Health Science*, *55*, 249–258.

Kim, S., Jung, D., Kho, Y., & Choi, K. (2014). Effects of benzophenone-3 exposure on endocrine disruption and reproduction of Japanese medaka (*Oryzias latipes*)—A two generation exposure study. *Aquatic Toxicology, 155*, 244–252.

Kim, K., Owens, G., Kwon, S., So, K., Lee, D., & Ok, Y. (2011). Occurrence and environmental fate of veterinary antibiotics in the terrestrial environment. *Water, Air, & Soil Pollution, 214* (1), 163–174.

Kime, D. E. (1999). A strategy for assessing the effects of xenobiotics on fish reproduction. *Science of the Total Environment, 225*, 3–11.

Kimura, K., Hara, H., & Watanabe, Y. (2007). Elimination of selected acidic pharmaceuticals from municipal wastewater by an activated sludge system and membrane bioreactors. *Environmental science & technology, 41*(10), 3708–3714.

Kinney, C. A., Furlong, E. T., Werner, S. L., & Cahill, J. D. (2006). Presence and distribution of wastewater-derived pharmaceuticals in soil irrigated with reclaimed water. *Environmental Toxicology and Chemistry: An International Journal, 25*, 317–326.

Kivits, T., Broers, H. P., Beeltje, H., Van Vliet, M., & Griffioen, J. (2018). Presence and fate of veterinary antibiotics in age-dated groundwater in areas with intensive livestock farming. *Environmental Pollution, 241*, 988–998.

Klavarioti, M., Mantzavinos, D., & Kassinos, D. (2009). Removal of residual pharmaceuticals from aqueous systems by advanced oxidation processes. *Environment International, 35*, 402–417.

Knepper, T. (2004). Analysis and fate of insect repellents. *Water Science and Technology: A Journal of the International Association on Water Pollution Research, 50*, 301–308.

Kolpin, D. W., Furlong, E. T., Meyer, M. T., Thurman, E. M., Zaugg, S. D., Barber, L. B., & Buxton, H. T. (2002). Pharmaceuticals, hormones, and other organic wastewater contaminants in United States streams, 1999 – 2000: A national reconnaissance. *Environmental Science & Technology, 36*, 1202–1211.

Kolpin, D. W., Skopec, M., Meyer, M. T., Furlong, E. T., & Zaugg, S. D. (2004). Urban contribution of pharmaceuticals and other organic wastewater contaminants to streams during differing flow conditions. *Science of the Total Environment, 328*, 119–130.

Kostich, M. S., Batt, A. L., & Lazorchak, J. M. (2014). Concentrations of prioritized pharmaceuticals in effluents from 50 large wastewater treatment plants in the United States and implications for risk estimation. *Environmental Pollution, 184*, 354–359.

Kovalakova, P., Cizmas, L., McDonald, T. J., Marsalek, B., Feng, M. B., & Sharma, V. K. (2020). Occurrence and toxicity of antibiotics in the aquatic environment: A review. *Chemosphere, 251*.

Kunz, P. Y., Gries, T., & Fent, K. (2006). The ultraviolet filter 3-benzylidene camphor adversely affects reproduction in fathead minnow (*Pimephales promelas*). *Toxicological Sciences, 93*, 311–321.

Lambropoulou, D. A., & Nollet, L. M. (2014). *Transformation products of emerging contaminants in the environment: Analysis, processes, occurrence, effects and risks*. John Wiley & Sons.

Li, X. D., Yu, H. X., Xu, S. S., & Hua, R. M. (2013). Uptake of three sulfonamides from contaminated soil by pakchoi cabbage. *Ecotoxicology and Environmental Safety, 92*, 297–302.

Limbu, S. M., Chen, L. Q., Zhang, M. L., & Du, Z. Y. (2021). A global analysis on the systemic effects of antibiotics in cultured fish and their potential human health risk: a review. *Reviews in Aquaculture, 13*(2), 1015–1059.

Lin, T., Yu, S., & Chen, W. (2016). Occurrence, removal and risk assessment of pharmaceutical and personal care products (PPCPs) in an advanced drinking water treatment plant (ADWTP) around Taihu Lake in China. *Chemosphere, 152*, 1–9.

Liu, F., Ying, G.-G., Yang, J.-F., Zhou, L.-J., Tao, R., Wang, L., Zhang, L.-J., & Peng, P.-A. (2010). Dissipation of sulfamethoxazole, trimethoprim and tylosin in a soil under aerobic and anoxic conditions. *Environmental Chemistry*, *7*, 370–376.

Loos, R., Locoro, G., Comero, S., Contini, S., Schwesig, D., Werres, F., Balsaa, P., Gans, O., Weiss, S., & Blaha, L. (2010). Pan-European survey on the occurrence of selected polar organic persistent pollutants in ground water. *Water Research*, *44*, 4115–4126.

Luo, Y., Guo, W., Ngo, H. H., Nghiem, L. D., Hai, F. I., Zhang, J., Liang, S., & Wang, X. C. (2014). A review on the occurrence of micropollutants in the aquatic environment and their fate and removal during wastewater treatment. *Science of the Total Environment*, *473*, 619–641.

Lyu, J., Yang, L. S., Zhang, L., Ye, B. X., & Wang, L. (2020). Antibiotics in soil and water in China-a systematic review and source analysis. *Environmental Pollution*, *266*.

Ma, J., Lin, H., Sun, W., Wang, Q., Yu, Q., Zhao, Y., & Fu, J. (2014). Soil microbial systems respond differentially to tetracycline, sulfamonomethoxine, and ciprofloxacin entering soil under pot experimental conditions alone and in combination. *Environmental Science and Pollution Research*, *21*, 7436–7448.

Madrid, Y., & Zayas, Z. P. (2007). Water sampling: Traditional methods and new approaches in water sampling strategy. *Trends in Analytical Chemistry*, *26*, 293–299.

Mailler, R., Gasperi, J., Coquet, Y., Bulete, A., Vulliet, E., Deshayes, S., Zedek, S., Mirande-Bret, C., Eudes, V., Bressy, A., Caupos, E., Moilleron, R., Chebbo, G., & Rocher, V. (2016). Removal of a wide range of emerging pollutants from wastewater treatment plant discharges by micro-grain activated carbon in fluidized bed as tertiary treatment at large pilot scale. *Science of the Total Environment*, *542*, 983–996.

Mailler, R., Gasperi, J., Coquet, Y., Deshayes, S., Zedek, S., Cren-Olive, C., Cartiser, N., Eudes, V., Bressy, A., Caupos, E., Moilleron, R., Chebbo, G., & Rocher, V. (2015). Study of a large scale powdered activated carbon pilot: Removals of a wide range of emerging and priority micropollutants from wastewater treatment plant effluents. *Water Research*, *72*, 315–330.

Martínez, J. L. (2008). Antibiotics and antibiotic resistance genes in natural environments. *Science (New York, N.Y.)*, *321*, 365–367.

McClellan, K., & Halden, R. U. (2010). Pharmaceuticals and personal care products in archived United States biosolids from the 2001 EPA national sewage sludge survey. *Water Research*, *44*, 658–668.

Merel, S., & Snyder, S. A. (2016). Critical assessment of the ubiquitous occurrence and fate of the insect repellent N, N-diethyl-m-toluamide in water. *Environment International*, *96*, 98–117.

Mirali, M., Ambrose, S. J., Wood, S. A., Vandenberg, A., & Purves, R. W. (2014). Development of a fast extraction method and optimization of liquid chromatography–mass spectrometry for the analysis of phenolic compounds in lentil seed coats. *Journal of Chromatography B*, *969*, 149–161.

Mompelat, S., Le Bot, B., & Thomas, O. (2009). Occurrence and fate of pharmaceutical products and by-products, from resource to drinking water. *Environment International*, *35*, 803–814.

Morthorst, J. E., Lister, A., Bjerregaard, P., & Van Der Kraak, G. (2013). Ibuprofen reduces zebrafish PGE2 levels but steroid hormone levels and reproductive parameters are not affected. *Comparative Biochemistry and Physiology Part C: Toxicology & Pharmacology*, *157*, 251–257.

Musson, S. E., & Townsend, T. G. (2009). Pharmaceutical compound content of municipal solid waste. *Journal of Hazardous Materials*, *162*, 730–735.

Muthanna, T., & Plosz, B. (2008). The impact of hospital sewage discharge on the assessment of environmental risk posed by priority pharmaceuticals: Hydrodynamic modelling and measurements. In *International Conference on Urban Drainage* (pp. 1–10). Edinburgh, Scotland, United Kingdom.

Nghiem, L. D., Schäfer, A. I., & Elimelech, M. (2004). Removal of natural hormones by nanofiltration membranes: Measurement, modeling, and mechanisms. *Environmental Science & Technology, 38*, 1888–1896.

Novo, M., Verdú, I., Trigo, D., & Martínez-Guitarte, J. (2018). Endocrine disruptors in soil: Effects of bisphenol A on gene expression of the earthworm *Eisenia fetida. Ecotoxicology and Environmental Safety, 150*, 159–167.

Oluwole, A. O., Omotola, E. O., & Olatunji, O. S. (2020). Pharmaceuticals and personal care products in water and wastewater: A review of treatment processes and use of photocatalyst immobilized on functionalized carbon in AOP degradation. *BMC Chemistry, 14*.

Oros, D. R., Jarman, W. M., Lowe, T., David, N., Lowe, S., & Davis, J. A. (2003). Surveillance for previously unmonitored organic contaminants in the San Francisco Estuary. *Marine Pollution Bulletin, 46*, 1102–1110.

Ostermann, A., Siemens, J., Welp, G., Xue, Q., Lin, X., Liu, X., & Amelung, W. (2013). Leaching of veterinary antibiotics in calcareous Chinese croplands. *Chemosphere, 91*, 928–934.

Overturf, M. D., Anderson, J. C., Pandelides, Z., Beyger, L., & Holdway, D. A. (2015). Pharmaceuticals and personal care products: A critical review of the impacts on fish reproduction. *Critical Reviews in Toxicology, 45*, 469–491.

Pan, B., Ning, P., & Xing, B. (2009). Part V—Sorption of pharmaceuticals and personal care products. *Environmental Science and Pollution Research, 16*, 106–116.

Papageorgiou, A., Voutsa, D., & Papadakis, N. (2014). Occurrence and fate of ozonation by-products at a full-scale drinking water treatment plant. *Science of the Total Environment, 481*, 392–400.

Papich, M., & Riviera, J. (2009). *Sulfonamides and potentiated sulfonamides. Veterinary pharmacology and therapeutics* (pp. 835–863). United States: Wiley-Blackwell.

Patel, M., Kumar, R., Kishor, K., Mlsna, T., Pittman, C. U., Jr, & Mohan, D. (2019). Pharmaceuticals of emerging concern in aquatic systems: Chemistry, occurrence, effects, and removal methods. *Chemical Reviews, 119*, 3510–3673.

Pazda, M., Kumirska, J., Stepnowski, P., & Mulkiewicz, E. (2019). Antibiotic resistance genes identified in wastewater treatment plant systems - A review. *Science of the Total Environment, 697*.

Pérez-Fernández, V., Rocca, L. M., Tomai, P., Fanali, S., & Gentili, A. (2017). Recent advancements and future trends in environmental analysis: Sample preparation, liquid chromatography and mass spectrometry. *Analytica Chimica Acta, 983*, 9–41.

Pérez-Fernández, V., Rocca, L. M., Tomai, P., Fanali, S., & Gentili, A. (2017). Recent advancements and future trends in environmental analysis: Sample preparation, liquid chromatography and mass spectrometry. *Analytica chimica acta, 983*, 9–41.

Perez-Lemus, N., Lopez-Serna, R., Perez-Elvira, S. I., & Barrado, E. (2019). Analytical methodologies for the determination of pharmaceuticals and personal care products (PPCPs) in sewage sludge: A critical review. *Analytica Chimica Acta, 1083*, 19–40.

Petrovic, M., Farré, M., De Alda, M. L., Perez, S., Postigo, C., Köck, M., Radjenovic, J., Gros, M., & Barcelo, D. (2010). Recent trends in the liquid chromatography–mass spectrometry analysis of organic contaminants in environmental samples. *Journal of Chromatography A, 1217*(25), 4004–4017.

Petrović, M., Gonzalez, S., & Barceló, D. (2003). Analysis and removal of emerging contaminants in wastewater and drinking water. *Trends in Analytical Chemistry*, *22*, 685–696.

Poté, J., Ceccherini, M. T., Rosselli, W., Wildi, W., Simonet, P., & Vogel, T. M. (2003). Fate and transport of antibiotic resistance genes in saturated soil columns. *European Journal of Soil Biology*, *39*, 65–71.

Price, O. R., Hughes, G. O., Roche, N. L., & Mason, P. J. (2010). Improving emissions estimates of home and personal care products ingredients for use in EU risk assessments. *Integrated Environmental Assessment and Management*, *6*, 677–684.

Petrović, M., Hernando, M. D., Díaz-Cruz, M. S., & Barceló, D. (2005). Liquid chromatography–tandem mass spectrometry for the analysis of pharmaceutical residues in environmental samples: a review. *Journal of Chromatography A*, *1067*(1-2), 1–14.

Płotka-Wasylka, J., Szczepańska, N., de La Guardia, M., & Namieśnik, J. (2016). Modern trends in solid phase extraction: New sorbent media. *Trends in Analytical Chemistry*, *77*, 23–43.

Quednow, K., & Püttmann, W. (2009). Temporal concentration changes of DEET, TCEP, terbutryn, and nonylphenols in freshwater streams of Hesse, Germany: Possible influence of mandatory regulations and voluntary environmental agreements. *Environmental Science and Pollution Research*, *16*, 630–640.

Radjenović, J., Petrović, M., Ventura, F., & Barceló, D. (2008). Rejection of pharmaceuticals in nanofiltration and reverse osmosis membrane drinking water treatment. *Water Research*, *42*, 3601–3610.

Rajapaksha, A. U., Vithanage, M., Lim, J. E., Ahmed, M. B. M., Zhang, M., Lee, S. S., & Ok, Y. S. (2014). Invasive plant-derived biochar inhibits sulfamethazine uptake by lettuce in soil. *Chemosphere*, *111*, 500–504.

Reddersen, K., Heberer, T., & Dünnbier, U. (2002). Identification and significance of phenazone drugs and their metabolites in ground-and drinking water. *Chemosphere*, *49*, 539–544.

Reinthaler, F. F., Galler, H., Feierl, G., Haas, D., Leitner, E., Mascher, F., Melkes, A., Posch, J., Pertschy, B., & Winter, I. (2013). Resistance patterns of *Escherichia coli* isolated from sewage sludge in comparison with those isolated from human patients in 2000 and 2009. *Journal of Water and Health*, *11*, 13–20.

Riaz, L., Mahmood, T., Khalid, A., Rashid, A., Siddique, M. B. A., Kamal, A., & Coyne, M. S. (2018). Fluoroquinolones (FQs) in the environment: A review on their abundance, sorption and toxicity in soil. *Chemosphere*, *191*, 704–720.

Rosi-Marshall, E. J., & Kelly, J. J. (2015). Antibiotic stewardship should consider environmental fate of antibiotics. *Environmental Science & Technology*, *49*(9), 5257–5258.

Ruiz-Angel, M., García-Alvarez-Coque, M., Berthod, A., & Carda-Broch, S. (2014). Are analysts doing method validation in liquid chromatography? *Journal of Chromatography A*, *1353*, 2–9.

Śliwka-Kaszyńska, Magdalena, Kot-Wasik, Agata, & Namieśnik, Jacek (2003). Preservation and storage of water samples. *Critical Reviews in Environmental Science and Technology*, *33*(1). doi:10.1080/10643380390814442.

Sacher, F., Lange, F. T., Brauch, H.-J., & Blankenhorn, I. (2001). Pharmaceuticals in groundwaters: Analytical methods and results of a monitoring program in Baden-Württemberg, Germany. *Journal of Chromatography A*, *938*, 199–210.

Salgado, R., Marques, R., Noronha, J., Carvalho, G., Oehmen, A., & Reis, M. (2012). Assessing the removal of pharmaceuticals and personal care products in a full-scale activated sludge plant. *Environmental Science and Pollution Research*, *19*, 1818–1827.

Sarmah, A. K., Meyer, M. T., & Boxall, A. B. (2006). A global perspective on the use, sales, exposure pathways, occurrence, fate and effects of veterinary antibiotics (VAs) in the environment. *Chemosphere*, *65*, 725–759.

Sasu, S., Kümmerer, K., & Kranert, M. (2012). Assessment of pharmaceutical waste management at selected hospitals and homes in Ghana. *Waste Management & Research, 30*, 625–630.

Schäfer, A. I., Akanyeti, I., & Semião, A. J. (2011). Micropollutant sorption to membrane polymers: A review of mechanisms for estrogens. *Advances in Colloid and Interface Science, 164*, 100–117.

Scheurer, M., Storck, F. R., Brauch, H.-J., & Lange, F. T. (2010). Performance of conventional multi-barrier drinking water treatment plants for the removal of four artificial sweeteners. *Water Research, 44*, 3573–3584.

Schmidt, B., Ebert, J., Lamshöft, M., Thiede, B., Schumacher-Buffel, R., Ji, R., Corvini, P. F.-X., & Schäffer, A. (2008). Fate in soil of 14C-sulfadiazine residues contained in the manure of young pigs treated with a veterinary antibiotic. *Journal of Environmental Science and Health, Part B, 43*, 8–20.

Schwarzbauer, J., & Heim, S. (2005). Lipophilic organic contaminants in the Rhine river, Germany. *Water Research, 39*, 4735–4748.

Schwarzbauer, J., Heim, S., Brinker, S., & Littke, R. (2002). Occurrence and alteration of organic contaminants in seepage and leakage water from a waste deposit landfill. *Water Research, 36*, 2275–2287.

Seifrtová, M., Nováková, L., Lino, C., Pena, A., & Solich, P. (2009). An overview of analytical methodologies for the determination of antibiotics in environmental waters. *Analytica Chimica Acta, 649*, 158–179.

Shan, J., Yang, P., Rahman, M. M., Shang, X., & Yan, X. (2018). Tetracycline and sulfamethazine alter dissimilatory nitrate reduction processes and increase N2O release in rice fields. *Environmental Pollution, 242*, 788–796.

Shenker, M., Harush, D., Ben-Ari, J., & Chefetz, B. (2011). Uptake of carbamazepine by cucumber plants—A case study related to irrigation with reclaimed wastewater. *Chemosphere, 82*, 905–910.

Siddiqui, M. R., AlOthman, Z. A., & Rahman, N. (2017). Analytical techniques in pharmaceutical analysis: A review. *Arabian Journal of Chemistry, 10*, S1409–S1421.

Slack, R., Gronow, J., & Voulvoulis, N. (2005). Household hazardous waste in municipal landfills: Contaminants in leachate. *Science of the Total Environment, 337*, 119–137.

Smallwood, A. W., DeBord, K. E., & Lowry, L. K. (1992). N, N′ Diethyl-m-Toluamide (m-DET): Analysis of an insect repellent in human urine and serum by high-performance liquid chromatography. *Journal of Analytical Toxicology, 16*, 10–13.

Song, L., Li, L., Yang, S., Lan, J., He, H., McElmurry, S. P., & Zhao, Y. (2016). Sulfamethoxazole, tetracycline and oxytetracycline and related antibiotic resistance genes in a large-scale landfill, China. *Science of the Total Environment, 551*, 9–15.

Stackelberg, P. E., Furlong, E. T., Meyer, M. T., Zaugg, S. D., Henderson, A. K., & Reissman, D. B. (2004). Persistence of pharmaceutical compounds and other organic wastewater contaminants in a conventional drinking-water-treatment plant. *Science of the Total Environment, 329*, 99–113.

Stackelberg, P. E., Gibs, J., Furlong, E. T., Meyer, M. T., Zaugg, S. D., & Lippincott, R. L. (2007). Efficiency of conventional drinking-water-treatment processes in removal of pharmaceuticals and other organic compounds. *Science of the Total Environment, 377*, 255–272.

Sui, Q., Cao, X., Lu, S., Zhao, W., Qiu, Z., & Yu, G. (2015). Occurrence, sources and fate of pharmaceuticals and personal care products in the groundwater: A review. *Emerging Contaminants, 1*, 14–24.

Sui, Q., Huang, J., Deng, S., Yu, G., & Fan, Q. (2010). Occurrence and removal of pharmaceuticals, caffeine and DEET in wastewater treatment plants of Beijing, China. *Water Research, 44*, 417–426.

Sun, L., Zha, J., & Wang, Z. (2009). Effects of binary mixtures of estrogen and antiestrogens on Japanese medaka (*Oryzias latipes*). *Aquatic Toxicology*, *93*, 83–89.

Swale, D. R., & Bloomquist, J. R. (2019). Is DEET a dangerous neurotoxicant? *Pest Management Science*, *75*, 2068–2070.

Söffker, M., & Tyler, C. R. (2012). Endocrine disrupting chemicals and sexual behaviors in fish—A critical review on effects and possible consequences. *Critical Reviews in Toxicology*, *42*, 653–668.

Tasho, R. P., & Cho, J. Y. (2016). Veterinary antibiotics in animal waste, its distribution in soil and uptake by plants: A review. *Science of the Total Environment*, *563*, 366–376.

Teijon, G., Candela, L., Tamoh, K., Molina-Díaz, A., & Fernández-Alba, A. (2010). Occurrence of emerging contaminants, priority substances (2008/105/CE) and heavy metals in treated wastewater and groundwater at Depurbaix facility (Barcelona, Spain). *Science of the Total Environment*, *408*, 3584–3595.

Ternes, T. A., Joss, A., & Siegrist, H. (2004). Peer reviewed: Scrutinizing pharmaceuticals and personal care products in wastewater treatment. *Environmental Science & Technology*, *38*, 392A–399A.

Ternes, T. A., Stumpf, M., Mueller, J., Haberer, K., Wilken, R-D., & Servos, M. (1999). Behavior and occurrence of estrogens in municipal sewage treatment plants—I. Investigations in Germany, Canada and Brazil. *Science of the Total Environment*, *225*(1-2), 81–90.

Thiele-Bruhn, S. (2003). Pharmaceutical antibiotic compounds in soils–a review. *Journal of plant nutrition and soil science*, *166*(2), 145–167.

Thiele-Bruhn, S., Seibicke, T., Schulten, H., & Leinweber, P. (2004). Technical reports–Organic compounds in the environment-sorption of sulfonamide pharmaceutical antibiotics on whole soils and particle-size fractions. *Journal of Environmental Quality*, *33*, 1331–1342.

Tian, G., Wang, W., Zong, L., Kang, Y., & Wang, A. (2016). A functionalized hybrid silicate adsorbent derived from naturally abundant low-grade palygorskite clay for highly efficient removal of hazardous antibiotics. *Chemical Engineering Journal*, *293*, 376–385.

Tisch, M., Schmezer, P., Faulde, M., Groh, A., & Maier, H. (2002). Genotoxicity studies on permethrin, DEET and diazinon in primary human nasal mucosal cells. *European Archives of Oto-Rhino-Laryngology*, *259*, 150–153.

Tobiszewski, M., Mechlińska, A., & Namieśnik, J. (2010). Green analytical chemistry—Theory and practice. *Chemical Society Reviews*, *39*, 2869–2878.

Togola, A., & Budzinski, H. (2007). Analytical development for analysis of pharmaceuticals in water samples by SPE and GC–MS. *Analytical and Bioanalytical Chemistry*, *388*, 627–635.

United States Environmental Protection Agency. (1998). *Reregistration eligibility decision (RED) DEET*. Washington, DC.

Vane, J., & Botting, R. (1998). Anti-inflammatory drugs and their mechanism of action. *Inflammation Research*, *47*, 78–87.

Veltri, J. C., Osimitz, T. G., Bradford, D. C., & Page, B. C. (1994). Retrospective analysis of calls to poison control centers resulting from exposure to the insect repellent N, N-diethyl-m-toluamide (DEET) from 1985–1989. *Journal of Toxicology: Clinical Toxicology*, *32*, 1–16.

Wang, D., Sui, Q., Lu, S.-G., Zhao, W.-T., Qiu, Z.-F., Miao, Z.-W., & Yu, G. (2014). Occurrence and removal of six pharmaceuticals and personal care products in a wastewater treatment plant employing anaerobic/anoxic/aerobic and UV processes in Shanghai, China. *Environmental Science and Pollution Research, 21*, 4276–4285.

Watanabe, N., Bergamaschi, B. A., Loftin, K. A., Meyer, M. T., & Harter, T. (2010). Use and environmental occurrence of antibiotics in freestall dairy farms with manured forage fields. *Environmental Science & Technology, 44*, 6591–6600.

Weeks, J., Guiney, P., & Nikiforov, A. (2012). Assessment of the environmental fate and ecotoxicity of N, N-diethyl-m-toluamide (DEET). *Integrated Environmental Assessment and Management, 8*, 120–134.

Wegst-Uhrich, S. R., Navarro, D. A. G., Zimmerman, L., & Aga, D. S. (2014). Assessing antibiotic sorption in soil: A literature review and new case studies on sulfonamides and macrolides. *Chemistry Central Journal, 8*.

Weigel, S., Berger, U., Jensen, E., Kallenborn, R., Thoresen, H., & Hühnerfuss, H. (2004). Determination of selected pharmaceuticals and caffeine in sewage and seawater from Tromsø/Norway with emphasis on ibuprofen and its metabolites. *Chemosphere, 56*, 583–592.

Weisbrod, C. J., Kunz, P. Y., Zenker, A. K., & Fent, K. (2007). Effects of the UV filter benzophenone-2 on reproduction in fish. *Toxicology and Applied Pharmacology, 225*, 255–266.

Weng, S., Sun, P., Ben, W., Huang, C.-H., Lee, L. T., & Blatchley, E. R., III (2014). The presence of pharmaceuticals and personal care products in swimming pools. *Environmental Science & Technology Letters, 1*, 495–498.

World Health Organization. (2014). Antimicrobial resistance global report on surveillance: 2014 summary *(No. WHO/HSE/PED/AIP/2014.2)*. World Health Organization.

Xu, Y., Yu, X., Xu, B., Peng, D., & Guo, X. (2020). Sorption of pharmaceuticals and personal care products on soil and soil components: Influencing factors and mechanisms. *Science of The Total Environment*, 141891.

Xu, L. Y., Zhang, H., Xiong, P., Zhu, Q. Q., Liao, C. Y., & Jiang, G. B. (2021). Occurrence, fate, and risk assessment of typical tetracycline antibiotics in the aquatic environment: A review. *Science of the Total Environment, 753*.

Yang, X., & Peppard, T. (1994). Solid-phase microextraction for flavor analysis. *Journal of Agricultural and Food Chemistry, 42*, 1925–1930.

Yang, X., Flowers, R. C., Weinberg, H. S., & Singer, P. C. (2011). Occurrence and removal of pharmaceuticals and personal care products (PPCPs) in an advanced wastewater reclamation plant. *Water research, 45*(16), 5218–5228.

Yang, Y., Ok, Y. S., Kim, K.-H., Kwon, E. E., & Tsang, Y. F. (2017). Occurrences and removal of pharmaceuticals and personal care products (PPCPs) in drinking water and water/sewage treatment plants: A review. *Science of the Total Environment, 596*, 303–320.

Yi, C. H., Qin, W., & Wen, X. H. (2020). Renovated filter filled with poly-3-hydroxybutyrateco-hydroxyvalerate and granular activated carbon for simultaneous removal of nitrate and PPCPs from the secondary effluent. *Science of the Total Environment, 749*.

Yoon, Y., Westerhoff, P., Snyder, S. A., & Wert, E. C. (2006). Nanofiltration and ultrafiltration of endocrine disrupting compounds, pharmaceuticals and personal care products. *Journal of Membrane Science, 270*, 88–100.

Yu, X., Sui, Q., Lyu, S., Zhao, W., Liu, J., Cai, Z., Yu, G., & Barcelo, D. (2020). Municipal solid waste landfills: An underestimated source of pharmaceutical and personal care products in the water environment. *Environmental Science & Technology, 54*, 9757–9768.

Zhang, G., Liu, X., Sun, K., He, Q., Qian, T., & Yan, Y. (2013). Interactions of simazine, metsulfuron-methyl, and tetracycline with biochars and soil as a function of molecular structure. *Journal of Soils and Sediments*, *13*, 1600–1610.

Zhou, J. L., Liu, R., Wilding, A., & Hibberd, A. (2007). Sorption of selected endocrine disrupting chemicals to different aquatic colloids. *Environmental science & technology*, *41*(1), 206–213.

Chapter 3

Fate and transport of pharmaceuticals and personal care products in soils and groundwater

Hao Chen and Linta Akhtar
Department of Agriculture, University of Arkansas at Pine Bluff, Pine Bluff, AR, United States

Chapter Outline

3.1 Overview

Pharmaceutical and personal care products (PPCPs) such as antiinflammatory drugs, blood lipid-lowering agents, antibiotics, and hormones have been frequently detected in soil and groundwater systems (Santos et al., 2010). As indicated in the previous chapter, PPCPs in soils and groundwater usually fall into the following broad categories: (1) over-the-counter and prescription pharmaceuticals; (2) personal care products such as surfactants (present in dishwashing liquids), caffeine, artificial sweeteners (ASs), and antimicrobial chemicals such as triclosan commonly found in soaps; (3) insect repellent such as *N*,*N*-diethyl-*meta*-toluamide (DEET).

Emerging Contaminants in Soil and Groundwater Systems. DOI: https://doi.org/10.1016/B978-0-12-824088-5.00004-5

The most frequently reported PPCPs, namely, DEET, caffeine, carbamazepine, and ibuprofen, are insect repellent, stimulant, anticonvulsant, and antiinflammatory, respectively. Antibiotics comprise approximately half of the top frequently reported PPCPs in the environment, therefore the fate and transport of antibiotics have been one of the top priorities of PPCP studies. The high detection rate of antibiotics is largely related to their high consumptions (Kuppusamy et al., 2018). Van Boeckel et al. (2015) reported that the global antibiotic consumption by livestock was about 63,151 tons in 2010 and would rise by 67% between 2010 and 2030, due to increased demands for food production and intensive farming systems.

PPCPs typically entered the environment system through three main sources: manufacture of PPCPs, PPCP consumption, and the disposal of out-of-date PPCPs in landfills (Ebele et al., 2017; Fick et al., 2009). Postuse PPCPs enter the environment through different routes and the major sources are via biosolid application, reclaimed water irrigation, and landfill leaching. Details on the sources and occurrences of PPCPs in soils and groundwater can be found in Chapter 2, Occurrences and Impacts of Pharmaceuticals and Personal Care Products in Soils and Groundwater. Public concerns of PPCPs begin decades ago with the detection of the PPCPs in surface water and groundwater worldwide (Daughton, 2001; Daughton & Ternes, 1999). In recent years, studies on the fate and transport of PPCPs in soil and groundwater systems are more focused on hot zones such as agricultural lands with reclaimed water irrigation, landfills, and deep groundwater aquifers. The goals of this chapter are to summarize the knowledge and recent development on (1) the effect of PPCP properties on their fate and transport in soil and groundwater systems, (2) interactions of PPCPs with different soil components, and (3) fate and transport of PPCPs in the environment under different systems.

3.2 Interaction with soil components

The fate and transport of PPCPs in soils and groundwater are strongly influenced by the interaction strength between PPCPs and soil components, which are determined by physiochemical properties of PPCPs (e.g., molecular structure, hydrophobicity, polarity, polarizability, and spatial configuration) and their surrounding environment [e.g., soil type, soil pH, coexisting ions, and soil organic matter (SOM)] (Schwarzenbach et al., 2005; Zhi et al., 2019). Hydrophobic PPCPs interact with soils mainly through hydrophobic interactions, for example, $\pi-\pi$ bonds or alignment of aromatic rings, while polar PPCPs interact mostly through H-bonds or electrostatic interaction (Ahmed et al., 2015). Several models have been developed for molecular mechanic calculations of the interactions between PPCPs and SOM (Ahmed et al., 2015).

3.2.1 Hydrophobicity, polarity, and ionization of pharmaceutical and personal care products

Most of the PPCPs are organic compounds showing various degrees of hydrophobicity. Hydrophilic compounds with high solubility often move as solutes and are shown high mobility in soil and groundwater systems. Hydrophobic compounds tend to be adsorbed by soil solids especially the hydrophobic part of the soil matrix. For example, Conde-Cid et al. (2020) studied the transport of three sulfonamide antibiotics and their interaction with soils, suggesting that antibiotic molecules with higher hydrophobic also have higher retention (sulfadiazine < sulfamethazine < sulfachloropyridazine) in soils. However, the transport of hydrophobic PPCPs can also be promoted through colloids or dissolved organic matter facilitated transport. PPCPs often constituent with heteroatoms and nonsymmetric structures therefore shown polarity. Polarity is a critical factor to determine the attraction strength between PPCPs and soil components especially SOM. Rasamimanana et al. (2017) studied the sorption of anthropogenic organic molecules to Callovian-Oxfordian claystone and reported a linear correlation between adsorption data and dipole moment of adsorbates. Xing et al. (1994) also reported that polarity can be used to predict the sorption of nonionic organic contaminants to the soil.

Some PPCPs can be partially charged in soil solution as cationic and anionic species based on the degree of protonation or deprotonation, which varies as a function of solution pH. In general, cationic PPCPs species show higher retention than negatively charged species due to the majority of the soil surface is negatively charged. Schaffer et al. (2012) studied the sorption and transport of ionizable pharmaceuticals onto natural sandy aquifer sediments and observed positively charged atenolol species shown as strong sorption to soil matrix and as increasing Ca concentrations atenolol sorption decreased dramatically, indicating that cation exchange plays a major role in the sorption process. In contrast, negatively charged anionic compounds like naproxen and phenobarbital are only weakly adsorbed on the soil matrix (Schaffer et al., 2012).

3.2.2 Sorption of pharmaceutical and personal care products to soil matrix

Sorption refers to "the accumulation of a substance or material at an interface between the solid surface and the bathing solution" (Figueroa et al., 2004; MacKay & Canterbury, 2005; Sparks, 2003; Wu et al., 2019). Adsorption of PPCPs means the transference and accumulation of PPCPs from the fluid phase to the solid phase of soil through physical and/or chemical interactions (Chen et al., 2013a,b; Ruthven, 1984). Sorption of PPCPs onto soil matrix has a fundamental impact on their mobility, degradation, bioavailability, and ecotoxicity. Physical adsorption typically involves intermolecular interactions, such as electrostatic forces, hydrogen bonding, Van der Waals forces, and hydrophobic interactions (Chen et al., 2013a,b; Zhang et al., 2016). Chemisorption involves chemical interactions with the transfer of electrons between PPCPs and the adsorbents, such as soil minerals or SOM. Factors controlling the sorption of PPCPs in soils include the properties of PPCPs, the properties of soil components, soil porous structure, environmental factors, and the interaction dynamics. Batch and fixed-bed (upflow or downflow) are common techniques used to characterize the sorption behaviors and dynamics of PPCPs in soils. Besides, instruments analysis such as Fourier-transform infrared spectroscopy (FTIR), nuclear magnetic resonance, and X-ray photoelectron spectroscopy (XPS) is also used to identify the interactions between PPCPs' functional group and soil components.

Due to the properties of PPCPs vary widely from one group to another, their interaction with soil also differs significantly (Fig. 3.1). For example, the montmorillonite adsorbs tetracycline mainly through electrostatic attraction; however, its adsorption of sulfamethoxazole is dominated by hydrophobic interaction. On the other hand, since the intrinsic differences of soil components, the sorption mechanism of antibiotics groups to different soil components also differs largely (Fig. 3.1). For instance, fluoroquinolone group antibiotics sorption onto silica-based mineral mainly through cation exchange due to amine group are positively charged in neutral pH; however, when fluoroquinolone group interacts with iron-containing minerals, surface complexation will be dominant sorption mechanism since the carboxylic group in fluoroquinolone can form strong complexation with irons. When multivalent cation coexisted in the system, fluoroquinolones can also adsorb onto the soil surface through the cation bridging mechanism. Similarly, tetracycline with several functional groups is also shown different dominant sorption mechanisms when interacting with different soil components, such as cation exchange and surface complexation.

3.2.2.1 The controlling factors of pharmaceutical and personal care products sorption

3.2.2.1.1 Properties of soil

Soil properties such as soil texture, cation exchange capacity (CEC) or anion exchange capacity (AEC), mineral components, and clay content are all impact soil PPCPs adsorption capacities. Soil texture determines the specific surface area of soils, thus influences the sorption capacities of soils. In general, soils with finer textures have higher sorption capacities than soils with coarser textures. Karnjanapiboonwong et al. (2010) studied sorption of estrone, 17β-estradiol, estriol, 17α-ethynylestradiol, triclosan, and caffeine in sandy loam and a silt loam soil and reported that silt loam soil has better

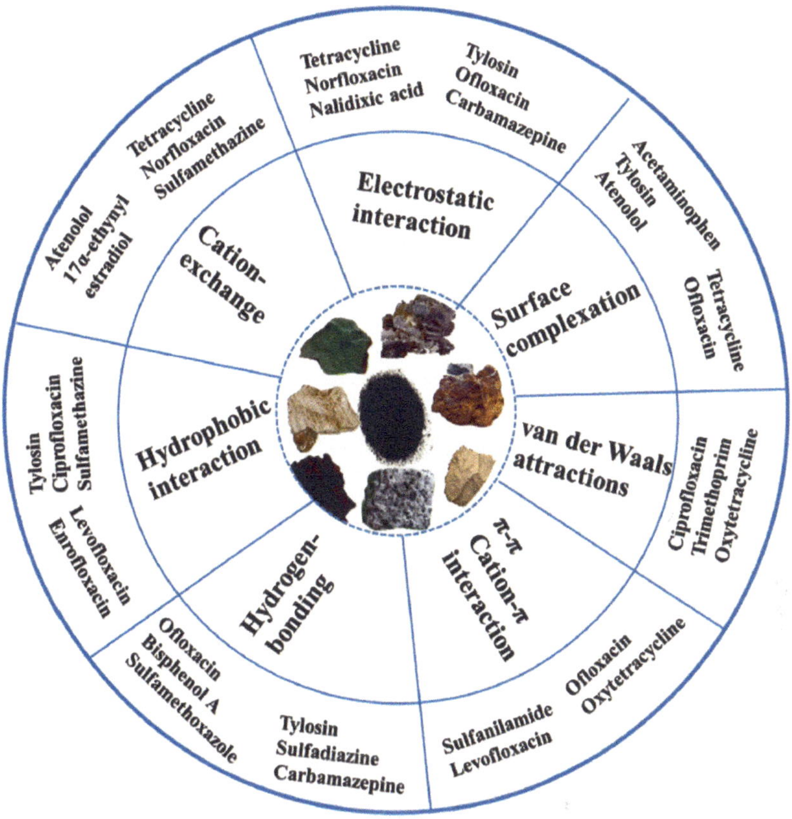

FIGURE 3.1 Sorption mechanisms of pharmaceutical and personal care products (PPCPs) to soils and components (e.g., SOM, kaolinite, montmorillonite, limestone, goethite). Various PPCPs can be adsorbed to soil and soil components via different mechanisms. The *dotted circle* inside represents the main components of soil, and each sector includes the sorption mechanism of various PPCPs to soil and soil components. *From Xu, Y., Yu, X., Xu, B., Peng, D., & Guo, X. (c. 2020). Sorption of pharmaceuticals and personal care products on soil and soil components: Influencing factors and mechanisms.* Science of the Total Environment.

sorption than sandy loam for all selected PPCPs except caffeinc. Soil CEC or AEC is often correlated to the sorption capacity of charged PPCPs since sorption on the exchange site is considered the dominant sorption mechanisms for ionizable PPPs (Sassman & Lee, 2005; Xu et al., 2020). Metal oxides especially iron oxides or aluminum oxides can provide additional positive charged attraction sites or complexation sites. Chen et al. (2013a, 2013b) reported that iron oxides on the sand surface are the main minerals on the sand for ciprofloxacin sorption through surface complexation mechanism. Clay or colloids are considered major reactive components in soil due to their large surface area and high CEC. Rabølle and Spliid (2000) study the sorption of tylosin in different soils and reported that tylosin adsorption correlated positively with the clay content of the soil.

3.2.2.1.2 Soil organic matter

SOM plays a key role in PPCPs sorption, due to their surface functional groups and large surface area. SOM exists in soil as dissolved soil organic matter (DOM) or organic matter bound in the soil matrix. In general, SOM in the soil matrix can promote PPCPs sorption due to they provide extra sorption sites or promote PPCPs hydrophobic partition on the soil surface.

However, if soil minerals showed high sorption capacity to PPCPs, SOM on the mineral surface can act as shielding agents and reduce PPCPs sorption. Yao et al. (2016) compared oxytetracycline sorption on the soil before and after SOM removal and observed OTC adsorption capacity of the red soil increased significantly when the organic matter was removed. This probably due to the OTC surface sorption sites shielded by SOM is exposed and increased OTC sorption since OTC is known for its high affinity to adsorbing soil minerals and clay (Kong et al., 2012).

Besides, DOM may compete with PPCPS for surface sorption sites or complex with PPCPs and promote their mobility. Oh et al. (2016) investigate that the effect of DOM on ibuprofen sorption, while citrate, and urea was used as DOM. As citrate concentration increased, the sorption coefficiency K_d value decreased but urea did not interrupt the ibuprofen sorption because citrate has three carboxyl functional groups which can attach easily to ibuprofen and hinder its sorption onto sediment. Kulshrestha et al. (2004) studied tetracyclines sorption to montmorillonite as a function of DOM (humic acid) concentration. They reported at a lower concentration of dissolved humic acid (1 mg/L), increased sorption was observed, while a higher concentration of humic acid (10 mg/L) resulted in an opposite effect. This phenomenon may be explained by hydrophobic partitioning of oxytetracycline between clay-associated humic acid in the solid phase and "free" humic acids in the solution phase. Since clay's sorption for humic acids has a limit, if a large amount of humic acid was added to the system, the concentration of humic acids in the solution phase may be higher than those associated with clay. Consequently, more oxytetracycline will remain in the solution phase and sorption on clay will decrease. This phenomenon suggests that soil-bound PPCPs may be mobilized at high DOM concentration, such as after liquid manure application.

3.2.2.1.3 pH

pH is a key environmental factor that determines the speciation of ionizable PPCPs and the dominant surface change of the soil matrix thus controls the attractions between them. Ionizable PPCPs show different forms such as cation, anion, zwitterion, or neutral species in the soil solution depending on their acid dissociation constant (pK_a) and the pH value of soil solution. pK_a values of PPCPs are associated with functional groups, for example, ciprofloxacin has two pK_as: pK_{a1} = 6.2 associated with carboxylic group and pK_{a2} = 8.8 associated with amine group, which means at pH = 6.2 50% of carboxylic groups are deprotonated and at pH = 8.8 50% of the amine group are protonated. Base on the pK_a values, the speciation of ciprofloxacin can be calculated: at solution pH = 5.6, ciprofloxacin showed 80% cationic and 20% zwitterion species; at pH = 9.5, ciprofloxacin showed 83.4% anionic and 16.6% zwitterion. Surface charge of soil minerals is also determined by soil solution pH base on point of zero charge (PZC) of soil. For example, silica-based clean sand with PZC = 5.1 shows positive charge when solution below 5.1 and negative charge goethite with PZC = 9.1, will show positive charge when pH below 9.1. (Chen et al., 2013a,b; Dabizha & Kersten, 2020).

Since sorption is controlled by both PPCPs species and surface charge of the soil matrix, three sorption trends are observed with the increase of pH: bell-shaped relationship, decrease in sorption as pH increases and increase in sorption as pH increases (Xu et al., 2020). Bell-shaped sorption pattern can be explained by at the peak sorption pH, the opposite charge between PPCPs species and soil surface sorption site is maximized, either further increase or decrease pH sorption decreased. For example, sorption of tetracycline to humic acid found increased as pH increased from 2.5, reaching a maximum around pH 4.3, and then decreasing as proton activity decreased further (Gu et al., 2007). At pH < 4.3, zwitterionic and cationic tetracycline species apparently competed with protons for binding sites, therefore when proton concentration increase as pH decrease tetracycline sorption decrease; as pH increased above 4.3, as the fractions of cationic and zwitterionic tetracycline species decreased and humic acid also became progressively deprotonated, tethacycline sorption decreases (Gu et al., 2007). Sorption decreases as pH increases mainly due to with pH increase majority PPCPs species turn negative and become repulsive to negatively charged soil surface if pH is above PZC of soil. A large portion of soil particles have relatively low PZC, therefore, sorption decreases as pH increase is commonly observed in PPCPs sorption studies. However, sorption may also increase with pH increase if the tested soil particles have high PZC, such as certain iron or aluminum-containing minerals. Guo et al. (2014) studied the sorption of tylosin on goethite at pH = 3.95, 7.25, and 9.45 and observed the sorption increased with pH increase. Compare to pH = 3.95 and 7.25, at pH = 9.5 tylosin has the highest number of natural species (99%) and the lowest amount of positively charged species; in the meanwhile, the positive charge sites of goethite also decreased to the lowest at pH = 9.45. Therefore, at pH = 9.45 the repulsive force caused by the positive charge from both tylosin and goethite is reduced to the minimum, tylosin had the highest attraction to the goethite surface.

3.2.2.1.4 Coexisting cations

Coexisting cations can promote or inhibit PPCPs sorption depending on cation's property and concentration (Table 3.1). Cations, especially monovalent cations such as Na^+ or K^+ in the system, can change PPCP sorption in different ways. If PPCPs are adsorbed on soil surface through cation exchange then cations in soil solution can act as competitors and reduce PPCPs sorption. In this case, PPCPs sorption decreases with cation concentration increase. Gu and Karthikeyan (2008) studied tetracycline sorption on humic acid and observed tetracycline sorption decreased with NaCl concentration increase from 0.01 to 0.1 mol/L ($pH = 6.2 \pm 0.1$). Cations especially divalent cations (Ca^{2+}, Mg^{2+}) or trivalent cations (Al^{3+}, Fe^{3+}) can act as cation bridges to promote PPCPs sorption. However, these cations can still inhibit PPCPs sorption when cation concentration increased to a certain range. Chen et al. (2013a,b) study the effect of coexisted Ca^{2+} or Cu^{2+} on ciprofloxacin sorption onto the sand and reported for sand coated with small amount of Fe or Al oxides, Ca^{2+} and Cu^{2+} at low concentration can both promote ciprofloxacin sorption due to complexation and cation bridging effects, but with concentration increase, Ca^{2+} or Cu^{2+} cation will compete with Ca-CIP or Cu-CIP complexation and reduce CIP sorption.

3.2.2.2 Adsorption kinetics and equilibrium

For porous sorbent such as soil, sorption kinetics mainly have three steps: (1) mass is transferred from bulk liquid phase through boundary film to solid external surface; (2) mass is transferred into the solid pores by in series or in parallel liquid phase diffusion and surface diffusion; (3) consist in the kinetics of mass transfer from liquid phase onto solid surface inside pores (Schwaab et al., 2017). SBatch kinetic study evaluates the adsorption rate, which can be limited by the nature of mass transfer in the system. Several models based on various assumptions have been proposed to examine mechanisms and the potential rate-determining factors. For the adsorption kinetics of PPCPs, commonly used models include pseudo-first-order model, pseudo-second-order model, and Elovich model (Chauhan et al., 2020; Ho & McKay, 1999; Zhuo et al., 2017). Governing equations of the (1) pseudo-first-order, (2) pseudo-second-order, and (3) Elovich models can be written as (Yao et al., 2016):

$$\frac{dq_t}{dt} = k_1(q_e - q_t) \text{ pseudo-first-order} \tag{3.1}$$

$$\frac{dq_t}{dt} = k_2(q_e - q_t)^2 \text{ pseudo-second-order} \tag{3.2}$$

$$\frac{dq_t}{dt} = \alpha \exp(-\beta q_t) \text{ Elovich} \tag{3.3}$$

TABLE 3.1 Effect of ionic strength on the sorption of pharmaceutical and personal care products (Xu et al., 2020).

Result	Sorbate	Sorbent	References
Bell-shaped relationship	Sulfamethoxazole	Soil	Qin et al. (2016)
	Sulfadiazine	Soil	Qin et al. (2016)
Increase in sorption as ionic strength increase	Sulfapyridine	Soil	Srinivasan et al. (2013)
	Sulfamethazine	Humic acid	Zhao et al. (2018)
	Tetracycline	Montmorillonite, kaolinite	Kolz et al. (2005) Zhao et al. (2011)
	Ciprofloxacin	Magnetite	Rakshita et al. (2013)
	Citalopram	Soil	Call et al. (2019)
Decrease in sorption as ionic strength increases	Sulfapyridine	Montmorillonite	Avisar et al. (2010)
	Sulfamethoxazole	Humic acid	Liu et al. (2017b)
	Sulfamethazine	Humic acid, soil	Guo et al. (2016) Pavlovic et al. (2014)
	Sulfadimethoxine	Soil	Bialk-Bielinska et al. (2012)
	Sulfaguanidine	Soil	Bialk-Bielinska et al. (2012)
	Tetracycline	Palygorskite, humic acid, goethite, ferrihydrite	Chang et al. (2009b), Christl et al. (2016), Zhou et al. (2016)
	Oxytetracycline	Nano-hydroxyapatite	Tian et al. (2016)
	Tylosin	Kaolinite, montmorillonite, humic acid goethite	Guo et al. (2016), Guo et al. (2014), Shen et al. (2018)
	Clarithromycin	Humic acid	Christl et al. (2016)
	Ciprofloxacin	Sediments, humic acid	Liu et al. (2017a,b)
	Norfloxacin	Humic acid	Li et al. (2016)

(*Continued*)

TABLE 3.1 (Continued)

Result	Sorbate	Sorbent	References
	Ofloxacin	Kaolinite	Vieira et al. (2017)
	Irbesartan	Soil	Call et al. (2019)
	Fexofenadine	Soil	Call et al. (2019)
	Nalidixic acid	Goethite	Li et al. (2017)
Negligible effect of ionic strength on sorption	Sulfapyridine	Graphite	Ji et al. (2009)
	Niflumic acid	Goethite	Li et al. (2017)

where q_t and q_e are the amounts of sorbate (e.g., PPCPs) adsorbed at time t and at equilibrium, respectively, and k_1 and k_2 are the pseudo-first-order and pseudo-second-order apparent adsorption rate constants, respectively. The constants α and β are the initial adsorption rate and the desorption constant, respectively.

The pseudo-first-order equation assumes that the rate of adsorption of adsorbate is directly proportional to the difference in saturation concentration and the amount of solid uptake with time, which is generally applicable over the initial stage of an adsorption process (Simonin, 2016). The pseudo-second-order kinetic model assumes that the removal from a solution is due to physicochemical interactions between the two phases (Ho & McKay, 1999). Elovich model assumes that the surface is energetically heterogeneous (Largitte & Pasquier, 2016).

Dynamic adsorption equilibrium is established when adsorption and desorption rates are equivalent (Grathwohl, 2012). In this condition, the overall mass transfer of solute is zero since the adsorption rate and desorption rate are the same in the system (Burris et al., 1998). For a given condition, adsorption isotherms relate the amount of solute adsorbed in the solid phase and its equilibrium concentration in the liquid phase (Ayawei et al., 2017). Models used to characterize the sorption isotherm include Langmuir and Freundlich models, etc. (Ayawei et al., 2017). Governing equations of Langmuir and Freundlich models can be written as:

$$Q_e = \frac{S_{max} K C_e}{1 + K C_e} \quad \text{Langmuir equation} \tag{3.4}$$

where C_e is the equilibrium concentration (mg/L), Q_e is the amount of SMX adsorbed at equilibration (mg/g), S_{max} is the maximum sorption value (mg/g$^-$), and K (L/mg) is the binding affinity of the sorption process.

$$Q_e = K_f C_e^{\,n} \quad \text{Freundlich equation} \tag{3.5}$$

where K_f ($L^n\ mg^{1-n}/g$) and n (dimensionless) are Freundlich affinity coefficient and linearity constant, respectively.

Langmuir model (Eq. 3.4) assumes that sorption occurs on homogeneous monolayer sorption without interactions among the solutes (Langmuir, 1918). The Freundlich model (Eq. 3.5) is empirical and often used to describe chemisorption onto the heterogeneous surface. The linear regression analysis has been used to define the best fitting adsorption models, verify the consistency of theoretical assumptions of the adsorption isotherm model, and analyze the adsorption system (Ayawei et al., 2017). Chen et al. (2013a, 2013b) used isotherm models to study the sorption capacities of sand with and without Fe/Al coatings and reported CIP sorption data fit better with Langmuir model ($R^2 = 0.9986–0.9998$), suggestive of monolayer sorption. The maximum sorption capacity derived from the best-fit value of the Langmuir model for coated sand was 50 mg/kg, which was 10 times greater than that of clean sand of 5 mg/kg. Moreover, the affinity of binding sites calculated from the Langmuir model on the coated sand (1.95 L/μg) was 6 times greater than that on the clean sand (0.33 L/μg). Based on the parameters calculated from isotherm models, they concluded coated sand with additional Fe/Al oxides on the surface not only provided more sorption sites but also had higher sorption affinity (Chen et al., 2013a,b).

3.2.2.3 *Continuous fixed-bed adsorption*

The batch experiment setting does not reflex the dynamic interactions between PPCPs and soil matrix. To eliminate these limitations, soil column was used as media to study sorption influenced transport behavior of PPCPs (Lewis & Sjöstrom, 2010; Magga et al., 2008). In a fixed-bed transport, the soil is first packed in a column and the solute containing a certain amount of PPCPs will pass through the packed column to mimic a transport event. Breakthrough curves (BTCs) generated through column transport experiments can be used to understand the interaction between PPCPs and soil matrix in a dynamic condition. Advection-Dispersion Equation is the fundamental equations for quantifying the transport of contaminants in the subsurface and analyzing advective and dispersive/diffusive transport parameters (Patel, 2019). The governing equation can be written as (Chen et al., 2011):

$$R\frac{\partial C_w}{\partial t} = D\frac{\partial^2 C_w}{\partial z^2} - v\frac{\partial C_w}{\partial z} - kC_w \tag{3.6}$$

where C_w is PPCP concentration in pore water; R is the retardation factor, which reflects the magnitude of equilibrium reactions; D is the dispersion coefficient; z is the coordinate parallel to flow; v is the velocity of pore water; and k is the irreversible, first-order kinetic reaction rate constant.

Various environmental factors may impact the PPCPs transport behavior such as the characteristics of porous media, saturation condition, flow rate, and water chemistry. Chen et al. (2013a,b) studied the transport of ciprofloxacin in sand media and reported that sand with metal oxides deposit on its surface has significantly higher ciprofloxacin retention ability than clean sand. They also compared the effects of coexisted Ca or Cu on ciprofloxacin mobility and reported that Cu can significantly promote ciprofloxacin mobility. Although divalent cations have influences on CIP transport in sand porous media, monovalent cation potassium has a limited effect on ciprofloxacin mobility in sand columns (Chen et al., 2011).

3.2.2.4 Pharmaceutical and personal care products sorption and degradation

PPCPs often undergo degradation after sorption. Besides sorption, many other factors may influence the PPCPs degradation rate such as their intrinsic properties, initial PPCPs concentration, microbial composition, aerobic condition, and soil texture. Xu et al. (Sassman & Lee, 2005; Xu et al., 2020) studied adsorption and degradation of six selected PPCPs, including clofibric acid, ibuprofen, naproxen, triclosan, diclofenac, and bisphenol using four US agricultural soils associated with reclaimed wastewater reuse; they stated that PPCPs degradation in soils are influenced by the SOM and clay contents. Yu et al. (2013) studied sorption and degradation bisphenol A, carbamazepine, gemfibrozil, octylphenol, and triclosan using three different soils and reported degradation of PPCPs in soils was influenced by the soil organic carbon contents, microbial activity, and soil type.

Zhang et al. (Chen et al., 2013a,b; Zhang et al., 2016) studied adsorption and degradation of metronidazole, tinidazole, caffeine, and chloramphenicol using two agricultural soils and one sediment from Changjiang River. Degradation of selected PPCPs in soils generally followed first-order exponential decay kinetics, and half-lives ranging from 0.97 to 10.21 days mainly due to microbial activity. They also reported that degradation rates decreased with increasing initial PPCPs concentrations, suggesting that microbial activity may be inhibited by high PPCPs loading concentrations. Conkle et al. (2012) studied the ibuprofen, DEET, and gemfibrozil degradations in selected soils under aerobic and anaerobic conditions and reported that degradation of the selected PPCPs was enhanced under aerobic conditions as compared to anaerobic conditions.

3.2.3 Pharmaceutical and personal care product interaction with plant roots

PPCPs can be accumulated in the root zone and translocate to plant tissue after entered surface soil. A detectable amount of PPCPs in root extracts has been reported in several studies indicating PPCPs may accumulate in or on roots of various types of plants (Figueroa et al., 2004; MacKay & Canterbury, 2005; Sparks, 2003; Wu et al., 2019). However, most studies with extracting PPCPs from whole roots cannot discriminate sorption to the root exterior from uptake into the root (Miller et al., 2016). Root concentration factor, the ratio of the chemical concentration in roots to that in the exposure medium (typically that for bulk soil rather than for soil pore water), is often used to quantify the chemical root accumulation rate (Miller et al., 2016). PPCPs accumulation in the root zone depends on the properties of PPCPs, plant species, soil properties, humidity, and temperature exposure time (Boonsaner & Hawker, 2015; Dodgen et al., 2015; Mathews et al., 2014).

Several properties of PPCPS are affecting their interactions with plants in the root zone (Fig. 3.2). PPCPs with higher lipophilicity allow more rapid diffusion across lipid bilayers of roots (Sterling, 1994). Partitioning to lipids and membrane permeability can be estimated from n-octanol/water partition coefficient or phospholipid − water partition coefficients depending on PPCPs speciation (Briggs et al., 1982; Topp et al., 1986). Hyland et al. (2015) studied nine pharmaceuticals and flame retardants in roots of lettuce and strawberry grown in soil and found that the root concentration factors are positively correlated with the pH-adjusted octanol-water partition coefficients for the target contaminants.

PPCPs uptake by roots is facilitated by passive uptake through protein channels or protein-mediated energy-dependent uptake (Miller et al., 2016).

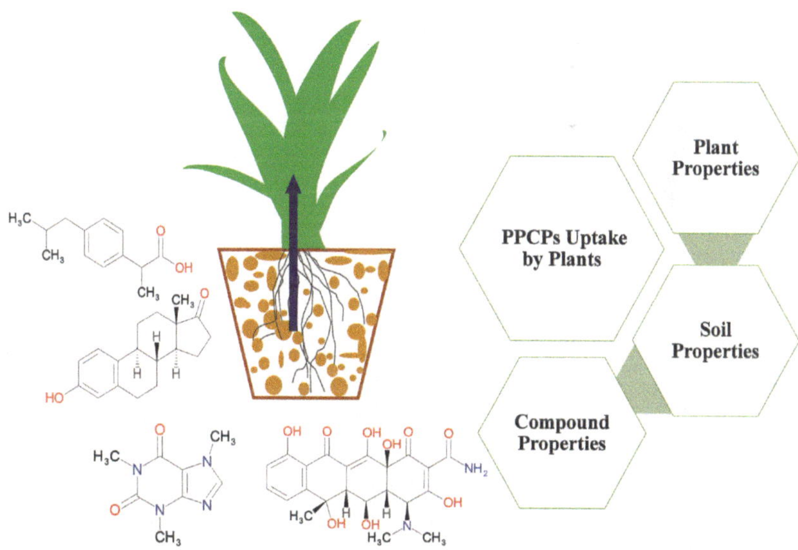

FIGURE 3.2 Factors that influence pharmaceutical and personal care products uptake by plants.

Plant proteins can mediate the uptake of organic compounds into root cells against concentration gradients (Näsholm et al., 2009). Due to the low selectivity of organic nitrogen transporters, PPCPs with structures similar to the natural organic compounds may use the same nitrogen transporters for plant uptake (Eggen & Lillo, 2012). Eggen et al. (2011) studied uptake and translocation of metformin, ciprofloxacin, and narasin in forage and crop plants and reported that uptake of metformin is higher than ciprofloxacin and narasin for tested plants, probably due to the effect of potential transporters.

3.3 Pharmaceutical and personal care products field transport studies and modeling

Field studies of PPCPs transport and modeling are challenging due to a lack of geological information (Lapworth et al., 2012). Most onsite PPCPs transport and modeling studies were conducted during aquifer recharge management characterized by the bank infiltration process.

Depending on the site-specific geological settings and data availability, various models have been developed to understand transport behavior. Barkow et al. (2020) investigated the temporal and spatial behaviors of selected PPCPs during bank filtration in dependence of relevant biogeochemical conditions with a 2D reactive transport model using MODFLOW (United States Geological Survey modular finite-difference flow model) and PHT3D (a multicomponent transport model for 3D reactive transport in saturated porous media) under the user interface ORTI3D (open reactive transport interface 3D). The 2D model was also used to simulate oxygen-, nitrate-, and temperature-dependent fate and attenuation of selected PPCPs during the bank filtration process. Data used for model stimulating including water levels, temperatures, and chemical parameters were regularly collected in the surface water and groundwater observation wells over 1.5 years (Barkow et al., 2020). Based on the maximum degradation rates stimulated by the models, valsartan was found degraded under oxic-conditions, whereas carbamazepine and sulfamethoxazole were degraded under anoxic conditions. Their study highlighted the influence of seasonal variations in oxygen supply and temperature on PPCPs' fate in surface water infiltrating into an aquifer.

Nham et al. (2015) conducted a field-scale experiment to investigate the transport behavior of 16 organic trace pollutants including PPCPs during soil aquifer treatment. In the modeling section, nonreactive 1D models were used for solute transport modeling, MODFLOW was used for modeling the groundwater flow, and MT3DMS (modular transport, 3D, multispecies model) was used as the multispecies transport simulator (Nham et al., 2015). Their study highlights the necessity to obtain intrinsic attenuation parameters at each site and the understanding of the complex interplay of various hydraulic, microbiological, and physicochemical factors, as attenuation is unique for each site (Nham et al., 2015).

Henzler et al. (2014) studied the fate and transport PPCPs under riverbank filtration using a 2D numerical flow and transport model to estimate linear distribution coefficients and first-order degradation rate constants of the PPCPs. Modeling was carried out in two steps. In the first step, a 2D vertical cross-sectional groundwater flow and nonreactive transport model were developed. Oxygen isotope signatures and hydraulic head data were used in model calibration. In the second step, the nonreactive and reactive transport simulations were carried out by the transport simulator MT3DMS (Zheng & Wang, 1999). For groundwater flow, the USGS groundwater flow simulation software MODFLOW was used (Harbaugh et al., 2000). Based on the stimulation results, the simplified model assuming the first-order degradation was incapable of reproducing the BTCs for clindamycin, sulfamethoxazole, diclofenac, and phenazone, as significant redox and/or temperature dependencies appear to exist.

Overall, due to site-specific microbial composition and redox condition, the land application data and site-specific geochemical information are critical for determining the fate and transport of PPCPs in the contaminated land due to the site-specific attenuation patterns in different soil systems (Henzler et al., 2014; Nham et al., 2015).

3.4 Typical systems of pharmaceutical and personal care product fate and transport research

3.4.1 Landfills

It is common to discharge solid waste containing PPCPs into landfills during domestic waste treatment. Some of the PPCPs with high persistency and high mobility can enter the leachates of landfills, and become risks to the surrounding environment. Municipal landfills as potential sources of PPCPs have been studied to understand the migration patterns of PPCPs and to evaluate their potential risks to ecosystems (Clarke et al., 2015; Wang et al., 2020; Wu et al., 2021). Besides the properties of PPCPs, characteristics of landfill sites (i.e., landfill ages), sociological factors (i.e., economic development), aquifer properties (i.e., permeability), redox potential, and seepage velocity all play critical roles in determining the fate and transport of PPCPs in soils and their migration to groundwater (Yu et al., 2013).

Musson and Townsend (2009) reported the sum of concentrations of 22 pharmaceutical ingredients was 8.1 mg/kg in a landfill located in Florida. Yu et al. (2020) reported that a total of 172 PPCPs were reported in landfill leachates globally over the last two decades. These PPCPs belong to various therapeutic classes, including antibiotics, antiinflammatories, stimulants, beta-blockers, etc. Yu et al. (2020) studied 50 relevant papers and summarized the top reported PPCPs including DEET, caffeine, carbamazepine, ibuprofen, diclofenac, and sulfamethoxazole, etc. (Fig. 3.3).

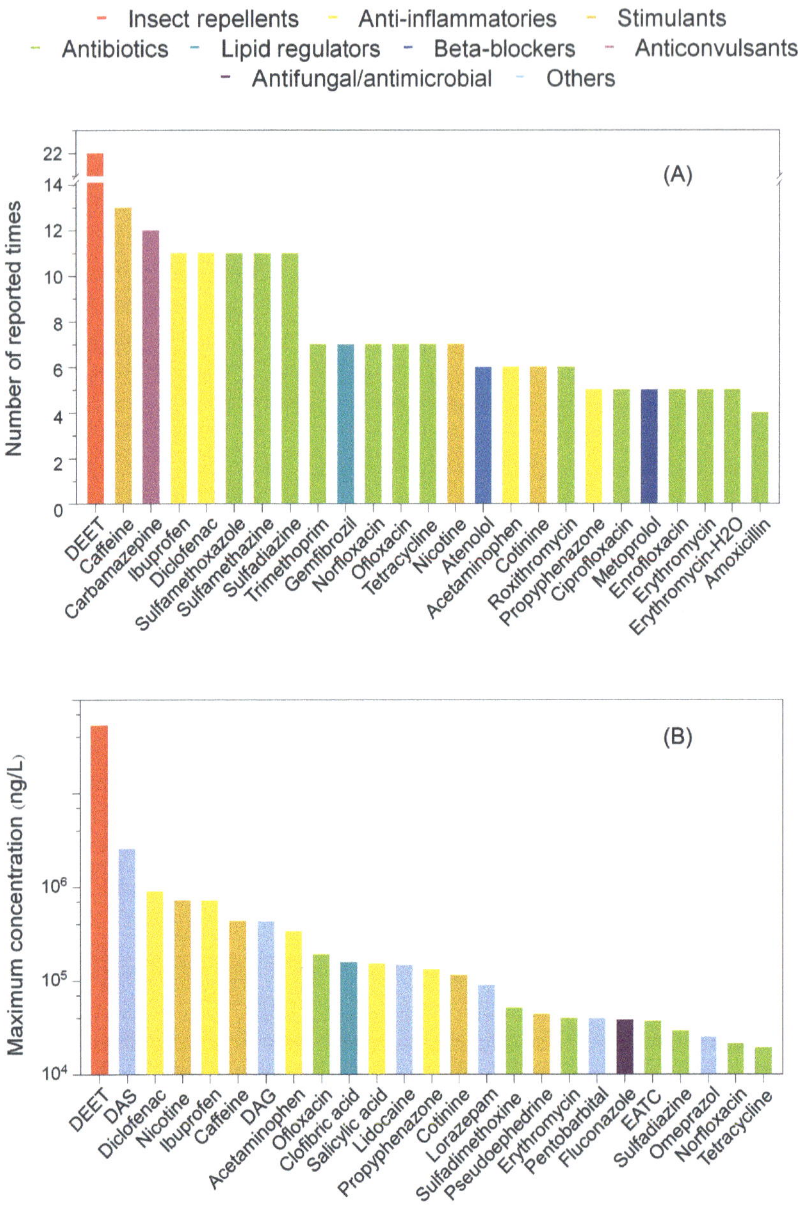

FIGURE 3.3 Top 25 pharmaceutical and personal care products sorted by the number of reported times (A) and by concentrations (maximum concentrations) (B) in landfill leachates. *DEET*, *N,N*-Diethyl-*meta*-toluamide; *DAS*, diacetone sorbose; *DAG*, diacetone alpha-keto-gulonic acid; *EATC*, 4-epianhydrotetracycline. *From Yu, X., Sui, Q., Lyu, S., Zhao, W., Liu, J., Cai, Z., Yu, G., & Barcelo, D. (c. 2020). Municipal solid waste landfills: An underestimated source of pharmaceutical and personal care products in the water environment.* Environmental Science and Technology, 54*(16), 9757–9768. https://doi.org/10.1021/acs.est.0c00565.*

High concentrations of pharmaceuticals in municipal solid waste were also detected in the nearby environments, especially nearby groundwater systems (Lapworth et al., 2012). Some of the PPCPs can exist in soils and groundwater after years of the landfill closure (Barnes et al., 2008; Lapworth et al., 2012). Han et al. (2016) analyzed scientific papers related to groundwater contamination caused by landfills in China and concluded groundwater contamination appeared after 5 years of the initial landfill stage and peaked some years afterward. Based on their study, groundwater contamination mainly appeared within 1000 m of a landfill. Yu et al. (2020) reported 14 most frequently reported PPCPs in the groundwater adjacent to landfills based on recent studies worldwide. Among them, DEET, ibuprofen, and diclofenac are detected most frequently (Fig. 3.4). As expected, the PPCPs with a high concentration in landfills show a higher detection rate in nearby groundwater; however, the fate and transport pathways of PPCPs from landfills to groundwater are still unclear.

3.4.2 Septic system

Septic systems are widely used especially in the rural area to treat conventional wastewater pollutants such as nutrients and pathogens with the goal to prevent water contamination (Wilhelm et al., 1994). However, some of the chemicals, especially PPCPs, cannot be completely removed from the septic system. In general, the removal efficiencies of PPCPs in septic systems depend on the physicochemical properties (e.g., water solubility) and environmental conditions. The remaining PPCPs in the septic tank effluent can cause pollution to soil, migrate to groundwater, and harm environmental health (Fig. 3.5) (Yang et al., 2017).

Yang et al. (2017) constructed two in situ conventional drain fields (drip dispersal and gravel trench) and an advanced drain field of septic systems to investigate the fate and transport of micropollutants in shallow groundwater. In their study, PPCP concentrations in septic tank effluent, unsaturated soil-water (0.31–1.07 m deep), and groundwater (3.1–3.4 m deep) samples were analyzed to calculate the removal efficiency of the septic tank and the transport of PPCPs from the drain fields to groundwater. In their study, PPCPs showed similar behavior in different septic tank systems. In the drainage zone, the most frequently detected pharmaceuticals in soil-water were sulfamethoxazole (85%) and acetaminophen (83%), whereas the least detected were carbamazepine (23%) and ibuprofen (6%) (Yang et al., 2017). Median concentrations of detected PPCPs in the soil-water were 0.8–20 ng/L. Benzene sulfonamide, a transformation product of sulfamethoxazole, was also present (<8–15 ng/L) in the soil-water. Among the personal care products, triclocarban (42%), ethylparaben (33%), and methylparaben (45%) were frequently detected, whereas butylparaben, heptyl paraben, and propylparaben were found in <20% of the soil-water samples (Yang et al., 2017).

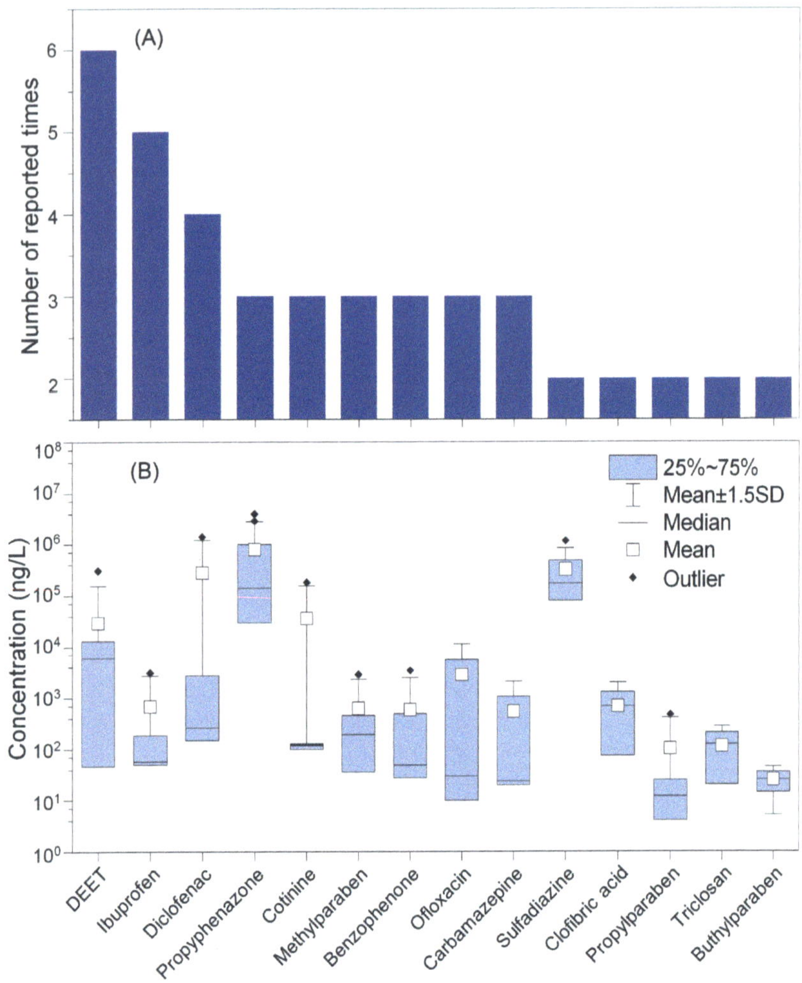

FIGURE 3.4 Numbers of reported times (A) and concentrations of selected PPCPs (B) in the groundwater adjacent to the landfills. *From Yu, X., Sui, Q., Lyu, S., Zhao, W., Liu, J., Cai, Z., Yu, G., & Barcelo, D. (c. 2020). Municipal solid waste landfills: An underestimated source of pharmaceutical and personal care products in the water environment.* Environmental Science and Technology, 54*(16), 9757–9768. https://doi.org/10.1021/acs.est.0c00565.*

The main attenuation processes of PPCPs attenuation in the environment are proposed to be dilution, dispersion, sorption, and microbial degradation (Yang et al., 2017).

3.4.3 Agriculture area

Agriculture areas are considered to be PPCP hot zones due to the applications of manure and wastewater treatment plant–related products such as

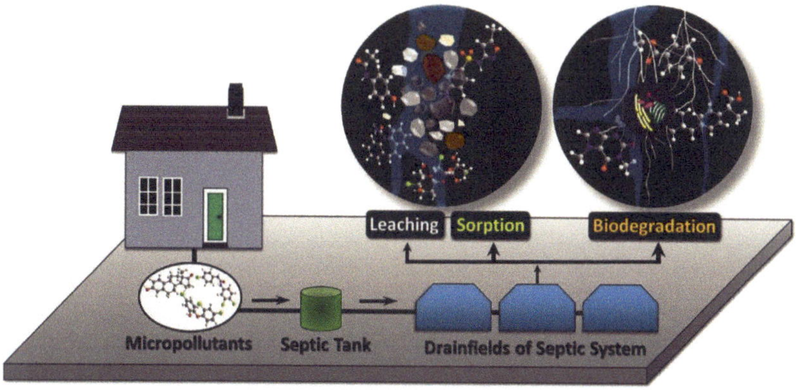

FIGURE 3.5 Septic systems may contribute pharmaceutical and personal care products to soil and groundwater. *From Yang, Y., Toor, G., Wilson, P., & Williams, C. (c. 2016). Septic systems as hot-spots of pollutants in the environment: Fate and mass balance of micropollutants in septic drainfields.* Science of the Total Environment, 566–567, *1535–1544. https://doi.org/10.1016/j.scitotenv.2016.06.043.*

biosolid or reclaimed water. PPCPs introduced to agricultural land can cause soil and groundwater pollutions.

The study conducted by Lee et al. (2019) compared PPCPs, including 33 pharmaceuticals and the 5 ASs in groundwater in the rural agricultural and rural nonagricultural areas in South Korea. A total of 31 target compounds were detected in agricultural groundwater at concentrations from not detected (ND) to 49.3 ng/L for PPCPs, ND to 1340 ng/L for ASs, and ND to 116 ng/L for pesticides (Lee et al., 2019). Four target compounds (two PPCPs and two ASs) were detected in rural nonagricultural groundwater in the range of 0.085–5.74 ng/L for PPCPs and 5.64–1330 ng/L for ASs (Fig. 3.6) (Lee et al., 2019). Among the target compounds, ASs, especially acesulfame, were predominantly detected in both agricultural (mean: 32.9 ng/L) and nonagricultural (mean: 536 ng/L) groundwater, but different occurrences of patterns were observed according to the sources of contamination (Fig. 3.7).

3.4.3.1 Biosolids or manure land application

Biosolids from water treatment plants are rich in nutrients and commonly used in crops as fertilizer in agriculture production. Due to the low removal efficiency of PPCPs in the wastewater treatment plant, biosolids may contain large amounts of PPCPs. Similarly, land application of manure as fertilizer can also introduce PPCPs to agricultural land, especially manure from animal farms containing veterinary antibiotics (Kemper et al., 2008). Therefore biosolids and manure application in agricultural land are identified sources of PPCPs into environments.

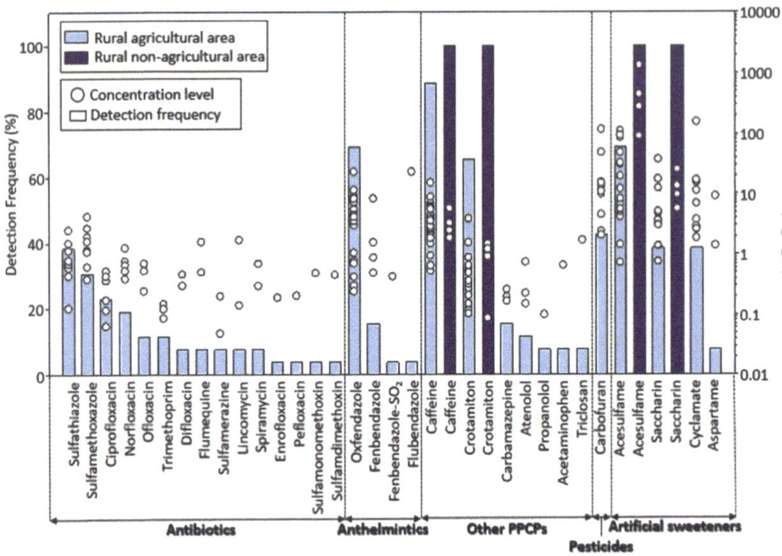

FIGURE 3.6 Concentrations and detection frequencies of pharmaceutical and personal care products, artificial sweeteners, and pesticides in groundwater samples from rural agricultural and rural nonagricultural areas. *Vertical bar*, detection frequency (%); *circles*, concentration (ng/L). *From Lee, H., Kim, K., Hamm, S., Kim, M., Kim, H., & Oh, J. (c. 2019). Occurrence and distribution of pharmaceutical and personal care products, artificial sweeteners, and pesticides in groundwater from an agricultural area in Korea.* Science of the Total Environment, 659, *168–176. https://doi.org/10.1016/j.scitotenv.2018.12.258.*

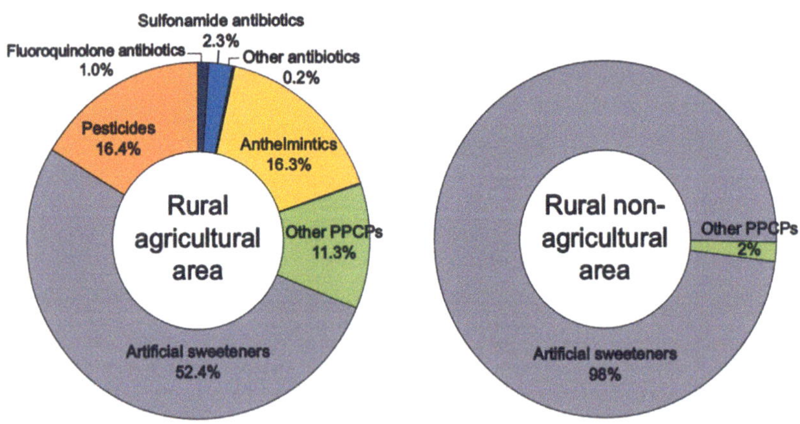

FIGURE 3.7 Comparison of the distribution of the six groups of groundwater pollutants between rural agricultural and rural nonagricultural areas. *From Lee, H., Kim, K., Hamm, S., Kim, M., Kim, H., & Oh, J. (c. 2019). Occurrence and distribution of pharmaceutical and personal care products, artificial sweeteners, and pesticides in groundwater from an agricultural area in Korea.* Science of the Total Environment, 659, *168–176. https://doi.org/10.1016/j.scitotenv.2018.12.258.*

To study the fate and transport of hormones introduced with biosolids, Gottschall et al. (2013) conducted a field study in agricultural land with the application of dewatered municipal biosolids at a rate of 22 mg dry weight per ha. Pre- and postapplication hormones concentrations were measured in tile drainage water, groundwater (2, 4, 6 m depth), surface soil cores, and dewatered municipal biosolids aggregates incorporated in the soil (~0.2 m depth) for a period of roughly 1-year postapplication. Hormones were detected upto 1-year postapplication in soil and in dewatered municipal biosolids aggregates. Hormones (androsterone, desogestrel, estrone) were detected briefly in tile water samples (22 days and ~2 months postapplication), at 2–34 ng/L concentrations but not detectable in groundwater. Based on this study, the transitory contamination from hormones is unlikely to result in any significant exposure risk in subsurface drainage and groundwater (Gottschall et al., 2013). However, some high mobility antibiotics can travel long distances downwards. Kivits et al. (2018) reported that sulfonamides were detected at all measured depths down to 23 m below surface level due to the manure application to agricultural land.

For manure land application, Conde-Cid et al. (2018) studied the presence of tetracyclines (tetracycline, oxytetracycline, chlortetracycline, and doxycycline) and sulfonamides (sulfadiazine, sulfamethazine, sulfachlorpyridazine, and sulfamethoxypyridazine) in manures, soils, and crops from Galicia (Spain), where a high number of cattle, pig, and poultry farms exist. The presence of antibiotics was detected in 42% of the manures, 17% of the soils, and 44% of crop samples, with maximum concentrations of 106.0 mg/kg for individual antibiotics in manures and 0.6 mg/kg in soils and plants (Conde-Cid et al., 2018). Based on their study, several antibiotics can simultaneously present in manures and plants. Pig slurries showed the highest antibiotic concentrations, as well as the highest number of different antibiotics. Antibiotics were detected in 71% of grass and corn samples and 33% of wheat grain samples, while they were not detected in potato samples (Conde-Cid et al., 2018).

Yeom et al. (2017) studied antibiotics occurrence in agricultural land using fertilizer produced from livestock excretions. Surface soil samples were collected at depths of 0, 7, and 15 cm from farmland in Incheon (South Korea) to test the concentrations of tetracycline, oxytetracycline, and sulfathiazole. In the surface soil, three compounds were detected: tetracycline (17.74 μg/kg), oxytetracycline (0.78 μg/kg), and sulfathiazole (0.23 μg/kg) (Yeom et al., 2017). However, except for sulfathiazole, tetracycline and oxytetracycline were not detected in the deeper samples (7 and 15 cm). Overall, tetracycline was shown strong adsorption to the organic matter and metals in soil therefore its mobility reduced.

3.4.3.2 Reclaimed water irrigation

Irrigation with reclaimed water is an important method to solve the issue of water scarcity. It is considered the best way to recycle the reclaimed water since it reduces nutrient discharges into surface water and benefits the soil and plants by reuse the remaining nutrients in reclaimed water (Anderson, 2003). Land application of reclaimed water is also beneficial for recharging groundwater aquifers and avoiding direct pollutants discharged to surface waters. However, nonregulated organic wastewater pollutants, such as PPCPs, can be introduced to agricultural land through reclaimed water irrigation.

Kibuye et al. (2019) conducted a 14-month study (October 2016 through December 2017) at a spray-irrigation site where effluent has been spray-irrigated since the early 1980s using 13 groundwater monitoring wells to evaluate the fate and potential risks of selected PPCPs. Sulfamethoxazole (40%) and caffeine (32%) are the most frequently detected compounds in groundwater samples. Average PPCPs concentrations in the groundwater samples were found to be nearly two orders of magnitude lower than concentrations in the effluents (Kibuye et al., 2019).

Ma et al. (2018) investigated 11 PPCPs in groundwater near irrigated farmland with a vadose zone (upto 16 m) in Beijing (China) and observed a unique bell-shape pattern vertical distribution in the vadose zone soils with wastewater irrigation, peaking at 3–6 m depth in meters below ground level (Fig. 3.8) (Ma et al., 2018). Redundancy analysis indicated that available

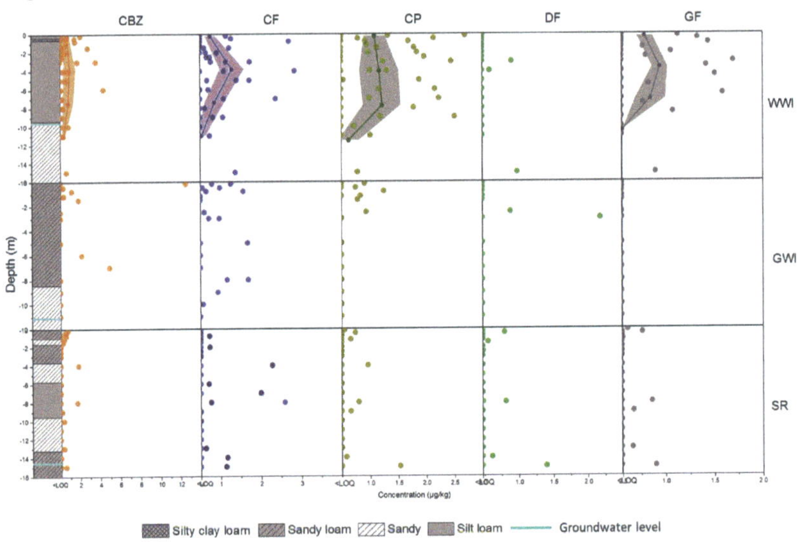

FIGURE 3.8 Vertical distribution of five pharmaceutical and personal care products (PPCPs) against sampling depth in boreholes with lithology and water table presented. *Point trend lines* illustrate the bell-shape pattern of the target PPCPs in the vadose zone soils irrigated with wastewater, estimated by the mean concentrations in the vadose zone soils at 0–2.5, 3–5.5, 6–8.5, and 9–11 m below ground level. The *shaded areas* designate mean ± standard error. *From Ma, L., Liu, Y., Zhang, J., Yang, Q., Li, G., & Zhang, D. (c. 2018). Impacts of irrigation water sources and geochemical conditions on vertical distribution of pharmaceutical and personal care products (PPCPs) in the vadose zone soils.* Science of the Total Environment, 626, *1148–1156. https://doi.org/10.1016/j.scitotenv.2018.01.168.*

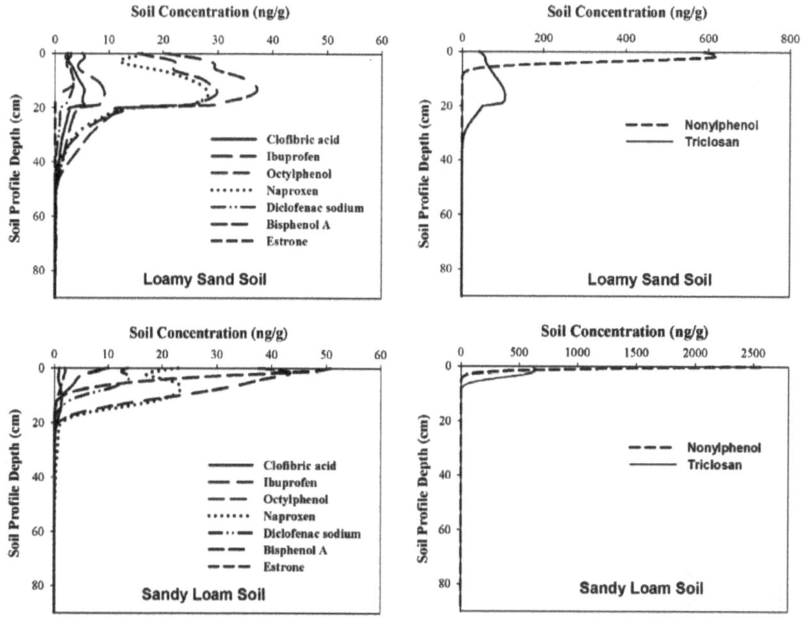

FIGURE 3.9 Pharmaceutical and personal care product distributions in the 90 cm soil profile, following 10 years of simulated reclaimed water irrigation of turfgrass fields. *From Chen, W., Xu, J., Lu, S., Jiao, W., Wu, L., & Chang, A. C. (c. 2013). Fates and transport of PPCPs in soil receiving reclaimed water irrigation.* Chemosphere, *93(10), 2621–2630. https://doi.org/10.1016/j.chemosphere.2013.09.088.*

potassium and silt content are highly correlated with PPCPs migration in the subsurface (Ma et al., 2018). Chen et al. (2013a) and Ruthven (1984) also studied selected PPCPs migration in agricultural land after 10 years of irrigation. They found that the properties of the soils affected PPCPs' distributions in soil and the overall vertical distributions of selected PPCPs were highly related to the adsorption coefficiency (Chen et al., 2013b). The sandy loam soil exhibited high affinity to the PPCPs with PPCPs is mostly accumulated with within surface 0–20 cm soil layer; however, in the loamy sand soil PPCPs were detected at soil profile depth beyond 40 cm due to relatively lower sorption capacity (Fig. 3.9) (Chen et al., 2013b). Recently models also developed to simulate the fate of PPCPs during wastewater reuse for agriculture irrigation using published research data (Shahriar et al., 2021).

3.4.4 Groundwater

Due to the long history of PPCPs productions and applications, PPCPs have traveled through soil profiles and entered into groundwater. PPCPs detections in shallow groundwater have been reported globally in the last two decades

(Bartelt-Hunt et al., 2011; Chen et al., 2018; Kuroda et al., 2012). Deep groundwater contamination of PPCPs also has been reported lately indicating large vertical distributions of PPCPs. Kuroda et al. (2012) assessed the occurrences of PPCPs in the unconfined and confined groundwater aquifers upto 260 m below surface. They found that PPCPs were more frequently detected in unconfined aquifers (66% detection rate, 21/32 groundwaters) than in confined aquifers (44% detection rate, 7/16 samples), and only caffeine was detected in the two groundwaters of more than 200 m. The authors suggested that PPCP occurrences in the confined aquifer may be due to groundwater contamination at the recharge zone, either by sewer exfiltration or by leakage from vault toilets before 1994 (Kuroda et al., 2012).

Kobayashi et al. (2021) studied PPCPs in groundwater after a series of earthquakes occurred in Kumamoto, which caused serious damage to underground sewerage networks. The detection frequency of caffeine, carbamazepine, crotamiton, ibuprofen, and anthropogenic gadolinium ranged from 29% to 45%. The concentrations of the pharmaceuticals in the groundwater were lower than those reported in previous studies. Based on their research, the

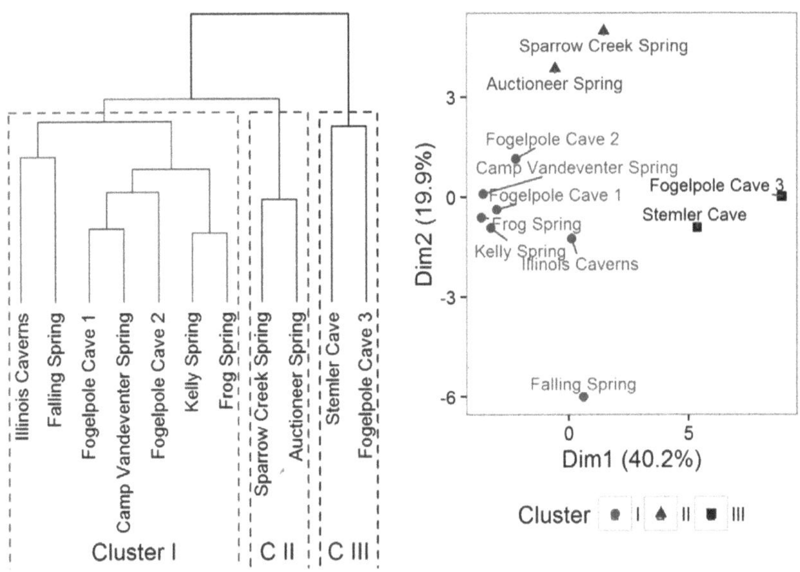

FIGURE 3.10 Hierarchical clustering (*left panel*) and k-means clustering (*right panel*) of 11 karst sites. k-means clustering is visually depicted through a biplot of the data showing the two principal components of the dataset. Clustering determined that sites fell into three subsets that had similar relationships between PPCP sum and other factors (e.g., ions, stream flow, etc.). *From Dodgen, L., Kelly, W., Panno, S., Taylor, S., Armstrong, D., Wiles, K., Zhang, Y., & Zheng, W. (c. 2017). Characterizing pharmaceutical, personal care product, and hormone contamination in a karst aquifer of southwestern Illinois, USA, using water quality and stream flow parameters.* Science of the Total Environment, 578, *281–289. https://doi.org/10.1016/j.scitotenv.2016.10.103.*

effect of the earthquakes on sewer exfiltration and groundwater contamination was small, probably because the damaged sewers were quickly repaired (Kobayashi et al., 2021).

Karst aquifers are drinking water sources for 25% of the global population; however, karst areas are vulnerable to water contamination due to the geological setting (Stevanović, 2018). Dodgen et al. (2017) studied PPCPs in karst aquifers over a 17-month period, 58 water samples were collected from 13 sites in the Salem Plateau, a karst region in southwestern Illinois, United States. Hormones were detected in only 23% of samples, with concentrations of 2.2–9.1 ng/L. In contrast, other PPCPs were identified in 89% of groundwater samples. The two most commonly detected PPCPs were the antimicrobial triclocarban, in 81% of samples, and the cardiovascular drug gemfibrozil, in 57%. Analytical results indicated that sites with similar PPCP contamination profiles can be grouped into three clusters (Fig. 3.10): PPCPs contamination in Cluster I sites was related to stream height, manganese, boron, and heterotrophic bacteria; Cluster II sites were characterized by groundwater temperature, specific conductivity, sodium, and calcium; Cluster III sites were characterized by dissolved oxygen and barium.1

3.5 Conclusion and perspectives

This chapter reviewed PPCPs fate and transport in soil and groundwater system. PPCPs' physicochemical properties, such as hydrophobicity and polarity, are critical for understand their interactions with soil components and degradation potential.

The major interaction between PPCPs and soil matrix is sorption. The sorption process is generally influenced by intrinsic properties of PPCPs and soil components: different PPCPs group can show completely different retention behavior to soil due to their structural differences; similarly, different soil components showed different sorption capacity to the same group of PPCPs due to the variations of surface properties, such as surface area, surface charge, surface functional groups, and surface hydrophobic or hydrophilic properties. Environmental properties such as pH value and coexisting cations affect the sorption process by changing the properties of both sorbents and sorbates (e.g., PPCPs speciation, soil surface charge, and the speciation of soil functional groups). The sorption mechanism of PPCPs includes electrostatic interaction, cationic bridging, hydrogen bonding, and surface complexation. Batch experiment and column transport coupled with modeling methods are often used to quantify the sorption capacities and sorption affinities. Recently instrument analysis is also used to characterize the interactions between PPCPs and soil at the molecular level (e.g., FTIR, SEM, XPS, etc.). Most PPCP transport and modeling field studies were conducted during aquifer recharge management characterized by the bank infiltration process.

Soil microbial and plants also impact the fate of PPCPs. PPCPs degradation is mainly caused by microbial activity, with aerobic microbial more effective than anaerobic microbial. PPCPs degradation rates are influenced by soil textural, soil aerobic status, and PPCPs property. For antibiotics, high initial concentration can inhibit their degradation due to their antibiotic properties. The plant system has a significant influence on the fate of PPCPs, PPCPs can accumulate in the root zone or translate to the upper part of plants. PPCPs speciation and hydrophobic properties influence the plant uptake potential. Plant uptake PPCPs through protein transport channels with passive or energy-demand pathways.

PPCPs as synthetic compounds are mainly detected in several hot zones such as landfills, septic systems, agriculture areas, and groundwater. For PPCPs in the landfills, characteristics of landfill sites (i.e., landfill ages), sociological factors (i.e., economic development), and aquifer properties (i.e., permeability, redox potential, and seepage velocity) all play critical roles in determining the fate and transport of PPCPs in soils. Agriculture land became PPCPs hot zones due to the applications of manure and wastewater treatment plant—related products such as biosolid or reclaimed water. Due to repeated applications, PPCPs accumulation and downward migration are proven from recent studies. The vertical distributions of PPCPs is determined by their sorption behavior in soil matrix. PPCP transport to groundwater is reported mostly in shallow aquifer. Factors influence PPCPs distribution in groundwater, which include stream height, groundwater temperature, groundwater solution chemistry, dissolved oxygen, and microorganisms.

It is recommended that expanding research efforts on the fate and transport of PPCPs in soils and groundwater should be made in the following aspects: (1) There is a great need to conduct additional large-scale field studies since only a few studies have focused on PPCP vertical distributions in the subsurface. (2) Future investigations on biodegradation or chemical oxidation processes of PPCPs in soils and groundwater should be promoted. (3) Environmental fate and transport of certain groups of PPCPs such as fragrances, antioxidants, and prescription drugs are less explored than that of antibiotics. (4) The modeling techniques should be improved to better represent the key factors that control PPCPs retention and transport in the complex multidimensional system. (5) Plant roots interactions with PPCPs in the rhizosphere are still unclear. Research on PPCP translocation or plant uptake mechanisms is needed, especially under different soil conditions.

References

Ahmed, A. A., Thiele-Bruhn, S., Aziz, S. G., Hilal, R. H., Elroby, S. A., Al-Youbi, A. O., Leinweber, P., & Kühn, O. (2015). Interaction of polar and nonpolar organic pollutants with soil organic matter: Sorption experiments and molecular dynamics simulation. *Science of the Total Environment*, *508*, 276–287. Available from https://doi.org/10.1016/j.scitotenv.2014.11.087.

Anderson, J. (2003). The environmental benefits of water recycling and reuse. *Water Supply, 3* (4), 1–10. Available from https://doi.org/10.2166/ws.2003.0041.

Avisar, D., Primor, O., Gozlan, I., & Mamane, H. (2010). Sorption of sulfonamides and Tetracyclines to montmorillonite clay. *Water Air Soil Pollut, 209*, 439–450.

Ayawei, N., Ebelegi, A. N., & Wankasi, D. (2017). Modelling and interpretation of adsorption isotherms. *Journal of Chemistry, 2017*. Available from https://doi.org/10.1155/2017/3039817.

Barkow, I. S., Oswald, S. E., Lensing, H. J., & Munz, M. (2020). Seasonal dynamics modifies fate of oxygen, nitrate, and organic micropollutants during bank filtration—temperature-dependent reactive transport modeling of field data. *Environmental Science and Pollution Research*. Available from https://doi.org/10.1007/s11356-020-11002-9.

Barnes, K. K., Kolpin, D. W., Furlong, E. T., Zaugg, S. D., Meyer, M. T., & Barber, L. B. (2008). A national reconnaissance of pharmaceuticals and other organic wastewater contaminants in the United States—(I) Groundwater. *Science of the Total Environment, 402*(2–3), 192–200. Available from https://doi.org/10.1016/j.scitotenv.2008.04.028.

Bartelt-Hunt, S., Snow, D. D., Damon-Powell, T., & Miesbach, D. (2011). Occurrence of steroid hormones and antibiotics in shallow groundwater impacted by livestock waste control facilities. *Journal of Contaminant Hydrology, 123*(3–4), 94–103. Available from https://doi.org/10.1016/j.jconhyd.2010.12.010.

Bialk-Bielinska, A., Maszkowska, J., Mrozik, W., Bielawska, A., Kolodziejska, M., Palavinskas, R., et al. (2012). Sulfadimethoxine and sulfaguanidine: their sorption potential on natural soils. *Chemosphere, 86*, 1059–1065.

Boonsaner, M., & Hawker, D. W. (2015). Transfer of oxytetracycline from swine manure to three different aquatic plants: Implications for human exposure. *Chemosphere, 122*, 176–182. Available from https://doi.org/10.1016/j.chemosphere.2014.11.045.

Briggs, G. G., Bromilow, R. H., & Evans, A. A. (1982). Relationships between lipophilicity and root uptake and translocation of non-ionised chemicals by barley. *Pesticide Science, 13*(5), 495–504. Available from https://doi.org/10.1002/ps.2780130506.

Burris, D. R., Allen-King, R. M., Manoranjan, V. S., Campbell, T. J., Loraine, G. A., & Deng, B. (1998). Chlorinated ethene reduction by cast iron: Sorption and mass transfer. *Journal of Environmental Engineering, 124*(10), 1012–1019. Available from https://doi.org/10.1061/(ASCE)0733-9372(1998)124:10(1012).

Call, J. J., Rakshit, S., & Essington, M. E. (2019). The adsorption of tylosin by montmorillonite and vermiculite: exchange selectivity and intercalation. *Soil Sci. Soc. Am. J., 83*, 584–596.

Chang, P.-H., Li, Z., Yu, T.-L., Munkhbayer, S., Kuo, T.-H., Hung, Y.-C., et al. (2009b). Sorptive removal of tetracycline from water by palygorskite. *J. Hazard. Mater., 165*, 148–155.

Chauhan, M., Saini, V. K., & Suthar, S. (2020). Removal of pharmaceuticals and personal care products (PPCPs) from water by adsorption on aluminum pillared clay. *Journal of Porous Materials, 27*(2), 383–393. Available from https://doi.org/10.1007/s10934-019-00817-8.

Chen, H., Gao, B., Li, H., & Ma, L. Q. (2011). Effects of pH and ionic strength on sulfamethoxazole and ciprofloxacin transport in saturated porous media. *Journal of Contaminant Hydrology, 126*(1–2), 29–36. Available from https://doi.org/10.1016/j.jconhyd.2011.06.002.

Chen, H., Ma, L. Q., Gao, B., & Gu, C. (2013a). Effects of Cu and Ca cations and Fe/Al coating on ciprofloxacin sorption onto sand media. *Journal of Hazardous Materials, 252–253*, 375–381. Available from https://doi.org/10.1016/j.jhazmat.2013.03.014.

Chen, L., Lang, H., Liu, F., Jin, S., & Yan, T. (2018). Presence of antibiotics in shallow groundwater in the northern and southwestern regions of China. *Groundwater, 56*(3), 451–457. Available from https://doi.org/10.1111/gwat.12596.

Chen, W., Xu, J., Lu, S., Jiao, W., Wu, L., & Chang, A. C. (2013b). Fates and transport of PPCPs in soil receiving reclaimed water irrigation. *Chemosphere, 93*(10), 2621–2630. Available from https://doi.org/10.1016/j.chemosphere.2013.09.088.

Christl, I., Ruiz, M., Schinidt, J. R., & Pedersen, J. A. (2016). Clarithromycin and tetracycline binding to soil humic acid in the absence and presence of calcium. *Environ. Sci. Technol., 50*, 9933–9942.

Clarke, B. O., Anumol, T., Barlaz, M., & Snyder, S. A. (2015). Investigating landfill leachate as a source of trace organic pollutants. *Chemosphere, 127*, 269–275. Available from https://doi.org/10.1016/j.chemosphere.2015.02.030.

Conde-Cid, M., Álvarez-Esmorís, C., Paradelo-Núñez, R., Nóvoa-Muñoz, J. C., Arias-Estévez, M., Álvarez-Rodríguez, E., Fernández-Sanjurjo, M. J., & Núñez-Delgado, A. (2018). Occurrence of tetracyclines and sulfonamides in manures, agricultural soils and crops from different areas in Galicia (NW Spain). *Journal of Cleaner Production, 197*, 491–500. Available from https://doi.org/10.1016/j.jclepro.2018.06.217.

Conde-Cid, M., Fernández-Calviño, D., Fernández-Sanjurjo, M., Núñez-Delgado, A., Álvarez-Rodríguez, E., & Arias-Estévez, M. (2020). Effects of pine bark amendment on the transport of sulfonamide antibiotics in soils. *Chemosphere, 248*, 126041. Available from https://doi.org/10.1016/j.chemosphere.2020.126041.

Conkle, J. L., Gan, J., & Anderson, M. A. (2012). Degradation and sorption of commonly detected PPCPs in wetland sediments under aerobic and anaerobic conditions. *Journal of Soils and Sediments, 12*(7), 1164–1173. Available from https://doi.org/10.1007/s11368-012-0535-8.

Dabizha, A., & Kersten, M. (2020). Exothermic adsorption of chromate by goethite. *Applied Geochemistry*, 123. Available from https://doi.org/10.1016/j.apgeochem.2020.104785.

Daughton, C. G. (2001). Pharmaceuticals and Personal Care Products in the Environment: Overarching Issues and Overview. In *Pharmaceuticals and Care Products in the Environment*, (791, pp. 2–38). American Chemical Society.

Daughton, C. G., & Ternes, T. A. (1999). Pharmaceuticals and personal care products in the environment: Agents of subtle change? *Environmental Health Perspectives, 107*(6), 907–938. Available from https://doi.org/10.1289/ehp.99107s6907.

Dodgen, L. K., Kelly, W. R., Panno, S. V., Taylor, S. J., Armstrong, D. L., Wiles, K. N., Zhang, Y., & Zheng, W. (2017). Characterizing pharmaceutical, personal care product, and hormone contamination in a karst aquifer of southwestern Illinois, USA, using water quality and stream flow parameters. *Science of the Total Environment, 578*, 281–289. Available from https://doi.org/10.1016/j.scitotenv.2016.10.103.

Dodgen, L. K., Ueda, A., Wu, X., Parker, D. R., & Gan, J. (2015). Effect of transpiration on plant accumulation and translocation of PPCP/EDCs. *Environmental Pollution, 198*, 144–153. Available from https://doi.org/10.1016/j.envpol.2015.01.002.

Ebele, A. J., Abou-Elwafa Abdallah, M., & Harrad, S. (2017). Pharmaceuticals and personal care products (PPCPs) in the freshwater aquatic environment. *Emerging Contaminants, 3*(1), 1–16. Available from https://doi.org/10.1016/j.emcon.2016.12.004.

Eggen, T., Asp, T. N., Grave, K., & Hormazabal, V. (2011). Uptake and translocation of metformin, ciprofloxacin and narasin in forage- and crop plants. *Chemosphere, 85*(1), 26–33. Available from https://doi.org/10.1016/j.chemosphere.2011.06.041.

Eggen, T., & Lillo, C. (2012). Antidiabetic II drug metformin in plants: Uptake and translocation to edible parts of cereals, oily seeds, beans, tomato, squash, carrots, and potatoes. *Journal of Agricultural and Food Chemistry*, *60*(28), 6929–6935. Available from https://doi.org/10.1021/jf301267c.

Fick, J., Söderström, H., Lindberg, R. H., Phan, C., Tysklind, M., & Larsson, D. G. J. (2009). Contamination of surface, ground, and drinking water from pharmaceutical production. *Environmental Toxicology and Chemistry*, *28*(12), 2522–2527. Available from https://doi.org/10.1897/09-073.1.

Figueroa, R. A., Leonard, A., & Mackay, A. A. (2004). Modeling tetracycline antibiotic sorption to clays. *Environmental Science and Technology*, *38*(2), 476–483. Available from https://doi.org/10.1021/es0342087.

Gottschall, N., Topp, E., Edwards, M., Payne, M., Kleywegt, S., Russell, P., & Lapen, D. R. (2013). Hormones, sterols, and fecal indicator bacteria in groundwater, soil, and subsurface drainage following a high single application of municipal biosolids to a field. *Chemosphere*, *91*(3), 275–286. Available from https://doi.org/10.1016/j.chemosphere.2012.10.108.

Grathwohl, P. (2012). *Diffusion in natural porous media: Contaminant transport, sorption/desorption and dissolution kinetics* (Vol. 1). Springer Science & Business Media.

Gu, C., & Karthikeyan, K. G. (2008). Sorption of the antibiotic tetracycline to humic-mineral complexes. *Journal of Environmental Quality*, *37*(2), 704–711. Available from https://doi.org/10.2134/jeq2007.0030.

Gu, C., Karthikeyan, K. G., Sibley, S. D., & Pedersen, J. A. (2007). Complexation of the antibiotic tetracycline with humic acid. *Chemosphere*, *66*(8), 1494–1501. Available from https://doi.org/10.1016/j.chemosphere.2006.08.028.

Guo, X., Tu, B., Ge, J., Yang, C., Song, X., & Dang, Z. (2016). Sorption of tylosin and sulfamethazine on solid humic acid. *J. Environ. Sci.*, *43*, 208–215.

Guo, X., Yang, C., Wu, Y., & Dang, Z. (2014). The influences of pH and ionic strength on the sorption of tylosin on goethite. *Environmental Science and Pollution Research*, *21*(4), 2572–2580. Available from https://doi.org/10.1007/s11356-013-2174-z.

Han, Z., Ma, H., Shi, G., He, L., Wei, L., & Shi, Q. (2016). A review of groundwater contamination near municipal solid waste landfill sites in China. *Science of the Total Environment*, *569–570*, 1255–1264. Available from https://doi.org/10.1016/j.scitotenv.2016.06.201.

Harbaugh, A.W., Banta, E.R., Hill, M.C., & McDonald, M.G. (2000). Modflow-2000, the United States geological survey modular ground-water model-user guide to modularization concepts and the ground-water flow process. Open-File Report. United States Geological Survey.

Henzler, A. F., Greskowiak, J., & Massmann, G. (2014). Modeling the fate of organic micropollutants during river bank filtration (Berlin, Germany). *Journal of Contaminant Hydrology*, *156*, 78–92. Available from https://doi.org/10.1016/j.jconhyd.2013.10.005.

Ho, Y. S., & McKay, G. (1999). Pseudo-second order model for sorption processes. *Process Biochemistry*, *34*(5), 451–465. Available from https://doi.org/10.1016/S0032-9592(98)00112-5.

Hyland, K. C., Blaine, A. C., & Higgins, C. P. (2015). Accumulation of contaminants of emerging concern in food crops-part 2: Plant distribution. *Environmental Toxicology and Chemistry*, *34*(10), 2222–2230. Available from https://doi.org/10.1002/etc.3068.

Ji, L. L., Chen, W., Zheng, S. R., Xu, Z. Y., & Zhu, D. Q. (2009). Adsorption of sulfonamide antibiotics to multiwalled carbon nanotubes. *Langmuir*, *25*, 11608–11613.

Karnjanapiboonwong, A., Morse, A. N., Maul, J. D., & Anderson, T. A. (2010). Sorption of estrogens, triclosan, and caffeine in a sandy loam and a silt loam soil. *Journal of Soils and Sediments, 10*(7), 1300–1307. Available from https://doi.org/10.1007/s11368-010-0223-5.

Kemper, N., Färber, H., Skutlarek, D., & Krieter, J. (2008). Analysis of antibiotic residues in liquid manure and leachate of dairy farms in Northern Germany. *Agricultural Water Management, 95*(11), 1288–1292. Available from https://doi.org/10.1016/j.agwat.2008.05.008.

Kibuye, F. A., Gall, H. E., Elkin, K. R., Ayers, B., Veith, T. L., Miller, M., Jacob, S., Hayden, K. R., Watson, J. E., & Elliott, H. A. (2019). Fate of pharmaceuticals in a spray-irrigation system: From wastewater to groundwater. *Science of the Total Environment, 654*, 197–208. Available from https://doi.org/10.1016/j.scitotenv.2018.10.442.

Kivits, T., Broers, H. P., Beeltje, H., van Vliet, M., & Griffioen, J. (2018). Presence and fate of veterinary antibiotics in age-dated groundwater in areas with intensive livestock farming. *Environmental Pollution, 241*, 988–998. Available from https://doi.org/10.1016/j.envpol.2018.05.085.

Kobayashi, J., Kuroda, K., Miyamoto, C., Uchiyama, Y., Sankoda, K., & Nakajima, D. (2021). Evaluating sewer exfiltration in groundwater by pharmaceutical tracers after the 2016 Kumamoto earthquakes, Japan. *Journal of Hazardous Materials, 411*. Available from https://doi.org/10.1016/j.jhazmat.2021.125183.

Kolz, A. C., Ong, S. K., & Moorman, T. B. (2005). Sorption of tylosin onto swine manure. *Chemosphere, 60*, 284–289.

Kong, W., Li, C., Dolhi, J. M., Li, S., He, J., & Qiao, M. (2012). Characteristics of oxytetracycline sorption and potential bioavailability in soils with various physical-chemical properties. *Chemosphere, 87*(5), 542–548. Available from https://doi.org/10.1016/j.chemosphere.2011.12.062.

Kulshrestha, P., Giese, R. F., & Aga, D. S. (2004). Investigating the molecular interactions of oxytetracycline in clay and organic matter: Insights on factors affecting its mobility in soil. *Environmental Science and Technology, 38*(15), 4097–4105. Available from https://doi.org/10.1021/es034856q.

Kuppusamy, S., Kakarla, D., Venkateswarlu, K., Megharaj, M., Yoon, Y. E., & Lee, Y. B. (2018). Veterinary antibiotics (VAs) contamination as a global agro-ecological issue: A critical view. *Agriculture, Ecosystems and Environment, 257*, 47–59. Available from https://doi.org/10.1016/j.agee.2018.01.026.

Kuroda, K., Murakami, M., Oguma, K., Muramatsu, Y., Takada, H., & Takizawa, S. (2012). Assessment of groundwater pollution in Tokyo using PPCPs as sewage markers. *Environmental Science and Technology, 46*(3), 1455–1464. Available from https://doi.org/10.1021/es202059g.

Langmuir, I. (1918). The adsorption of gases on plane surfaces of glass, mica and platinum. *Journal of the American Chemical Society, 40*(9), 1361–1403. Available from https://doi.org/10.1021/ja02242a004.

Lapworth, D. J., Baran, N., Stuart, M. E., & Ward, R. S. (2012). Emerging organic contaminants in groundwater: A review of sources, fate and occurrence. *Environmental Pollution, 163*, 287–303. Available from https://doi.org/10.1016/j.envpol.2011.12.034.

Largitte, L., & Pasquier, R. (2016). A review of the kinetics adsorption models and their application to the adsorption of lead by an activated carbon. *Chemical Engineering Research and Design, 109*, 495–504. Available from https://doi.org/10.1016/j.cherd.2016.02.006.

Lee, H. J., Kim, K. Y., Hamm, S. Y., Kim, M. S., Kim, H. K., & Oh, J. E. (2019). Occurrence and distribution of pharmaceutical and personal care products, artificial sweeteners, and pesticides in groundwater from an agricultural area in Korea. *Science of the Total Environment, 659*, 168–176. Available from https://doi.org/10.1016/j.scitotenv.2018.12.258.

Lewis, J., & Sjöstrom, J. (2010). Optimizing the experimental design of soil columns in saturated and unsaturated transport experiments. *Journal of Contaminant Hydrology, 115*(1–4), 1–13. Available from https://doi.org/10.1016/j.jconhyd.2010.04.001.

Li, F., Pan, B., Liang, N., Chang, Z., Zhou, Y., Wang, L., & Xing, B. (2017). Reactive mineral removal relative to soil organic matter heterogeneity and implications for organic contaminant sorption. *Environmental Pollution, 227*, 49–56.

Li, Y., Wang, H., Liu, X., Zhao, G., & Sun, Y. (2016). Dissipation kinetics of oxytetracycline, tetracycline, and chlortetracycline residues in soil. *Environ. Sci. Pollut. Res., 23*, 13822–13831.

Liu, Q., Li, M., Zhang, F., Yu, H., Zhang, Q., & Liu, X. (2017a). The removal of trimethoprim and sulfamethoxazole by a high infiltration rate artificial composite soil treatment system. *Frontiers of Environmental Science & Engineering, 11*(2), 12.

Liu, X., Lu, S., Liu, Y., Meng, W., & Zheng, B. (2017b). Adsorption of sulfamethoxazole (SMZ) and ciprofloxacin (CIP) by humic acid (HA): characteristics and mechanism. *RSC advances, 7*(80), 50449–50458.

Ma, L., Liu, Y., Zhang, J., Yang, Q., Li, G., & Zhang, D. (2018). Impacts of irrigation water sources and geochemical conditions on vertical distribution of pharmaceutical and personal care products (PPCPs) in the vadose zone soils. *Science of the Total Environment, 626*, 1148–1156. Available from https://doi.org/10.1016/j.scitotenv.2018.01.168.

MacKay, A. A., & Canterbury, B. (2005). Oxytetracycline sorption to organic matter by metal-bridging. *Journal of Environmental Quality, 34*(6), 1964–1971. Available from https://doi.org/10.2134/jeq2005.0014.

Magga, Z., Tzovolou, D. N., Theodoropoulou, M. A., Dalkarani, T., Pikios, K., & Tsakiroglou, C. D. (2008). Soil column experiments used as a means to assess transport, sorption, and biodegradation of pesticides in groundwater. *Journal of Environmental Science and Health - Part B Pesticides, Food Contaminants, and Agricultural Wastes, 43*(8), 732–741. Available from https://doi.org/10.1080/03601230802388868.

Mathews, S., Henderson, S., & Reinhold, D. (2014). Uptake and accumulation of antimicrobials, triclocarban and triclosan, by food crops in a hydroponic system. *Environmental Science and Pollution Research, 21*(9), 6025–6033. Available from https://doi.org/10.1007/s11356-013-2474-3.

Miller, E. L., Nason, S. L., Karthikeyan, K. G., & Pedersen, J. A. (2016). Root uptake of pharmaceuticals and personal care product ingredients. *Environmental Science and Technology, 50*(2), 525–541. Available from https://doi.org/10.1021/acs.est.5b01546.

Musson, S. E., & Townsend, T. G. (2009). Pharmaceutical compound content of municipal solid waste. *Journal of Hazardous Materials, 162*(2–3), 730–735. Available from https://doi.org/10.1016/j.jhazmat.2008.05.089.

Näsholm, T., Kielland, K., & Ganeteg, U. (2009). Uptake of organic nitrogen by plants. *New Phytologist, 182*(1), 31–48. Available from https://doi.org/10.1111/j.1469-8137.2008.02751.x.

Nham, H. T. T., Greskowiak, J., Nödler, K., Rahman, M. A., Spachos, T., Rusteberg, B., Massmann, G., Sauter, M., & Licha, T. (2015). Modeling the transport behavior of 16 emerging organic contaminants during soil aquifer treatment. *Science of the Total Environment, 514*, 450–458. Available from https://doi.org/10.1016/j.scitotenv.2015.01.096.

Oh, S., Shin, W. S., & Kim, H. T. (2016). Effects of pH, dissolved organic matter, and salinity on ibuprofen sorption on sediment. *Environmental Science and Pollution Research, 23*(22), 22882–22889. Available from https://doi.org/10.1007/s11356-016-7503-6.

Patel, H. (2019). Fixed-bed column adsorption study: a comprehensive review. *Applied Water Science*, *9*(3), 45. Available from https://doi.org/10.1007/s13201-019-0927-7.

Pavlovic, D. M., Curkovic, L., Blazek, D., & Zupan, J. (2014). The sorption of sulfamethazine on soil samples: isotherms and error analysis. *Sci. Total Environ.*, *497*, 543–552.

Qin, X., Liu, F., Zhao, L., Hou, H., Wang, G., Li, F., et al. (2016). Adsorption of levofloxacin to goethite: batch and column studies. *Environ. Eng. Sci.*, *33*, 235–241.

Rabølle, M., & Spliid, N. H. (2000). Sorption and mobility of metronidazole, olaquindox, oxytetracycline and tylosin in soil. *Chemosphere*, *40*(7), 715–722. Available from https://doi.org/10.1016/S0045-6535(99)00442-7.

Rakshita, S., Sarkar, D., Elzinga, E. J., Punamiya, P., & Datta, R. (2013). Mechanisms of ciprofloxacin removal by nano-sized magnetite. *J. Hazard. Mater.*, *246*, 221–226.

Rasamimanana, S., Lefèvre, G., & Dagnelie, R. V. H. (2017). Adsorption of polar organic molecules on sediments: Case-study on Callovian-Oxfordian claystone. *Chemosphere*, *181*, 296–303. Available from https://doi.org/10.1016/j.chemosphere.2017.04.086.

Ruthven, D. M. (1984). *Principles of adsorption and adsorption processes*. John Wiley & Sons.

Santos, L. H. M. L. M., Araújo, A. N., Fachini, A., Pena, A., Delerue-Matos, C., & Montenegro, M. C. B. S. M. (2010). Ecotoxicological aspects related to the presence of pharmaceuticals in the aquatic environment. *Journal of Hazardous Materials*, *175*(1–3), 45–95. Available from https://doi.org/10.1016/j.jhazmat.2009.10.100.

Sassman, S. A., & Lee, L. S. (2005). Sorption of three tetracyclines by several soils: Assessing the role of pH and cation exchange. *Environmental Science and Technology*, *39*(19), 7452–7459. Available from https://doi.org/10.1021/es0480217.

Schaffer, M., Boxberger, N., Börnick, H., Licha, T., & Worch, E. (2012). Sorption influenced transport of ionizable pharmaceuticals onto a natural sandy aquifer sediment at different pH. *Chemosphere*, *87*(5), 513–520. Available from https://doi.org/10.1016/j.chemosphere.2011.12.053.

Schwaab, M., Steffani, E., Barbosa-Coutinho, E., & Severo Júnior, J. B. (2017). Critical analysis of adsorption/diffusion modelling as a function of time square root. *Chemical Engineering Science*, *173*, 179–186. Available from https://doi.org/10.1016/j.ces.2017.07.037.

Schwarzenbach, R. P., Gschwend, P. M., & Imboden, D. M. (2005). *Environmental organic chemistry. Environmental organic chemistry* (pp. 1–1313). Wiley. Available from https://doi.org/10.1002/0471649643.

Shahriar, A., Tan, J., Sharma, P., Hanigan, D., Verburg, P., Pagilla, K., & Yang, Y. (2021). Modeling the fate and human health impacts of pharmaceuticals and personal care products in reclaimed wastewater irrigation for agriculture. *Environmental Pollution*, *276*, 116532.

Shen, G., Zhang, Y., Hu, S., Zhang, H., Yuan, Z., & Zhang, W. (2018). Adsorption and degradation of sulfadiazine and sulfamethoxazole in an agricultural soil system under an anaerobic condition: kinetics and environmental risks. *Chemosphere*, *194*, 266–274.

Simonin, J. P. (2016). On the comparison of pseudo-first order and pseudo-second order rate laws in the modeling of adsorption kinetics. *Chemical Engineering Journal*, *300*, 254–263. Available from https://doi.org/10.1016/j.cej.2016.04.079.

Sparks, D. L. (2003). *Environmental soil chemistry*. Elsevier.

Srinivasan, P., Sarmah, A. K., & Manley-Harris, M. (2013). Co-contaminants and factors affecting the sorption behaviour of two sulfonamides in pasture soils. *Environ. Pollut.*, *180*, 165–172.

Sterling, T. M. (1994). Mechanisms of herbicide absorption across plant membranes and accumulation in plant cells. *Weed Science*, 263–276. Available from https://doi.org/10.1017/s0043174500080383.

Stevanović, Z. (2018). *Global distribution and use of water from karst aquifers. Geological society special publication* (Vol. 466, pp. 217–236). Geological Society of London Issue 1. Available from https://doi.org/10.1144/SP466.17.

Tian, G., Wang, W., Zong, L., Kang, Y., & Wang, A. (2016). A functionalized hybrid silicate adsorbent derived from naturally abundant low-grade palygorskite clay for highly efficient removal of hazardous antibiotics. *Chem. Eng. J.*, *293*, 376–385.

Topp, E., Scheunert, I., Attar, A., & Korte, F. (1986). Factors affecting the uptake of 14C-labeled organic chemicals by plants from soil. *Ecotoxicology and Environmental Safety*, *11* (2), 219–228. Available from https://doi.org/10.1016/0147-6513(86)90066-7.

Van Boeckel, T. P., Brower, C., Gilbert, M., Grenfell, B. T., Levin, S. A., Robinson, T. P., Teillant, A., & Laxminarayan, R. (2015). Global trends in antimicrobial use in food animals. *Proceedings of the National Academy of Sciences of the United States of America*, *112*(18), 5649–5654. Available from https://doi.org/10.1073/pnas.1503141112.

Vieira, A. P., Rath, S., & Fostier, A. H. (2017). Sorption of Sulfachloropyridazine in Brazilian soils. *J. Braz. Chem. Soc.*, *28*, 158–167.

Wang, K., Reguyal, F., & Zhuang, T. (2020). Risk assessment and investigation of landfill leachate as a source of emerging organic contaminants to the surrounding environment: A case study of the largest landfill in Jinan City, China. *Environmental Science and Pollution Research*. Available from https://doi.org/10.1007/s11356-020-10093-8.

Wilhelm, S. R., Schiff, S. L., & Cherry, J. A. (1994). Biogeochemical evolution of domestic waste water in septic systems: 1. Conceptual Model. *Groundwater*, *32*(6), 905–916. Available from https://doi.org/10.1111/j.1745-6584.1994.tb00930.x.

Wu, D., Sui, Q., Yu, X., Zhao, W., Li, Q., Fatta-Kassinos, D., & Lyu, S. (2021). Identification of indicator PPCPs in landfill leachates and livestock wastewaters using multi-residue analysis of 70 PPCPs: Analytical method development and application in Yangtze River Delta, China. *Science of the Total Environment*, *753*. Available from https://doi.org/10.1016/j.scitotenv.2020.141653.

Wu, M., Zhao, S., Tang, M., Jing, R., Shao, Y., Liu, X., Dong, Y., Li, M., Liao, Q., Lv, G., Zhang, Q., Meng, Z., & Liu, A. (2019). Adsorption of sulfamethoxazole and tetracycline on montmorillonite in single and binary systems. *Colloids and Surfaces A: Physicochemical and Engineering Aspects*, *575*, 264–270. Available from https://doi.org/10.1016/j.colsurfa.2019.05.025.

Xing, B., McGill, W. B., & Dudas, M. J. (1994). Cross-correlation of polarity curves to predict partition coefficients of nonionic organic contaminants. *Environmental Science and Technology*, *28*(11), 1929–1933. Available from https://doi.org/10.1021/es00060a025.

Xu, Y., Yu, X., Xu, B., Peng, D., & Guo, X. (2020). Sorption of pharmaceuticals and personal care products on soil and soil components: Influencing factors and mechanisms. *Science of the Total Environment*.

Yang, Y. Y., Toor, G. S., Wilson, P. C., & Williams, C. F. (2017). Micropollutants in groundwater from septic systems: Transformations, transport mechanisms, and human health risk assessment. *Water Research*, *123*, 258–267. Available from https://doi.org/10.1016/j.watres.2017.06.054.

Yao, F., Zhaojun, L., & Xiaoqing, H. (2016). Impacts of soil organic matter, iron-aluminium oxides and pH on adsorption-desorption behaviors of oxytetracycline. *Research Journal of Biotechnology*, *11*(1), 121–131. Available from http://WWW.biotekjournal.net.

Yeom, J. R., Yoon, S. U., & Kim, C. G. (2017). Quantification of residual antibiotics in cow manure being spread over agricultural land and assessment of their behavioral effects on antibiotic resistant bacteria. *Chemosphere*, *182*, 771–780. Available from https://doi.org/10.1016/j.chemosphere.2017.05.084.

Yu, X., Sui, Q., Lyu, S., Zhao, W., Liu, J., Cai, Z., Yu, G., & Barcelo, D. (2020). Municipal solid waste landfills: An underestimated source of pharmaceutical and personal care products in the water environment. *Environmental Science and Technology*, *54*(16), 9757–9768. Available from https://doi.org/10.1021/acs.est.0c00565.

Yu, Y., Liu, Y., & Wu, L. (2013). Sorption and degradation of pharmaceuticals and personal care products (PPCPs) in soils. *Environmental Science and Pollution Research*, *20*(6), 4261–4267. Available from https://doi.org/10.1007/s11356-012-1442-7.

Zhang, P., Chen, Y. P., Wang, W., Shen, Y., & Guo, J. S. (2016). Surface plasmon resonance for water pollutant detection and water process analysis. *Trends in Analytical Chemistry*, *85*, 153–165. Available from https://doi.org/10.1016/j.trac.2016.09.003.

Zhao, Y., Geng, J., Wang, X., Gu, X., & Gao, S. (2011). Tetracycline adsorption on kaolinite: pH, metal cations and humic acid effects. *Ecotoxicology*, *20*, 1141–1147.

Zhao, F., Yang, L., Chen, L., Li, S., & Sun, L. (2018). Co-contamination of antibiotics and metals in peri-urban agricultural soils and source identification. *Environ. Sci. Pollut. Res.*, *25*, 34063–34075.

Zheng, C., & Wang, P. P. (1999). MT3DMS: A modular three-dimensional multi-speciestransport model for simulation of advection, dispersion, and chemical reactions ofcontaminants in ground-water systems. documentation and user's guide. In *InContractReport*. (SERDP-99-1,). Environmental Laboratory, U.S. Army Engineering Researchand Development.

Zhi, D., Yang, D., Zheng, Y., Yang, Y., He, Y., Luo, L., & Zhou, Y. (2019). Current progress in the adsorption, transport and biodegradation of antibiotics in soil. *Journal of Environmental Management*, 251. Available from https://doi.org/10.1016/j.jenvman.2019.109598.

Zhou, D., Thiele-Bruhn, S., Arenz-Leufen, M. G., Jacques, D., Lichtner, P., & Engelhardt, I. (2016). Impact of manure-related DOM on sulfonamide transport in arable soils. *J. Contam. Hydrol.*, *192*, 118–128.

Zhuo, N., Lan, Y., Yang, W., Yang, Z., Li, X., Zhou, X., Liu, Y., Shen, J., & Zhang, X. (2017). Adsorption of three selected pharmaceuticals and personal care products (PPCPs) onto MIL-101(Cr)/natural polymer composite beads. *Separation and Purification Technology*, *177*, 272–280. Available from https://doi.org/10.1016/j.seppur.2016.12.041.

Chapter 4

Occurrences and impacts of perfluorinated compounds in soils and groundwater

Jian Zhang, Shengsen Wang, Xianni Yang and Xiaozhi Wang
College of Environmental Science and Engineering, Yangzhou University, Yangzhou, P.R. China

Chapter Outline

4.1 Overview

4.1.1 Overview of perfluorinated compounds

Perfluorinated compounds (PFCs) or perfluoroalkyl substances (PFASs) are a class of emerging persistent organic pollutants (POPs) existing in the environment globally (Liu et al., 2019a, 2019b). Common PFCs include perfluoroalkyl carboxylic acids (PFCAs), perfluoroalkyl sulfonic acids (PFSAs), perfluoroalkyl sulfonamides, perfluoroalkyl phosphonic acids (PFPAs), etc. PFCs have a hydrophobic and oleophobic perfluoroalkyl chain with varying lengths and a hydrophilic functional group (e.g., sulfonate, carboxylate, sulfonamides, and phosphate). The energy of $C-F$ bond is extremely high

Emerging Contaminants in Soil and Groundwater Systems. DOI: https://doi.org/10.1016/B978-0-12-824088-5.00007-0

(~531.5 kJ/mol), making PFCs very stable and difficult to be decomposed (Liu et al., 2019a). These unique structural properties of PFCs result in their very high oil and water-repellency, thermal stability, and acid-base resistance (Liu et al., 2020b). Therefore PFCs have been extensively used in manufacturing and households such as coating, textiles, paper packaging, fire-fighting foams, polishes, and electroplates. This makes it difficult to identify their distribution in the environment and find the exposure pathways for people owing to their plenty uses and unique physicochemical properties (Lindstrom, Strynar, Delinsky, et al., 2011). PFCs are ubiquitous and globally detected in soils, water, and organisms, even in remote areas including the arctic. Moreover, PFCs are persistent in the environment and bioaccumulated in high trophic level organisms through biomagnification, posing serious ecological and health risks. Two well-known PFCs, perfluorooctanoic acid (PFOA) and perfluorooctane sulfonate (PFOS), are the predominantly final degradation products of related PFCs and contain an eight-carbon framework (Kovářová & Svobodová, 2008). In addition, PFOA and PFOS are also often manufactured directly for domestic and industrial uses (Liu et al., 2020b). Therefore the occurrences and impacts of PFOA and PFOS in the environment have received increasing attention and concern recently.

4.1.2 Categories of PFCs

4.1.2.1 PFCAs

PFCAs or perfluorocarboxylic acids are compounds of the formula $C_nF_{2n+1}CO_2H$, including perfluorononanoic acid (PFNA, $n=8$), PFOA ($n=7$), perfluoroheptanoic acid (PFHpA, $n=6$), perfluorohexanoic acid (PFHxA, $n=5$), perfluoropentanoic acid (PFPeA, $n=4$), and perfluorobutanoic acid (PFBA, $n=3$). The carboxylate group is the same part as one end of different PFCA molecules. The other end in a PFCA molecule includes a perfluorinated alkyl carbon-chain a different length, endowing its water and oil repellency, thermal stability, and surfactant properties (Lindstrom, Strynar, Delinsky, et al., 2011). These unique physical and chemical characteristics make PFCAs be applied widely in industrial and consumer uses. Molecular structures of some common PFCAs are shown in Fig. 4.1.

4.1.2.2 PFSAs

Generally, the molecules of PFSAs are composed of a sulfonate head and a tail of perfluorinated carbon-chain with various lengths. The best known PFSA type is PFOS. Moreover, the common molecular formula of PFSAs is $C_nF_{2n+1}SO_3H$ ($n=4-14$). The linear and/or branched carbon units with different formula and molecule structures are generally possible. The names and structures of some common perfluoroalkyl sulfonic acids are showed in Fig. 4.2. Taking PFOS as an example, it is ubiquitous in surface and drinking

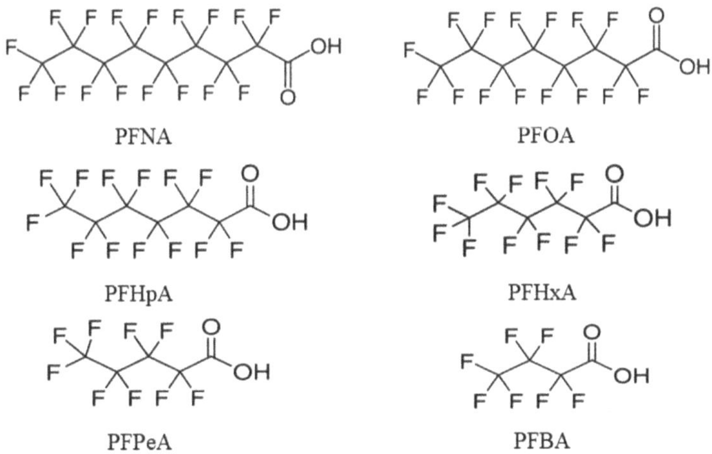

FIGURE 4.1 Molecular structures of some common perfluorinated carboxylic acids (PFCAs). *Modified from Ghisi, R., Vamerali, T., & Manzetti, S. (2019). Accumulation of perfluorinated alkyl substances (PFAS) in agricultural plants: A review. Environmental Research, 169, 326–341. https://doi.org/10.1016/j.envres.2018.10.023.*

water, even in remote areas of China, and the high contamination levels were observed in industrialized regions along the southeast coast than that in other regions in China (Gao et al., 2020). Moreover, PFOS in soils can be accumulated in some vegetables (e.g., lettuce, carrots, and tomato), posing potential health risks to human (Lal et al., 2020). Therefore more environmental media and food types should be monitored to determine the concentrations and distributions of PFSAs including PFOS.

4.1.2.3 Perfluoroalkyl sulfonamides

The main type of perfluoroalkyl sulfonamides is perfluorooctane sulfonamide (PFOSA). PFOSA is a synthetic fluorocarbon derivative having an eight-carbon chain and a terminal sulfonamide functional group. PFOSA is usually used to repel grease and water in food packaging. As a precursor, PFOSA in environment can gradually be degraded to form PFOS. A previous work indicated that the PFOS yield was 4.0% from *N*-ethyl perfluorooctane sulfonamide (EtFOSA) in aerobic soils via the biotransformation after 182 days (Mejia Avendaño & Liu, 2015). Moreover, it was reported that the hydrolysis of the sulfonamide bond can be a new biotransformation pathway, leading to PFOS (Liu et al., 2019b). Further, the precursors and intermediates of formed PFOS are various in soils and have different structures (e.g., linear and branched), causing their complex biotransformation processes (Liu et al., 2019b). Therefore further studies are still needed to investigate the occurrences, impacts, and degradation of perfluoroalkyl sulfonamides in soils and groundwater.

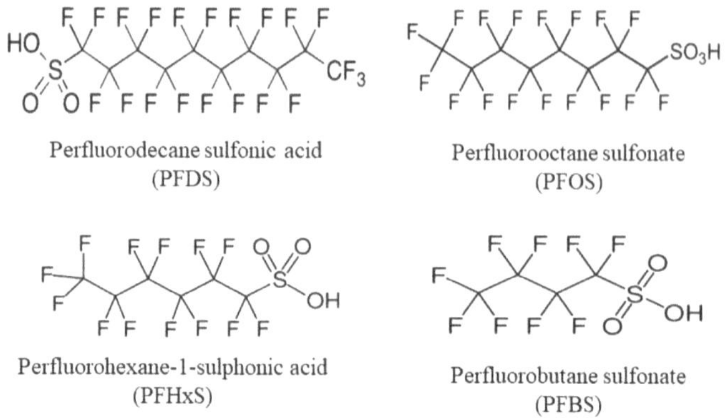

FIGURE 4.2 Molecular structures of some common perfluoroalkyl sulfonic acids (PFSAs). *Modified from Ghisi, R., Vamerali, T., & Manzetti, S. (2019). Accumulation of perfluorinated alkyl substances (PFAS) in agricultural plants: A review. Environmental Research, 169, 326–341. https://doi.org/10.1016/j.envres.2018.10.023.*

4.1.2.4 PFPAs

The molecules of PFPAs similarly have a perfluorinated carbon chain and a phosphate group. Compared with PFCAs and PFSAs, PFPAs own higher migration capability from emission sources into environmental compartments as a result of the di-anionic character (Huang et al., 2020). It is challenging to analyze the levels of PFPAs in environmental samples owing to this di-anionic character. A previous study reported the concentrations of three perfluorophosphonic acids (PFPAs) in fish, vegetables, and amended soil samples, including perfluorooctylphosphonic acid, perfluorohexylphosphonic acid, and perfluorodecylphosphonic acid (Zabaleta et al., 2014). As commercial surfactants, PFPAs have been used mostly in the industrial sector because of their favorable wetting and leveling properties (Ullah et al., 2011).

4.1.3 Chemical structure and properties

The chemical structure of PFCs mainly includes a hydrophobic tail perfluoroalkyl carbon-chain tail and a hydrophilic functional head (e.g., sulfonate, carboxylate, sulfonamides, and phosphate). PFCs are often divided into long-chain and short-chain ones. The former refers to PFCAs with seven and more fluorinated carbons and PFSAs with six and more fluorinated carbons (Liu et al., 2020b). Some common PFCs are listed in Table 4.1.

The physicochemical properties of PFCs are characterized with high stability, hydrophobicity, oleophobicity, and hydrophilicity, repelling water and oil and reducing surface tension. The high stability of PFCs can be attributed to the high carbon-fluorine bond energy. The one end of PFC molecules (i.e., perfluorinated alkyl carbon-chain) results in their hydrophobicity and oleophobicity. Moreover, the other end of PFC molecules, including carboxylic acid, sulfonic acid, or phosphate, leads to their hydrophilicity. PFCs are also persistent can exist in the environment for a long time, causing their long-range transport and spread. In addition, as two most important compounds of PFCs, PFOS, and PFOA are persistent in the environment, bioaccumulative, and can cause toxicity to organisms. They have been found in the water environment of various countries at the ng/L levels (Suja et al., 2009). With 48-hour acute toxicity, PFOS is more toxic than PFOA (21-day chronic toxicity). They can show strong synergistic toxicity to *Daphnia*

TABLE 4.1 The acronym and formula of some common perfluorinated compounds (Navarro et al., 2020).

Compound	Acronym	Formula
Perfluorobutanoic acid	PFBA	$CF_3(CF_2)_2COOH$
Perfluoropentanoic acid	PFPeA	$CF_3(CF_2)_3COOH$
Perfluorohexanoic acid	PFHxA	$CF_3(CF_2)_4COOH$
Perfluoroheptanoic acid	PFHpA	$CF_3(CF_2)_5COOH$
Perfluorooctanoic acid	PFOA	$CF_3(CF_2)_6COOH$
Perfluorononanoic acid	PFNA	$CF_3(CF_2)_7COOH$
Perfluorodecanoic acid	PFDA	$CF_3(CF_2)_8COOH$
Perfluoroundecanoic acid	PFUnDA	$CF_3(CF_2)_9COOH$
Perfluorododecanoic acid	PFDoDA	$CF_3(CF_2)_{10}COOH$
Perfluorotridecanoic acid	PFTrDA	$CF_3(CF_2)_{11}COOH$
Perfluorotetradecanoic acid	PFTeDA	$CF_3(CF_2)_{12}COOH$
Perfluorohexadecanoic acid	PFHxDA	$CF_3(CF_2)_{14}COOH$
Perfluorooctadecanoic acid	PFODA	$CF_3(CF_2)_{16}COOH$
Perfluorobutane sulfonic acid	PFBS	$CF_3(CF_2)_3SO_3H$
Perfluorohexane sulfonic acid	PFHxS	$CF_3(CF_2)_5SO_3H$
Perfluoroheptane sulfonic acid	PFHpS	$CF_3(CF_2)_6SO_3H$
Perfluorooctane sulfonic acid	PFOS	$CF_3(CF_2)_7SO_3H$
Perfluorodecane sulfonic acid	PFDS	$CF_3(CF_2)_9SO_3H$
Perfluorooctane sulfonamide	FOSA	$CF_3(CF_2)_7SO_2NH_2$

Source: From Navarro, I., de la Torre, A., Sanz, P., & Martínez, M. d. I. Á. (2020). Perfluoroalkyl acids (PFAAs): Distribution, trends and aquatic ecological risk assessment in surface water from Tagus River basin (Spain). *Environmental Pollution, 256*. https://doi.org/10.1016/j.envpol.2019.113511.

magna in acute and chronic toxicity of mixed PFOS and PFOA (Yang et al., 2019).

4.2 Applications and sources

4.2.1 Main applications

PFCs are extensively applied to various aspects in industry and in households, such as surfactants, coatings, fire-fighting foams, polishes, textiles, electroplates, and pesticides. For instance, PFOA and its sodium salts or ammonium salts are usually used as dispersants in the polymerization of polytetrafluoroethylene and in the production of fluororubber and also as raw materials and mineral concentrators for the preparation of water and oil repellents. Although the amount of PFOS used in some applications is usually very small (e.g., ~0.05% for the content of sulfluramid in cockroach-killing baits), the total amount of products involved and the seriousness of pollution are still quite large. Although PFOS has important applications in many industrial fields important to the national economy and people's livelihood, it is recognized as a POP with persistent pollution, bioaccumulation, and toxicity. Therefore PFOS and its salts were added to Annex B of the Stockholm Convention in 2009, and PFOA was planned to be eliminated globally by 2015 and has been listed in the candidate list of highly-concerned substances in legislation from the European Union (Fang et al., 2020).

4.2.2 Sources

PFCs in soils and groundwater are mainly from anthropogenic sources and barely natural. As shown in Fig. 4.3, some common anthropogenic sources include daily domestic consumption, emission of fluorine in industrial parks, fire-fighting products, industrial wastes, and wastewater treatment plant drainage (Stoiber et al., 2020). According to the ratio analysis, principal component analysis, and cluster analysis of sampling sites, the main sources, including industrial discharge, domestic wastewater, atmospheric inputs, and wastewater treatment plants, were identified qualitatively for PFCs in the Guanlan river from the rapidly urbanizing areas of Shenzhen, China (Wu et al., 2020). Moreover, the sources of PFCs in sediments from Dianchi Lake in China were apportioned quantitatively by using three multivariate factor analysis receptor models (PCA-MLR, PMF, and Unmix). The contribution percentages from the PFOS-cluster related sources were 43% for PCA-MLR, 48% for PMF, and 46% for Unmix and from the PFOA-cluster sources were 54% for PCA-MLR, 43% for PMF, and 44% for Unmix (Xu, Li, et al., 2013; Xu, Tian, et al., 2013). A prior study indicated that the highest total concentration of 16 PFCs in 68 soils from seven cities in the Pearl River Delta

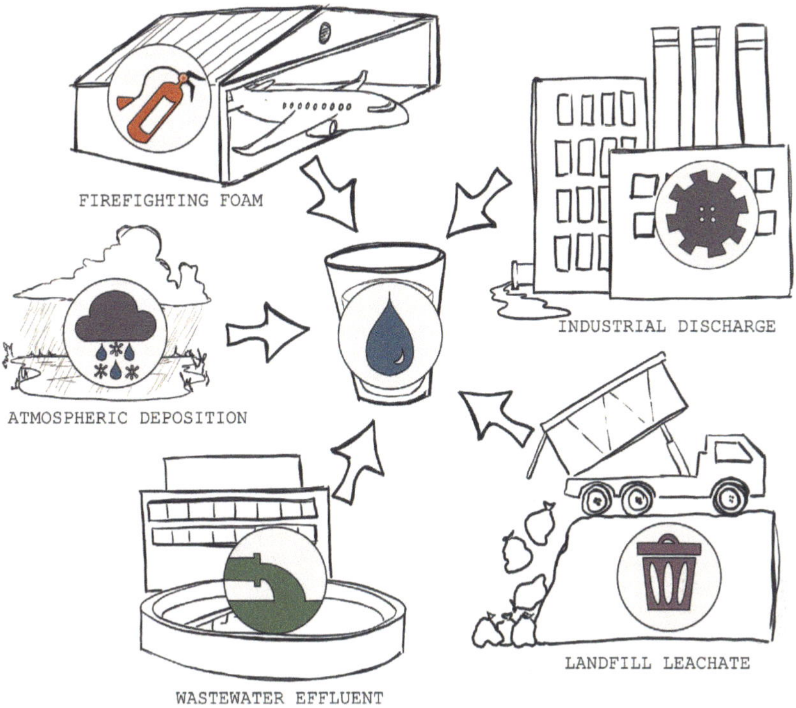

FIGURE 4.3 Some main sources that can contribute to the PFAS pollution of water and soil (Stoiber et al., 2020).

(PRD), China was found in the soil sampled from Dongguan with a large number of manufacturing industries, revealing the anthropogenic sources of PFCs.

In China, total industrial emission of PFOS was estimated to be ~70 t in 2010 based on PFOS equivalents from major industrial sources including textile treatment, metal plating, fire-fighting, and semiconductor industry (Fig. 4.4A). Higher industrial emission of PFOS equivalents was observed in the eastern region of China than those in central and western regions. Metal plating and textile treatment were dominant in the emissions in eastern China, while fire-fighting and metal plating contributed the larger portion of emissions of PFOS equivalents in western and central China (Wang et al., 2015). Moreover, in terms of domestic emissions, the three main economic belts (i.e., Pearl River Delta, Bohai coastal region, and Yangtze River Delta) emitted ~74% of the total emission of PFOS in the eastern and coastal regions of China (Fig. 4.4B). The emission from industrial sources was much higher than that from domestic sources (Wang et al., 2015). These results provide strong evidences that PFCs in the environment are mainly sourced from the emissions of human activities.

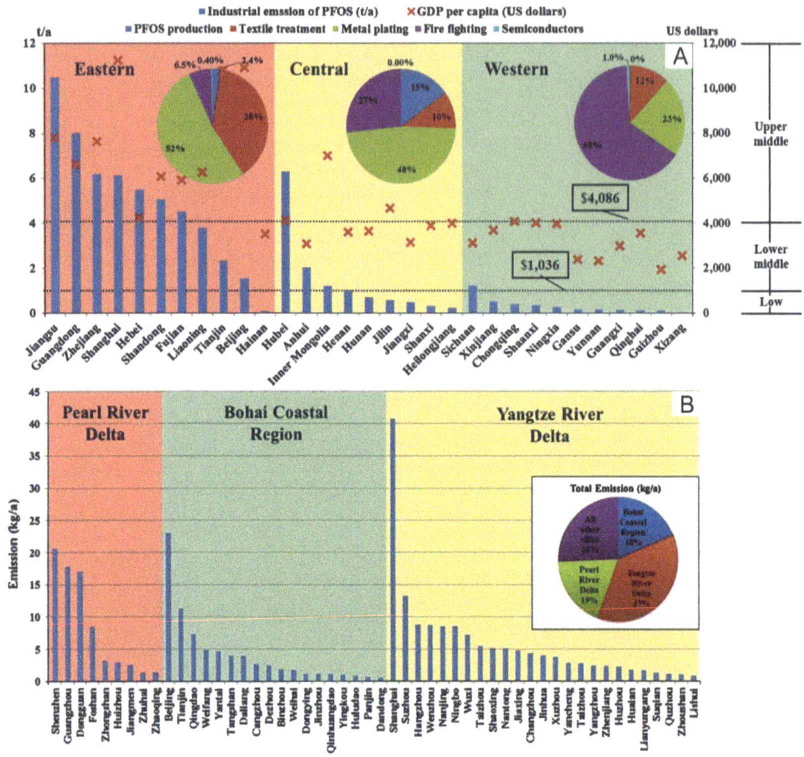

FIGURE 4.4 PFOS emissions from industrial sources (A) and domestic sources (B) in China in 2010 (Wang et al., 2015).

4.3 Occurrences

4.3.1 Occurrences in soils

The contents of total PFCs ($\sum$PFC), PFOS, and PFOA in soils across the globe are summarized in Table 4.2. PFCs in most of these soils are at the ppb level, but may range up to ppm levels at some contaminated sites (Brusseau et al., 2020). PFCs in these soils are mainly from irrigation water, atmospheric deposition, fire-training, emission of fluorochemical plants, etc. For instance, higher PFC levels in the skiing area were observed compared to the control area without skiing activities because PFCs are used ski products, such as ski waxes (Grønnestad et al., 2019). In addition, a prior study indicated that trace levels of PFCs were detected in soil cores at the depths of 120 cm, suggesting the potential movement of PFCs within the soil profile over time (Sepulvado et al., 2011). PFCs in soil can be accumulated in plants through the soil-to-plant transfer pathway. In general, PFOA is more easily absorbed by plants than PFOS, which can be attributed to the difference in functional groups of the two compounds. For the same type of PFCs, the short-chain ones are more likely to be accumulated in plant stems and leaves, while the long-chain molecules are more intended to be accumulated in the roots of plants.

The bioconcentration factor (BCF) of PFCs in plants is influenced to different extent according to the concentrations, functional group, and carbon chain length, plant species and variety, soil organic matter, etc. Because long-chain PFCs are gradually regulated, nonregulated short-chain PFCs (C4-C6) are increasingly used as alternatives. However, the short-chain PFCs are easier to be absorbed and translocated from the roots to the shoots, which may increase their exposures to human via the food chain. Therefore additional work should be carried out to investigate the accumulation and potential risks of short-chain PFCs in plants.

PFCs from various emission sources can be adsorbed in soils, which is affected by various factors including electrostatic interactions, hydrogen bonds, hydrophobic interactions, soil organic matter contents, and ligand exchange (Ahmed et al., 2020). Further, PFCs entering the topsoil can migrate along the soil profile into the vadose zone, thereby contaminating the groundwater. Generally, the concentrations of PFCs in the soil profile decreased with the depth, which is related to the source, soil type, and migration rate of PFCs in soil profile (Brusseau et al., 2020). According to the findings in the US air force aqueous film-forming foam impacted-site database, the contents of PFOS in the soil were somewhat high in the range of 15–25 m depth, indicating the large degree of migration of PFCs (Fig. 4.5A). Additionally, the depth distribution of total PFCs in soil was a function of chain length, indicating that the short-chain PFCs migrated faster than the long-chain PFCs (Fig. 4.5B) (Brusseau et al., 2020). These results suggest that PFCs can migrate to significant depths in soils, posing contamination risks to groundwater. Therefore additional attention should be paid to the transport and fate of PFCs in soils and groundwater.

4.3.2 Occurrences in groundwater

The concentration and distribution of PFCs in some groundwater sites are summarized in Table 4.3. The concentrations of PFCs in groundwater samples in most areas are at the ppb level. In particular, the contents of PFCs in the groundwater near fluorochemical plants are higher than those near the nonindustrial areas. The shorter-chain PFAAs show higher concentrations in groundwater than in soil samples, which can be ascribed to the different polarity and hydrophobicity of short- and long-chain molecules (Gao et al., 2020). Moreover, the sources of PFCs in groundwater also include the discharges from domestic wastewater, precipitation, and wastewater treatment plants in addition to industrial discharges (Wu et al., 2020). Previous studies have demonstrated that domestic sewage and atmospheric deposition may

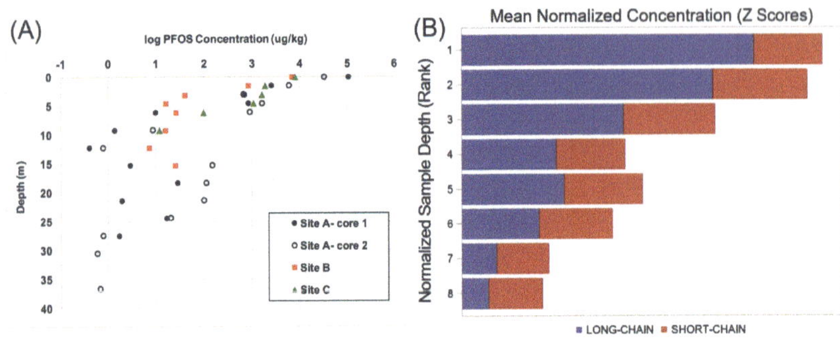

FIGURE 4.5 Occurrences of PFCs in soils: (A) Depth profiles of PFOS soil concentrations and (B) relationship between depth distribution of total PFAS in the soils and PFAS chain length. *Modified from Brusseau, M. L., Anderson, R. H., & Guo, B. (2020). PFAS concentrations in soils: Background levels versus contaminated sites. Science of the Total Environment, 740. https://doi.org/10.1016/j.scitotenv.2020.140017.*

contribute significantly to PFAS occurrence in groundwater in nonindustrial areas (Wei et al., 2018). Interestingly, the concentration of PFOS in groundwater near a fluorochemical industrial park was lower than that of the short-chain PFBS, which can be attributable to the increasing uses of PFBS as a replacement for the long-chain PFHxS and PFOS (Bao et al., 2020). The ability of PFCs to enter groundwater aquifers increases with the increase of their water solubility, which is negatively related to the chain-length of perfluoroalkyl carbon (Lindstrom, Strynar, & Libelo, 2011). In water environment and sediments, the long-chain PFCs (>8 carbons) present higher contents in sediments than in water, while the short-chain PFCs (≤ 8 carbons) are predominant in water bodies (Bai & Son, 2021).

The BCFs are calculated based on the following equation:

$$\mathrm{BCF} = \frac{C_{\mathrm{biota}}}{C_{\mathrm{water}}}$$

where C_{biota} is analyte concentration in the respective matrix and C_{water} is analyte concentration in the water. BCF (L/kg) ≥ 5000 indicates highly bioaccumulative, BCF ≥ 1000 indicates bioaccumulative, and BCF < 1000 indicates not bioaccumulative (Wilkinson et al., 2018). In terms of PFOS, its BCFs for *Potamogeton*, *Callitriche*, *Bithynia tentaculata*, and *Gammarus pluex* are 1330, 19,500, 440, and 590, respectively; while, their BCFs of PFOA are 243, 1140, 162, and 56, respectively (Wilkinson et al., 2018). Moreover, the PFCs are toxic to aquatic organisms e.g., *D. magna*, *Chlorella vulgaris*, zebrafish, and carp. For example, at certain concentrations, PFOS can exert lethal effect on zooplankton and fish and has suppression effect on the growth of phytoplankton. The toxic effect is positively correlated with its concentration. Yang et al. reported that PFOS was more toxic than PFOA to *D. magna* both in 48-hour acute toxicity and 21-day chronic toxicity, and

TABLE 4.2 Distribution of PFCs in soils.

Sample site	PFC category	Concentration range	References
Shandong Province, China	PFOS	0.06–1.23 ng/g	Zhang et al. (2019)
	PFOA	0.48–22.99 ng/g	
	∑PFCs ($n = 12$)	4.23–48.72 ng/g	
Zarqa catchment, Jordan	∑PFCs ($n = 6$)	0.34–0.97 ng/g	Shigei et al. (2020)
Shouguang vegetable base, China	∑PFCs ($n = 12$)	0.29–4.28 ng/g	Zhang et al. (2020b)
New York State, United States	∑PFCs ($n = 8$)	421–43261 ng/g	Zhang et al. (2020a)
Ten landfills, Ireland	PFOS	2.4–140 pg/g (downwind) and 3.7–2000 pg/g (upwind)	Harrad et al. (2020)
Ten landfills, Ireland	PFOA	0.15–5.8 ng/g (downwind) and 0.13–7.8 ng/g (upwind)	Harrad et al. (2020)
Yuqiao reservoir from Tianjin of China	∑PFCs ($n = 17$)	0.622–5.089 ng/g	Cao et al. (2019)
Hyung-san River, Gyeongju and Pohang, Korea	∑PFCs ($n = 19$)	2.5–8.8 ng/g	Seo et al. (2019)
Washington County, Ohio, United States	∑PFCs ($n = 6$)	150 ng/g	Zhu and Kannan (2019)
Hubei Province, China	∑PFCs ($n = 13$)	1.3–913 ng/g	Gao et al. (2019)
The Maltese Islands	PFOS	<LOQ–5.91 ng/g	Sammut et al. (2019)
The Maltese Islands	PFOA	<LOQ–0.58 ng/g	Sammut et al. (2019)
Antwerp, Belgium	PFOA	<LOD–1.5 ng/g	Groffen et al. (2019)
Antwerp, Belgium	PFOS	<LOD–8.9 ng/g	Groffen et al. (2019)
Fuxin, China	PFOA	1.2–6.3 ng/g	Bao et al. (2019)
Fuxin, China	PFOS	<LOD–1.4 ng/g	Bao et al. (2019)

(*Continued*)

TABLE 4.2 (Continued)

Sample site	PFC category	Concentration range	References
Natural forested mountain sites, China	PFOA	<0.9 to 9.0 pg/g	Wang et al. (2018)
Eastern Twin Cities, United States	∑PFCs ($n = 3$)	1.3–30 ng/g	Scher et al. (2018)
Huantai County, Shandong Province, China	∑PFCs ($n = 12$)	1.9–126 ng/g	Liu et al. (2017)
Global survey of soil	∑PFCAs (C4–C14)	0.029–14.3 ng/g	Rankin et al. (2016)
Global survey of soil	∑PFSAs (S6, S8, S10)	<LOQ–3.27 ng/g	Rankin et al. (2016)
Rural areas in eastern China	∑PFCs ($n = 17$)	0.34–65.8 ng/g	Chen et al. (2016)
Stockholm, Sweden	PFOS	2.18–8525 ng/g	Filipovic et al. (2015)
Stockholm, Sweden	PFOA	<0.12–287 ng/g	Filipovic et al. (2015)
Along Highway 10, China	PFOS	0.2–28.2 ng/g	Xiao et al. (2015)
Along Highway 10, China	PFOA	5.5–125.7 ng/g	Xiao et al. (2015)
Near the fluorochemistry industrial facilities in Jiangsu, China	PFOA	<0.20–9.03 ng/g	Shan et al. (2014)
	PFOS	< 0.10–0.34 ng/g	Shan et al. (2014)
	∑PFCs ($n = 11$)	0.75–28.8 ng/g	Shan et al. (2014)
Vicinity of Koshi River in Nepal	∑PFCs ($n = 16$)	<0.1–1.8 ng/g	Tan et al. (2014)
Huaihe watershed, China	∑PFCs ($n = 16$)	<LOQ–1.22 ng/g	Meng et al. (2013)
Huaihe watershed, China	PFOS	<LOQ–0.21 ng/g	Meng et al. (2013)

(*Continued*)

TABLE 4.2 (Continued)

Sample site	PFC category	Concentration range	References
Coast of Liaodong Bay, China	∑PFCs ($n=6$)	<LOQ–3.14 ng/g	Wang et al. (2013)
Birmingham, United Kingdom	PFOA	1.1 ng/g	Goosey & Harrad, 2012)
Birmingham, United Kingdom	PFOS	0.56 ng/g	Goosey & Harrad, 2012)
Vicinity of Guanting Reservoir, China	∑PFCs ($n=6$)	0.13–8.5 ng/g	Wang et al. (2011)
Suburban districts of Tianjin, China	PFOA	<0.1–0.51 ng/g	Pan et al. (2011)
Suburban districts of Tianjin, China	PFOS	0.023–2.36 ng/g	Pan et al. (2011)

mixtures of PFOA and PFOS had strong synergistic effect to *D. magna* in terms of acute and chronic toxicity (Yang et al., 2019). As a whole, PFCs have been widely detected in groundwater, exerted toxicological effect on aquatic organisms, and can be transferred to humans through drinking water and the food chain. Moreover, PFCs in groundwater can be transferred to surface water to affect the surface environment compartments (e.g., surface water and sediments). A previous study indicated that the concentrations of PFCs in groundwater were significantly lower than those in lake water and river water, which can be ascribed to the retention and adsorption of PFCs in soils (Kleywegt et al., 2020). Therefore more work is needed to investigate the occurrences and impacts of PFCs in groundwater.

4.4 Analytical techniques

4.4.1 Sampling

The schematic diagram of sampling and analysis of PFCs is shown in Fig. 4.6. The specific analytical techniques include sample collection, preconcentration, reconstitution, and quantitative analysis of PFCs in soil and groundwater samples.

4.4.1.1 Soil samples

To study the distribution of PFCs in soil samples, each sample should be collected using the five-point method, which uses a composite of five

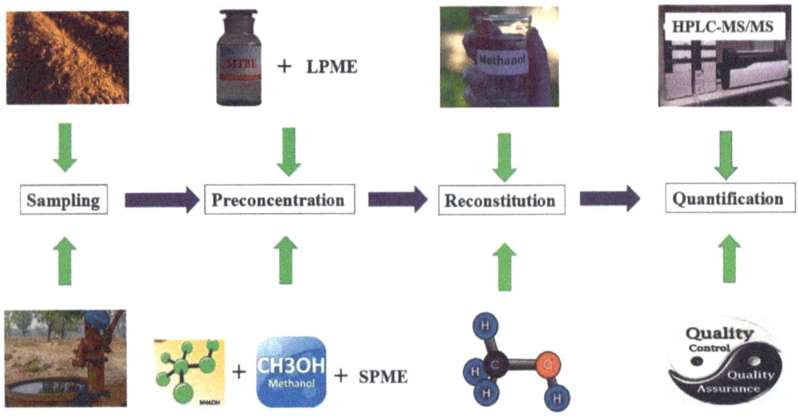

FIGURE 4.6 Schematic diagram of sampling and analysis in terms of the PFCs in soils and groundwater. *MTBE*, Methyl tertbutyl ether; *LPME*, liquid-phase microextraction; *SPME*, solid phase microextraction.

subsamples including the center and four corners of a square. In general, the soil samples are packed in polypropylene (PP) bags, which have been rinsed with methanol prior to use. Soil samples are collected using a clean stainless steel trowel that has been rinsed using methanol. Soil samples should be stored and frozen at −20°C on ice for transport to the laboratory (Wang et al., 2015). In addition, to ensure the accuracy of the experiment, three replicates should be collected in each sampling location.

4.4.1.2 Groundwater samples

For groundwater, the samples are generally collected from some fixed monitoring sites. Different depths of the monitoring wells and reference groundwater samples should also be considered (Li, Hu, et al., 2020; Li, Tang, et al., 2020). For liquid samples such as groundwater, 1.0 L water samples are sampled in PP plastic bottles that has been prerinsed with methanol. Each bottle should be rinsed with the water sample that will be collected at the sampling site before filling the bottle for further analysis. After that, the water samples are transported to the laboratory and pH values should be adjusted to 2.5 ± 0.5 to reduce the bacterial degradation (Guardian et al., 2020). Additionally, the reference or control water samples should also be collected.

4.4.2 Sample pretreatment

4.4.2.1 Groundwater samples

To analyze the PFC concentrations, groundwater samples (~500 mL each) should be firstly filtered using glass fiber filters (47 mm and 0.7 μm) to

TABLE 4.3 Distribution of PFCs in groundwater.

Sample site	PFC category	Concentration range	References
Ontario, Canada	∑PFCs ($n=10$)	3.8 ng/L	Kleywegt et al. (2020)
Maozhou river basin, Shenzhen city, south China	∑PFCs ($n=18$)	4.8–320.4 ng/L (wet reason) and 0.08–592.2 ng/L (dry reason)	Li, Hu, et al. (2020)
Yuqiao reservoir from Tianjin of China	∑PFCs ($n=17$)	1.426–17.138 ng/L	Cao et al. (2019)
Four areas, China	∑PFCs ($n=10$)	0.05–198.8 ng/L	Qiao et al. (2020)
Hubei Province, China	∑PFCs ($n=13$)	31.4–15656 ng/L	Gao et al. (2019)
The Maltese Islands	PFOA	<LOD–2.68 ng/L	Sammut et al. (2019)
Nonindustrial areas, Jiangsu of China	∑PFCs ($n=14$)	2.69–556 ng/L	Wei et al. (2018)
Shanghai, China	ΣPFCAs ($n=7$)	61.15–519.86 ng/L	Sun et al. (2018)
Rural areas in eastern China	∑PFCs ($n=17$)	5.3–615 ng/L	Chen et al. (2016)
A suburb of Tianjin, China	∑PFCs ($n=13$)	0.32–8.3 ng/L	Qi et al. (2016)
A suburb of Tianjin, China	PFBS	1.1 ng/L	Qi et al. (2016)
A suburb of Tianjin, China	PFOA	0.81 ng/L	Qi et al. (2016)
Stockholm, Sweden	∑PFCs ($n=4$)	738–51,000 ng/L	Filipovic et al. (2015)
Fluorochemical plants in Fuxin, China	∑PFCs ($n=8$)	37.0–37.6 μg/L	Bao et al. (2020)
Fluorochemical plants in Fuxin, China	PFBS	31.8–32.4 μg/L	Bao et al. (2020)
Hanoi and Ho Chi Minh City, Vietnam	∑PFCs ($n=7$)	<LOD–21 ng/L	Duong et al. (2015)

(*Continued*)

TABLE 4.3 (Continued)

Sample site	PFC category	Concentration range	References
Hanoi and Ho Chi Minh City, Vietnam	PFOS	<LOD–8.2 ng/L	Duong et al. (2015)
Central Tokyo, Japan	PFOA	<0.25–1800 ng/L	Kuroda et al. (2014)
Central Tokyo, Japan	PFOS	<0.25–990 ng/L	Kuroda et al. (2014)
The central part of the Netherlands	∑PFCs ($n = 8$)	2.8–4400 ng/L	Eschauzier et al. (2013)

remove large particles and biota. Then the samples are spiked with corresponding internal standard solutions. Further, the solid phase extraction (SPE) method can be used to extract the target analytes in water sample. For the SPE method, the brief description is as follows. The cartridges are firstly activated by 4 mL of 0.1% NH_4OH in methanol, 4 mL of methanol, and 4 mL of Milli-Q water sequentially, then the water samples are loaded onto the cartridges at the speed of 1–2 drops per second. Afterward, the cartridges are vacuum dried for 2 hours to eliminate water. Then, the cartridges with target PFCs are eluted using 4 mL of methanol and 4 mL of 0.1% NH_4OH in methanol with gravity, respectively. Then, the eluents are dried under a gentle stream of nitrogen and reconstituted in 500 μL methanol (Li, Hu, et al., 2020; Li, Tang, et al., 2020). Finally, the reconstituted solution should be filtered through a 0.22 μm membrane before instrumental analysis (Qu et al., 2019).

4.4.2.2 Soil samples

Soil samples should be firstly freeze-dried, homogenized and passed through a 2.0 mm sieve. Then, 2 g of homogenized soil is weighed and transferred into a 50 mL PP tube, and 100 μL of 50 ng/mL internal standard mixture corresponding to target analytes are added to the tube. Then the sample is wetted by adding 2 mL of Milli-Q water prior to vortexing for 2 minutes. Then 2 mL of 0.25 M Na_2CO_3 and 1 mL of 0.5 M tetrabutylammonium hydrogen sulfate solutions are added for extraction by vortexing for 5 minutes. Subsequently, 5 mL of methyl tertbutyl ether (MTBE) is added into the tube and the mixture is shaken for 20 minutes. The tube is centrifuged for 30 minutes at 3500 rpm/minute, then the supernatant is transferred into a 10-mL PP tube. The remnant aqueous mixture is performed the aforementioned repeated process including that 5 mL of MTBE is added again, followed by shaking and centrifuging under the same conditions. The two suspensions are mixed in the 10-mL PP tube. The extract in MTBE solvent is dried under a gentle flow of nitrogen, and then reconstituted in 500 μL methanol. Finally, the mixture is passed through a 0.22 μm nylon filter, transferred into PP snap top vial, and stored at −18°C for the further analysis (Li, Hu, et al., 2020; Li, Tang, et al., 2020).

4.4.3 Instrumental analysis

High performance liquid chromatography tandem mass spectrometry (HPLC-MS-MS) is usually used to quantitatively analyze the concentrations of PFCs. Firstly, the target compounds are separated using a column that is operated in negative ion multiple reaction monitoring mode (Wang et al., 2019). The mobile phases are 10 mM ammonium acetate and acetonitrile, and 40% acetonitrile is used to initialize the column at a flow rate of 250 μL/min. In general, the column temperature should be set at 35°C or 40°C (Zhang et al., 2019). Finally, each measured compounds are identified and quantified using the tandem mass spectrometry. PFCs in the extracts are usually quantified via matrix-matched calibration curves (Bao et al., 2020).

4.4.4 Quality assurance and quality control

During the sampling, extraction and analysis, strict quality assurance and quality control procedures should be performed. To avoid external or cross contamination, field controls, procedural blanks, and recoveries should be implemented during the sampling, extracting and analytical processes (Li, Hu, et al., 2020; Li, Tang, et al., 2020). The polytetrafluorethylene and other fluorinated materials should be not used to avoid contamination during the sample collection, preparation, and analysis. All vessels used should be alternately rinsed twice with ultrapure water and methanol prior to use (Wang et al., 2019). The limit of detection (LOD) of each analyte is set as three folds of the signal-to-noise (S/N) ratio, while the limit of quantification (LOQ) is defined as 10 folds of the S/N ratio. The S/N ratio is computed by the instrumental quantitative software (Li, Hu, et al., 2020; Li, Tang, et al., 2020). Moreover, matrix spiked recoveries are used to estimate the precision and accuracy of the analytical procedure. The recoveries between 80% and 120% are normally considered acceptable (Bao et al., 2020). The reproducibility of duplicate samples also should be evaluated by calculating the RSD for target compound, and the acceptable RSD is generally lower than 20%. Finally, the procedural blank sample is conducted to check the possible laboratory contamination and interferences for every eight samples.

4.5 Risks and impacts

4.5.1 Risks of PFCs

4.5.1.1 Ecological risks

4.5.1.1.1 Toxicity in aquatic ecosystems

The aquatic risks of PFCs mainly depend on their environmental behaviors and concentrations in aquatic ecosystems, aquatic species, and characteristics of water quality. The distribution and environmental behaviors of PFCs in typical aquatic ecosystem are displayed in Fig. 4.7. The dissociated PFCs can be absorbed by particles in the atmosphere or water and can be transported through dry and wet deposition. The water-suspended solid, water-sediment, and suspended solid-sediment partitions are the mainly categories of PFC partition (Liu et al., 2019a). The partitions of PFCs in the water-suspended solid-sediment system play an important role in controlling the fate of the compounds in multimedia. For instance, increasing organic matter of sediments and the decreasing pH of pore water can facilitate the accumulation of PFCs in sediments (Ahrens et al., 2009). Moreover, the length of the carbon chain of PFCs obviously affects their partition behavior in the water-sediment interface.

The ecological risk assessment of aquatic ecosystems mainly evaluates water and sediments. Two main methods, the risk quotient (RQ) and the species sensitivity distribution (SSD), are commonly used in the process of ecological risk assessment (Liu et al., 2019a). The RQ method is the simplest

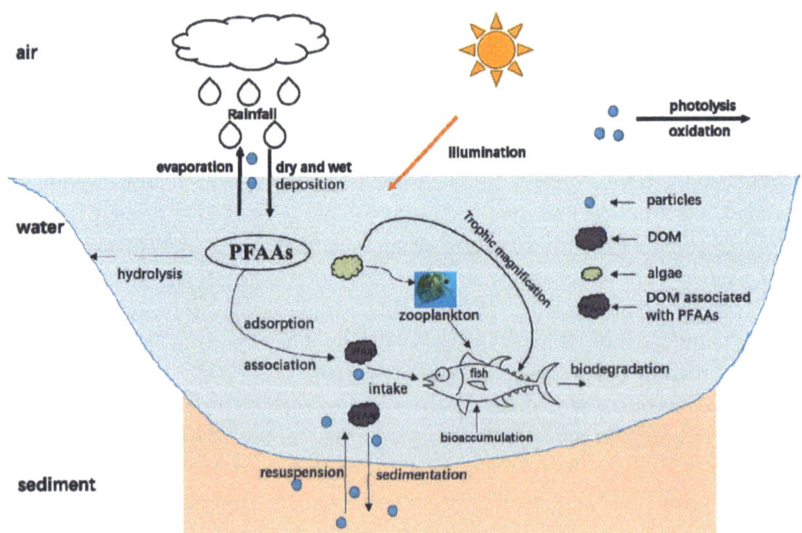

FIGURE 4.7 Distribution and environmental behaviors of PFCs in typical aquatic ecosystems (Liu et al., 2019a).

and most widely used method by calculating the ratio of pollutant concentrations in water or sediment to the environmental criteria. If the ratio is less than one, the water or sediment is nonrisky and the pollutants are relatively safe for the ecological environment; on the contrary, the greater the ratio is, the higher the risk (Liu et al., 2019a). According to the toxicity data of certain aquatic species, the SSD curve is plotted, and the ecological risk levels can be obtained in accordance with the environmental exposure and the SSD curve. For example, the HC_5, namely the hazardous concentration at 5% of species to protect 95% species, are 386.09 and 26.37 μg/L for PFOA and PFOS, respectively (Liu et al., 2019a), indicating that the toxicity of PFOS is higher than that of PFOA (Fig. 4.8).

In terms of the aquatic ecotoxicology, exposures of PFOS and PFOA at 1, 10, and 25 μM do not cause acute lethality of *D. magna*, however, hatching is reduced following exposure to both compounds (at 25 μM), and lifespan is decreased following exposure to PFOS at 25 μM. According to the lipid staining, PFAS exposure causes lipid accumulation, which can be attributable to the inhibition effect of PFAS on the genes involved on the absorption and catabolism of fatty acids as evidenced by the qRT-PCR analysis (Seyoum et al., 2020). Additionally, PFOS (25 μM) exposure can reduce the body length and the fecundity of *D. magna* (Seyoum et al., 2020). In addition, exposure to PFCs can affect the metabolic process of *D. magna* and affect the metabolites abundances. For instance, the sublethal PFOS

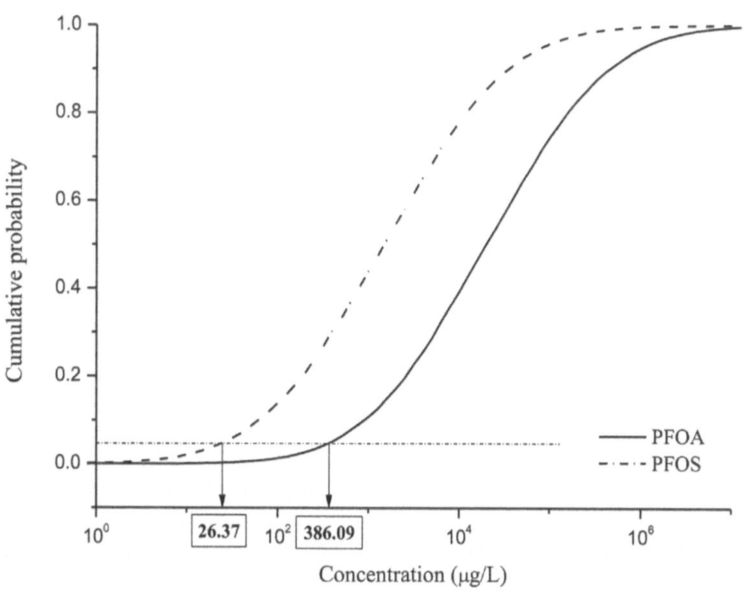

FIGURE 4.8 Comparison of species sensitivity distribution curves of PFOA and PFOS (Liu et al., 2019a).

exposure at 2 and 20 mg/L alters regulation of arginine, tyrosine, and adenosine monophosphate in *D. magna*, and the PFOS bioaccumulation in *D. magna* is time-dependent (Jeong & Simpson, 2020). The younger *D. magna* can bioaccumulate more PFOS during the monitoring process of 16 days.

Glutathione S-transferases (GSTs) enzyme activity in hepatopancreas of *Mytilus galloprovincialis* significantly increased when exposed to PFOS in low concentrations, however, this protective mechanism in the mussels was weakened at higher exposure concentration owing to the toxic effect of PFOS (Gizem & Hatice, 2018). Moreover, the PFCs can retard the development of zebrafish embryos, and the toxicity effect depends on the functional group, concentration, and carbon-chain length of PFCs (Zheng et al., 2012).

Perfluorotetradecanoic acid (PFTeDA), the longest carbon chain PFCs investigated, can pose ecological risks to four aquatic organisms (green algae, daphnid, fish, and mysid) during both dry and wet seasons. The risk induced by PFCs to the aquatic organisms obviously depends on the carbon chain length (Wu et al., 2020). In general, the longer the carbon chain length, the higher the risk. Due to biomagnification, PFCs may have higher risks to the high trophic mammals. Acute toxicity to *D. magna* decreases with decreasing carbon chain length of PFCs, and C4–C6 chemistry is thus less hazardous to daphnids than C7–C8 chemistry (Barmentlo et al., 2015).

4.5.1.1.2 Toxicity in terrestrial ecosystems

PFCs in soils can cause risks and toxicological effects on terrestrial organisms such as plants, microbes, and invertebrates due to their well-known persistence, toxicity, and bioaccumulative potential. Since they are frequently detected in soils, PFCs may have caused toxicological effects on terrestrial organisms. Therefore this section briefly summarizes some toxicological research of common PFCs on terrestrial organisms.

High phyto-toxicity of PFCs to lettuce and arugula was observed in a prior study, and it increases as a function of the concentration and decreases as a function of exposure time to the environment (Montagnolli et al., 2017). In addition, their toxicity to lettuce and green algae is similar and has good relations with the fluorinated carbon-chain length of the PFCs investigated. For instance, the EC_{50} of lettuce (Lactuca *sativa*) on 5-day root elongation follows an order of PFBA (4.19) > PFOA (1.80) > PFNA (0.85) > PFDA (0.27) > PFUnA (0.21) > PFDoA (0.14 mM) (Ding et al., 2012). In terms of PFOA, for the model plant species *Arabidopsis thaliana*, PFOA can trigger certain oxidative stress-associated events more strongly in shoots than in roots. Moreover, PFOA exposure can reduce the *Arabidopsis* biomass in a dose-related manner, alter the PFOA distribution, reconfigure the transcriptomes, and induce oxidative stress in a tissue-dependent manner (Fan et al., 2020). In addition, under a high exposure dose (50 mg/L) of PFOS, the chlorophyll levels of *Acorus calamus* and *Phragmites communis* decreased by

38.3% and 36.7%, respectively. Similarly, the soluble protein contents were reduced by 21.7% and 35.7% in *A. calamus* and *P. communis*, respectively. However, both the chlorophyll levels and soluble protein contents of both plants were increased at 1.0 mg/L PFOS (Qian et al., 2019). It was reported that the chlorophyll synthesis of *Eichhornia crassipes* and *Cyperus alternifolius* were promoted by low concentration (<0.1 mg/L) of PFOS; however, its synthesis was suppressed under high concentration (10 mg/L) of PFOS (Li, Hu, et al., 2020). Similarly, PFOS stress can cause the alteration of the membrane of plant leaves and the cell structure of the two wetland plants, and the incomplete mitochondrial contour of root cells was also observed (Li, Tang, et al., 2020). PFOS stress can also change the expression levels of genes involved in cell wall formation, the cell apoptosis pathway, phytochemicals synthesis, and metabolism in the plants (Li, Tang, et al., 2020). In addition, PFOS can cause the oxidative stress, which may be demonstrated by the elevated concentrations of malonaldehyde and hydrogen peroxide, and the reduced activities of superoxide dismutase (SOD), catalase (CAT), and peroxidase (POD) in *A. calamus* and *P. communis* at 50 mg/L PFOS for 48 days (Qian et al., 2019). For the same PFCs at the same exposure dose, the toxicological responses of different plants are different. For example, *A. calamuss* is more sensitive at low concentration (<10 mg/L) of PFOS than *P. communis*. Overall, the stress of PFCs can led to toxicological effects and the oxidative stress responses in plants, thereby generating ecotoxicological effect. However, the molecular mechanism of toxicity underlying plant exposure to PFCs has not been clearly revealed yet. More research on genetic and transcriptional analysis are encouraged.

Numerous studies have revealed that PFCs can influence the community structure, bacterial richness, and diversity of microorganisms in the soil. This influence extent is related to the carbon-chain length, functional groups, concentration of PFCs in the soil, moreover, it is also affected by soil properties and the type of soil microorganisms (Cai et al., 2019, 2020; Chen et al., 2020). For instance, sulfonic PFCs have higher toxicity to soil microorganisms than carboxylic ones. This is because the longer the carbon chain length, the greater the toxicity of PFCs (Cai et al., 2019). Soils with higher pH, organic matter, and available N, P, and K can weaken the toxicity of PFCs to bacteria because these soil conditions are favorable to improve bacterial activities.

Low concentration of PFOS exposure (e.g., 0.01 and 0.1 μgPFOS/g soil) can improve the activities of soil bacteria. When the exposure dose is 1.0 μgPFOS/g soil, the richness, diversity, and community structure of the bacterial community are enhanced. Conversely, high concentrations of PFOS (e.g., 10 μg/g) can undermine the normal functions of soil microorganisms (Qiao et al., 2018). In addition, short-chain PFCs (e.g., PFBS and PFHxS) can activate the urase and sucrose; while the long chain PFCs (e.g., PFOS) may inhibit these enzyme activities, reduce soil functional genes, destroy cellular structure, immune system and gene expression of soil bacteria (Qiao et al., 2018). This indicates that the PFCs with long carbon chains are more toxic to soil bacteria than the short-chain ones. The toxicity of PFOA to alkaline phosphatase is stronger than that to urease because of the lower binding free energy between PFOA and alkaline phosphatase (Chen et al., 2020). Furthermore, PFC tress can activate the glycosyl transferases in PFC-polluted soils, but severely subdue the glycoside hydrolases (Cai et al., 2020).

PFCs also have toxicological effect on invertebrates in soils such as *Eisenia fetida* (redworm), a model soil animal, commonly used for soil ecological risk assessment and early warning. The accumulation of PFCs into earthworms is a potential route of their entry into terrestrial food chains. It has been demonstrated that PFOS in soils can induce oxidative stresses, antioxidative defense system responses, and DNA damages to earthworms (Xu, Tian, et al., 2013). For instance, after being exposed to PFOS for 48 hours, the DNA of earthworm coelomocyte is typically damaged (Fig. 4.9B), whereas the DNA of control coelomocytes (i.e., deionized water) is undamaged (Fig. 4.9B).

PFCs in soils can change some biomarkers in *E. fetida*, such as SOD, CAT, POD, and GST, cellulase, and malondialdehyde (MDA). In addition, the toxicity of the same PFCs in different soils is different, and the sensitivity of different types of invertebrates to the same PFC in the same soil is also different (Princz et al., 2018). For instance, *Oppia nitens* is significantly more sensitive than *Folsomia candida* to the exposure of PFOS, regardless of soil types (Princz et al., 2018) (Fig. 4.10). A biomagnification factor (BMF) > 1 was detected for PFOS derived from ski waxes in earthworms in the soil in a skiing area, suggesting high risks of PFCs to high trophic mammals because of biomagnification (Grønnestad et al., 2019).

4.5.1.2 Health risks

The total concentration of PFASs ($\sum$PFASs, $n = 10$) in 169 groundwater samples from four regions in China (Shijiazhuang City, Du'an County,

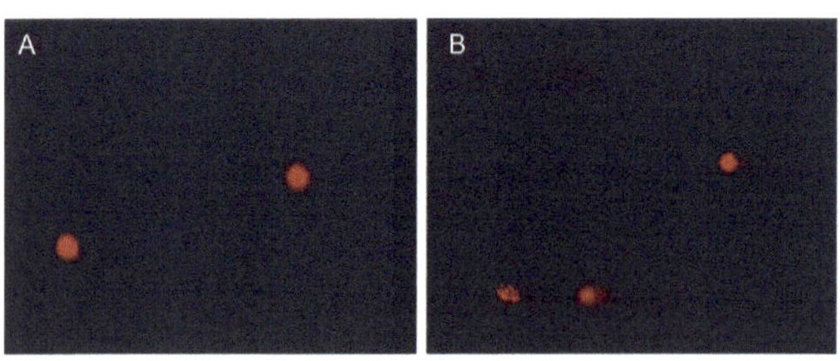

FIGURE 4.9 Typical comet figures of earthworm coelomocytes of deionized water control (A) and 2 $\mu g/cm^2$ PFOS (B) (Xu, Li, et al., 2013).

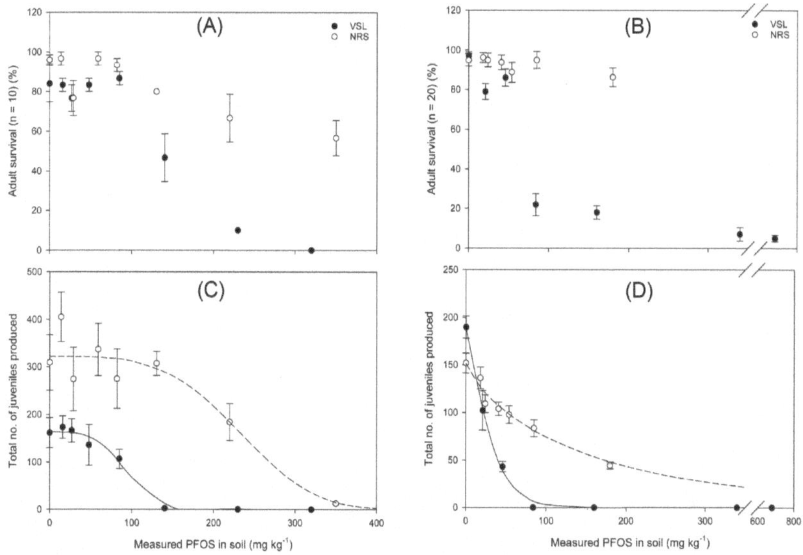

FIGURE 4.10 Mean (± standard error) adult survival of *Folsomia candida* (A) and *Oppia nitens* (B) and the number of their juveniles (C: *F. candida*; D: *O. nitens*) produced after 28 days exposure to PFOS in two soil types: a sandy loam (VSL) and a clay loam (NRS) soil. *Cited from Princz, J., Jatar, M., Lemieux, H., & Scroggins, R. (2018). Perfluorooctane sulfonate in surface soils: Effects on reproduction in the collembolan, Folsomia candida, and the oribatid mite, Oppia nitens. Chemosphere, 208, 757–763. https://doi.org/10.1016/j.chemosphere.2018.06.020.*

Lanzhou City, and Golmud City) ranges from 0.05 to 198.80 ng/L, from which the estimated health RQs of PFOA and PFOS are acceptable via the consumption of water (Qiao et al., 2018). The contents of PFCs in many edible food products have been measured, and the corresponding health risks through exposures to these food products have also been assessed, including chicken, duck, their eggs (Qi et al., 2019), fish (Meng et al., 2019), vegetables and fruits (Li et al., 2019), maize (Wen et al., 2013), potatoes (Clarke et al., 2010), lettuce and strawberry (Blaine et al., 2014), etc. Moreover, PFCs in urine (Wu et al., 2020), follicular fluid (Kim et al., 2020), breast milk, umbilical cord blood, fetal organs, and serum (Liu et al., 2020d) at various concentration have been determined.

In general, the health risks of PFCs to human are estimated by calculating the estimated daily intake (EDI) of PFCs through different exposure routes (e.g., eating food and drinking water) and the ratio of the EDI to the reference dosage recommended by USEPA, European Food Safety Authority (EFSA) or WHO. The obtained RQs of children are commonly higher than the adult's owing to the greater body weight of adult used in the assessment. Therefore the risk levels PFCs to residents are generally age dependent. The PFCs in infants mainly come from breast milk, thereby the feeding of breast milk is mainly the pathway of PFC exposure (Hu et al., 2021). In addition, based on the serum levels of PFOS and PFOA and the EDIs, the tolerable weekly intakes (TWIs) established are 13 and 6 ng/kg body weight per week for PFOS and for PFOA, respectively (EFSA Panel on Contaminants in the Food Chain (CONTAM) et al., 2018). According to this criterion, a considerable proportion of the population exceeds the proposed TWIs for both compounds (EFSA Panel on Contaminants in the Food Chain (CONTAM) et al., 2018). However, the uncertainty of health risk assessment may exist due to the various exposure pathways and the ubiquitous distribution of PFCs.

4.5.1.3 Epidemiological studies

Some epidemiological studies have indicated that PFCs have adverse effects on reproduction, development, immunity, endocrine, etc. (Grandjean et al., 2012; Granum et al., 2013; Petro et al., 2014). For example, Ballesteros et al. reported a positive association between PFHxS and PFOS exposures and thyroid stimulating hormone (TSH) levels measured in maternal blood, and PFNA exposure and TSH levels measured in the blood of boys with the age of more than 11 years old (Ballesteros et al., 2017). Prenatal exposures to higher PFOS and PFOA levels of pregnant women can cause larger number of episodes of cooccurrence of fever and coughing and fever and nasal discharge for their child at 1–4 years (Dalsager et al., 2016).

Exposure to PFCs can disrupt the endocrine hormones in humans. The positive associations of PFOS, PFOA, and/or their replacements (6:2 chlorinated polyfluorinated ether sulfonate, trade name F-53B) with progesterone, cortisol, and 11-deoxycortisol in newborns were observed (Liu et al., 2020a). Borghese et al. collected data from 1739 participants and found higher levels of PFHxS were associated with the development of preeclampsia, which may be affected by the fetal sex (Borghese et al., 2020). Ren et al. obtained an association of PFAS exposure with an increased risk of high one-hour plasma glucose (1 h-PG) among pregnant women, but could not found the association for fasting plasma glucose (Ren et al., 2020). Cui et al. found stronger inverse associations of semen PFASs with reproductive hormones of 651 adult men, and semen PFASs reflect exposures of reproductive system more directly than serum PFASs (Cui et al., 2020).

Epidemiological studies have indicated that PFOA and PFOS levels are associated with diabetes. For instance, PFASs may alter pancreatic development after the pancreatic domain emerges from the gut tube based on the human pluripotent stem cell pancreatic induction model and human pancreatic progenitor cell endocrine induction model (Liu et al., 2020c). It was found that several long-chain PFAS were associated with disrupted the homeostasis of thyroid hormones (TH) in Chinese thyroid peroxidase antibody-positive pregnant women (Aimuzi et al., 2020). However, uncertainty may exist due to the complex coexposure effects from other contaminants, the equivocal disturbance mechanisms of THs by PFCs, and the uncertain clinical significance of small changes in THs levels. Additionally, most of the studies on effects of PFCs on the reproduction, development, immunity, and endocrine of organism were cross-sectional, which can barely speculate the resultant conclusions.

4.6 Case studies

In a case from Yuqiao reservoir (YQR), a major drinking water source in Tianjin of China (Cao et al., 2019), the concentrations of 17 PFCs in surface water ($n = 17$), groundwater ($n = 9$), soils ($n = 17$), and sediment ($n = 13$) were detected. The occurrences, latent sources and health risks of 17 PFASs were also determined. The $\sum$PFCs concentrations of surface water, groundwater, soil, and sediment were 5.84−120.89 ng/L, 1.43−17.14 ng/L, 0.62−5.09 μg/kg dw, and 0.24−1.21 μg/kg dw, respectively. Short-chain PFBS exhibited high level and was detected frequently in groundwater because of its high solubility and mobility. PFCs with long chains were detected at lower detection rates. The levels of PFCs in sediments from downstream of rivers were slightly lower than that from upstream, which may be ascribed to the rapid economic development of the upstream regions where large amounts of wastewater were discharged into rivers. Interestingly, PFC concentrations in sediment cores from the 6−8 cm depth were obviously higher than that from the top layers 0−2 cm, which can be attributed to the limited production since 2006. RQs of PFOA and PFOS were far less than one, suggesting little adverse effect on the aquatic ecosystem and human health.

In the case of Gao et al. (2020), ecological and health risk assessment of PFOS in surface and drinking water resources in China was performed (Fig. 4.11). According to available monitoring data of PFOS in surface and drinking waters, PFOS was ubiquitous in China, whereas the overall level of contamination was low in comparison to that in other countries. Moreover, relatively high concentrations were observed in the southeast coastal area of China, which can be attributed to the abundant PFOS-related companies in the area. In terms of Chinese surface waters, when 5% of aquatic species exceeded the growth and development toxicity, the relevant probability was 0.65%, while the greatest probability (0.90%) was observed in Tai Lake in east China according to the joint probability curve method. However, most of the health risks of PFOS in China were acceptable, although the relative higher risk was found in east China.

Lal et al. selected three home-grown vegetables (lettuce, carrot, and tomato) to investigate the uptake and accumulation of PFOS via a greenhouse pot experiment (Lal et al., 2020). The BCF of PFOS in these vegetables were analyzed and the relative health risks to human through consuming these vegetables were assessed. Three treatments including the

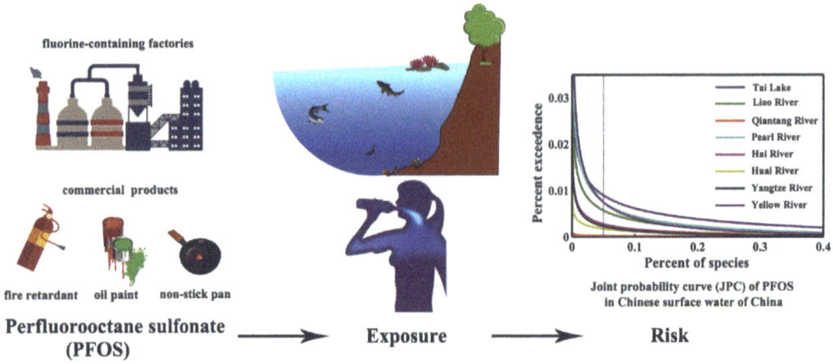

FIGURE 4.11 Exposure and risk assessment of PFOS in water by the joint probability curve (Gao et al., 2020).

control (no added PFOS), low PFOS-spiked mix soils, and high-level ones were used in the study. LC-MS results showed that the concentrations of PFOS were 232 ± 37, 3817 ± 331, and 41,947 ± 4603 ng/g for the control soil, the low-level treatment, and the high-level treatment, respectively. Moreover, increasing concentrations of PFOS in the potting mix soil caused the higher accumulation of PFOS in lettuce. Lower accumulation of PFOS in the edible roots was found than that in the leafy part of carrot, despite that roots have more contact with PFOS. Similarly, PFOS was accumulated more in the roots of tomato plants than that in the leafy shoots. Considerably high potential risk was found according to the concentrations of PFOS in these vegetables since the soil was spiked with high PFOS levels. However, such high concentrations of PFOS may not exist in the environment. Therefore there is no need to worry too much. More soil types and more plant species need to be included to assess the potential health risks of PFOS globally.

In a case study by Cai et al. (2019), toxic effects of PFCs on microorganisms in soils were evaluated using the microcalorimetry. The influence of functional group (carboxylic and sulfonic) and carbon chain length (C4, C8, and C10) of PFCs on the soil ecotoxicology were also investigated. Furthermore, the effects of soil properties [pH, OM, available nitrogen (N), phosphorus (P), and potassium (K)] on the toxicity of different PFCs were also examined in three typical soils from Heilongjiang, Fujian, and Sichuan in China. Results showed that sulfonic PFCs had higher toxic to soil bacteria than carboxylic ones. In addition, longer chain length of PFCs introduced greater toxicity to soil microorganisms. High soil pH values can improve the adsorption capability of soil to PFCs, weakening their toxicity to soil microorganisms. Favorable soil properties (sufficient OM, adequate available N, enough available P, and sufficient available K) can also attenuate the toxic effects of PFCs in soils.

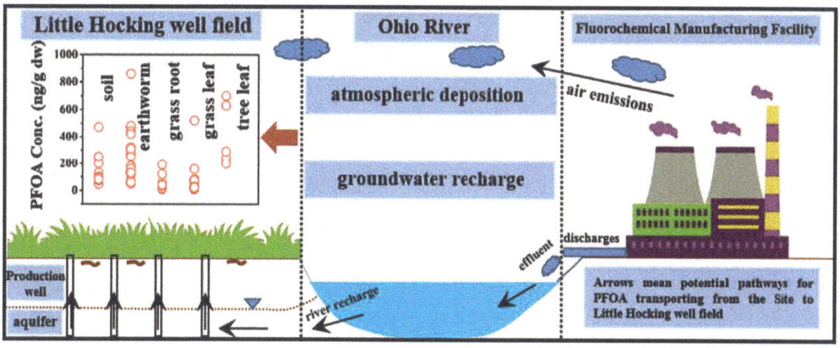

FIGURE 4.12 Distribution, transport pathways, and sources of PFCs in the Little Hocking well field in Washington County, Ohio, United States (Zhu and Kannan, 2019).

Another case study is from a contaminated site, the Little Hocking well field in Washington County, Ohio, United States (Zhu & Kannan, 2019). In this site, the distribution and partitioning of PFOA, PFHpA (C7), PFNA (C9), PFDA (C10), PFUnDA (C11), and PFDoDA (C12) in surface soil, earthworms, grass leaves, grass roots, and tree leaves were investigated (Fig. 4.12). The mean concentrations of $\sum$PFCAs ($n = 6$) were 150 ng/g for surface soil, 420 ng/g for earthworms, 61 ng/g for grass roots, 68 ng/g for grass leaves, and 430 ng/g for tree leaves on a dry weight basis. This site was contaminated by both atmospheric deposition and groundwater recharge based on spatial distribution of PFCAs. The biota-soil accumulation factors (BSAFs) of PFCAs for earthworm and grass root were positive associated with the length of perfluorocarbon chain. However, a negative association was found for both the translocation factor from grass root to shoot and the BSAF in tree leaves with the fluorocarbon chain length of PFCAs. This indicates that the short chain PFCAs are more readily translocated from the plant roots to shoots than the long chain ones.

4.7 Conclusion and perspectives

This chapter summarizes the properties, sources, occurrences, analytical methods, impacts, toxicity, and risks of PFCs in soils and groundwater. PFCs are highly stable, hydrophobic, oleophobic, hydrophilic, bioaccumulative, persistent, and difficultly degradable. The sources of PFCs are mainly anthropogenic, including daily domestic consumption, emission of fluorine in industrial parks, fire-fighting products, industrial wastes, and wastewater treatment plant drainage. The specific analytical techniques include sample collection, preconcentration, reconstitution, and quantitative analysis of PFCs in soil and groundwater samples. PFCs are widely detected in soil and groundwater, indicating that they have undergone significant migration from the sources to the receiving soils and groundwater. Therefore the occurrences and transfer of PFCs from the source zones (landfills, areas affected by fire foam, airports, areas near fluorine industrial parks, ski sites, etc.) to other areas in soils and groundwater need extra attention. PFCs can migrate to significant depths in soil profile, posing PFCs contamination to groundwater at most of sites. Thus the transport behaviors of PFCs in the vadose zone and groundwater are very important to their environmental fate. The next chapter of this book mainly focuses on the fate and transport of PFCs in soil and groundwater system under various conditions including the influence of some important factors such as pH, soil particulate size, saturated degree, and organic matter.

The ecological and health risks of PFCs are of great concern as they are highly toxic and bioaccumulative. However, the molecular mechanism of toxicity underlying plant and soil organism exposure to PFCs has not been clearly revealed yet. More research with genetic and transcriptional analyses is encouraged. Some epidemiological studies have indicated that PFCs have adverse effects on reproduction, development, immunity, endocrine, etc. However, these studies are cross-sectional and statistic-based, lacking direct toxicological mechanisms to organisms, which are critical and need more research investigations. Because the uses of perfluorinated alternatives, such as PFBA and PFBS, are encouraged, the toxicity of short-chain PFCs should be emphasized and further explored with additional research efforts.

References

Ahmed, M. B., Johir, M. A. H., McLaughlan, R., Nguyen, L. N., Xu, B., & Nghiem, L. D. (2020). Per- and polyfluoroalkyl substances in soil and sediments: Occurrence, fate, remediation and future outlook. *Science of the Total Environment, 748*. Available from https://doi.org/10.1016/j.scitotenv.2020.141251.

Ahrens, L., Yamashita, N., Yeung, L. W. Y., Taniyasu, S., Horii, Y., Lam, P. K. S., & Ebinghaus, R. (2009). Partitioning behavior of per- and polyfluoroalkyl compounds between pore water and sediment in two sediment cores from Tokyo Bay, Japan. *Environmental Science and Technology, 43*(18), 6969–6975. Available from https://doi.org/10.1021/es901213s.

Aimuzi, R., Luo, K., Huang, R., Huo, X., Nian, M., Ouyang, F., Du, Y., Feng, L., Wang, W., & Zhang, J. (2020). Perfluoroalkyl and polyfluroalkyl substances and maternal thyroid hormones in early pregnancy. *Environmental Pollution, 264*. Available from https://doi.org/10.1016/j.envpol.2020.114557.

Bai, X., & Son, Y. (2021). Perfluoroalkyl substances (PFAS) in surface water and sediments from two urban watersheds in Nevada, USA. *Science of the Total Environment, 751*. Available from https://doi.org/10.1016/j.scitotenv.2020.141622.

Ballesteros, V., Costa, O., Iñiguez, C., Fletcher, T., Ballester, F., & Lopez-Espinosa, M. J. (2017). Exposure to perfluoroalkyl substances and thyroid function in pregnant women and children: A systematic review of epidemiologic studies. *Environment International, 99*, 15–28. Available from https://doi.org/10.1016/j.envint.2016.10.015.

Bao, J., Li, C. L., Liu, Y., Wang, X., Yu, W. J., Liu, Z. Q., Shao, L. X., & Jin, Y. H. (2020). Bioaccumulation of perfluoroalkyl substances in greenhouse vegetables with long-term groundwater irrigation near fluorochemical plants in Fuxin, China. *Environmental Research, 188*. Available from https://doi.org/10.1016/j.envres.2020.109751.

Bao, J., Yu, W. J., Liu, Y., Wang, X., Jin, Y. H., & Dong, G. H. (2019). Perfluoroalkyl substances in groundwater and home-produced vegetables and eggs around a fluorochemical industrial park in China. *Ecotoxicology and Environmental Safety*, *171*, 199–205. Available from https://doi.org/10.1016/j.ecoenv.2018.12.086.

Barmentlo, S. H., Stel, J. M., Van Doorn, M., Eschauzier, C., De Voogt, P., & Kraak, M. H. S. (2015). Acute and chronic toxicity of short chained perfluoroalkyl substances to *Daphnia magna*. *Environmental Pollution*, *198*, 47–53. Available from https://doi.org/10.1016/j.envpol.2014.12.025.

Blaine, A. C., Rich, C. D., Sedlacko, E. M., Hyland, K. C., Stushnoff, C., Dickenson, E. R. V., & Higgins, C. P. (2014). Perfluoroalkyl acid uptake in lettuce (*Lactuca sativa*) and Strawberry (Fragaria ananassa) irrigated with reclaimed water. *Environmental Science and Technology*, *48*(24), 14361–14368. Available from https://doi.org/10.1021/es504150h.

Borghese, M. M., Walker, M., Helewa, M. E., Fraser, W. D., & Arbuckle, T. E. (2020). Association of perfluoroalkyl substances with gestational hypertension and preeclampsia in the MIREC study. *Environment International*, *141*. Available from https://doi.org/10.1016/j.envint.2020.105789.

Brusseau, M. L., Anderson, R. H., & Guo, B. (2020). PFAS concentrations in soils: Background levels versus contaminated sites. *Science of the Total Environment*, *740*. Available from https://doi.org/10.1016/j.scitotenv.2020.140017.

Cai, Y., Chen, H., Yuan, R., Wang, F., Chen, Z., & Zhou, B. (2019). Toxicity of perfluorinated compounds to soil microbial activity: Effect of carbon chain length, functional group and soil properties. *Science of the Total Environment*, *690*, 1162–1169. Available from https://doi.org/10.1016/j.scitotenv.2019.06.440.

Cai, Y., Chen, H., Yuan, R., Wang, F., Chen, Z., & Zhou, B. (2020). Metagenomic analysis of soil microbial community under PFOA and PFOS stress. *Environmental Research*, *188*. Available from https://doi.org/10.1016/j.envres.2020.109838.

Cao, X., Wang, C., Lu, Y., Zhang, M., Khan, K., Song, S., Wang, P., & Wang, C. (2019). Occurrence, sources and health risk of polyfluoroalkyl substances (PFASs) in soil, water and sediment from a drinking water source area. *Ecotoxicology and Environmental Safety*, *174*, 208–217. Available from https://doi.org/10.1016/j.ecoenv.2019.02.058.

Chen, S., Jiao, X. C., Gai, N., Li, X. J., Wang, X. C., Lu, G. H., Piao, H. T., Rao, Z., & Yang, Y. L. (2016). Perfluorinated compounds in soil, surface water, and groundwater from rural areas in eastern China. *Environmental Pollution*, *211*, 124–131. Available from https://doi.org/10.1016/j.envpol.2015.12.024.

Chen, H., Wang, Q., Cai, Y., Yuan, R., Wang, F., Zhou, B., & Chen, Z. (2020). Effect of perfluorooctanoic acid on microbial activity in wheat soil under different fertilization conditions. *Environmental Pollution*, 264. Available from https://doi.org/10.1016/j.envpol.2020.114784.

Clarke, D. B., Bailey, V. A., Routledge, A., Lloyd, A. S., Hird, S., Mortimer, D. N., & Gem, M. (2010). Dietary intake estimate for perfluorooctanesulphonic acid (PFOS) and other perfluorocompounds (PFCs) in UK retail foods following determination using standard addition LC–MS/MS. *Food Additives & Contaminants: Part A*, 530–545. Available from https://doi.org/10.1080/19440040903476590.

Cui, Q., Pan, Y., Wang, J., Liu, H., Yao, B., & Dai, J. (2020). Exposure to per- and polyfluoroalkyl substances (PFASs) in serum versus semen and their association with male reproductive hormones. *Environmental Pollution*, 266. Available from https://doi.org/10.1016/j.envpol.2020.115330.

Dalsager, L., Christensen, N., Husby, S., Kyhl, H., Nielsen, F., Høst, A., Grandjean, P., & Jensen, T. K. (2016). Association between prenatal exposure to perfluorinated compounds and symptoms of infections at age 1–4 years among 359 children in the Odense Child Cohort. *Environment International*, *96*, 58–64. Available from https://doi.org/10.1016/j.envint.2016.08.026.

Ding, G., Wouterse, M., Baerselman, R., & Peijnenburg, W. J. G. M. (2012). Toxicity of polyfluorinated and perfluorinated compounds to lettuce (*Lactuca sativa*) and green algae (Pseudokirchneriella subcapitata). *Archives of Environmental Contamination and Toxicology*, *62*(1), 49–55. Available from https://doi.org/10.1007/s00244-011-9684-9.

Duong, H. T., Kadokami, K., Shirasaka, H., Hidaka, R., Chau, H. T. C., Kong, L., Nguyen, T. Q., & Nguyen, T. T. (2015). Occurrence of perfluoroalkyl acids in environmental waters in Vietnam. *Chemosphere*, *122*, 115–124. Available from https://doi.org/10.1016/j.chemosphere.2014.11.023.

EFSA Panel on Contaminants in the Food Chain (CONTAM)., Knutsen, H. K., Alexander, J., Barregård, L., Bignami, M., Brüschweiler, B., Ceccatelli, S., Cottrill, B., Dinovi, M., Edler, L., Grasl-Kraupp, B., Hogstrand, C., Hoogenboom, L. (Ron), Nebbia, C. S., Oswald, I. P., Petersen, A., Rose, M., Roudot, A.-C., Vleminckx, C., & Schwerdtle, T. (2018). Risk to human health related to the presence of perfluorooctane sulfonic acid and perfluorooctanoic acid in food. *EFSA Journal*, *16*(12)e05194. Available from https://doi.org/10.2903/j.efsa.2018.5194.

Eschauzier, C., Raat, K. J., Stuyfzand, P. J., & De Voogt, P. (2013). Perfluorinated alkylated acids in groundwater and drinking water: Identification, origin and mobility. *Science of The Total Environment*, *458-460*, 477–485. Available from https://doi.org/10.1016/j.scitotenv.2013.04.066.

Fan, L., Tang, J., Zhang, D., Ma, M., Wang, Y., & Han, Y. (2020). Investigations on the phytotoxicity of perfluorooctanoic acid in *Arabidopsis thaliana*. *Environmental Science and Pollution Research*, *27*(1), 1131–1143. Available from https://doi.org/10.1007/s11356-019-07018-5.

Fang, S., Sha, B., Yin, H., Bian, Y., Yuan, B., & Cousins, I. T. (2020). Environment occurrence of perfluoroalkyl acids and associated human health risks near a major fluorochemical manufacturing park in southwest of China. *Journal of Hazardous Materials*, *396*. Available from https://doi.org/10.1016/j.jhazmat.2020.122617.

Filipovic, M., Woldegiorgis, A., Norström, K., Bibi, M., Lindberg, M., & Österås, A. H. (2015). Historical usage of aqueous film forming foam: A case study of the widespread distribution of perfluoroalkyl acids from a military airport to groundwater, lakes, soils and fish. *Chemosphere*, *129*, 39–45. Available from https://doi.org/10.1016/j.chemosphere.2014.09.005.

Gao, X., Liu, Z., Li, J., Wang, X., Cui, L., Ai, S., Zhao, S., & Xu, Q. (2020). Ecological and health risk assessment of perfluorooctane sulfonate in surface and drinking water resources in China. *Science of the Total Environment*, *738*. Available from https://doi.org/10.1016/j.scitotenv.2020.139914.

Gao, Y., Liang, Y., Gao, K., Wang, Y., Wang, C., Fu, J., Wang, Y., Jiang, G., & Jiang, Y. (2019). Levels, spatial distribution and isomer profiles of perfluoroalkyl acids in soil, groundwater and tap water around a manufactory in China. *Chemosphere*, *227*, 305–314. Available from https://doi.org/10.1016/j.chemosphere.2019.04.027.

Gizem, G., & Hatice, P. (2018). Effects of perfluorooctane sulfonate compounds on the biochemical activities in mussels (*Mytilus galloprovincialis*). *Ege Journal of Fisheries and Aquatic Sciences*, 417–422. Available from https://doi.org/10.12714/egejfas.2018.35.4.07.

Goosey, E., & Harrad, S. (2012). Perfluoroalkyl substances in UK indoor and outdoor air: Spatial and seasonal variation, and implications for human exposure. *Environment International*, *45*, 86–90. Available from https://doi.org/10.1016/j.envint.2012.04.007.

Grandjean, P., Andersen, E. W., Budtz-Jørgensen, E., Nielsen, F., Mølbak, K. R., Weihe, P., & Heilmann, C. (2012). Serum vaccine antibody concentrations in children exposed to perfluorinated compounds. *Journal of the American Medical Association*, *307*(4), 391–397. Available from https://doi.org/10.1001/jama.2011.2034.

Granum, B., Haug, L. S., Namork, E., Stølevik, S. B., Thomsen, C., Aaberge, I. S., Van Loveren, H., Løvik, M., & Nygaard, U. C. (2013). Pre-natal exposure to perfluoroalkyl substances may be associated with altered vaccine antibody levels and immune-related health outcomes in early childhood. *Journal of Immunotoxicology*, *10*(4), 373–379. Available from https://doi.org/10.3109/1547691X.2012.755580.

Groffen, T., Rijnders, J., Verbrigghe, N., Verbruggen, E., Prinsen, E., Eens, M., & Bervoets, L. (2019). Influence of soil physicochemical properties on the depth profiles of perfluoroalkylated acids (PFAAs) in soil along a distance gradient from a fluorochemical plant and associations with soil microbial parameters. *Chemosphere*, *236*. Available from https://doi.org/10.1016/j.chemosphere.2019.124407.

Grønnestad, R., Vázquez, B. P., Arukwe, A., Jaspers, V. L. B., Jenssen, B. M., Karimi, M., Lyche, J. L., & Krøkje, Å. (2019). Levels, patterns, and biomagnification potential of perfluoroalkyl substances in a terrestrial food chain in a Nordic Skiing Area. *Environmental Science and Technology*, *53*(22), 13390–13397. Available from https://doi.org/10.1021/acs.est.9b02533.

Guardian, M. G. E., Boongaling, E. G., Bernardo-Boongaling, V. R. R., Gamonchuang, J., Boontongto, T., Burakham, R., Arnnok, P., & Aga, D. S. (2020). Prevalence of per- and polyfluoroalkyl substances (PFASs) in drinking and source water from two Asian countries. *Chemosphere*, *256*. Available from https://doi.org/10.1016/j.chemosphere.2020.127115.

Harrad, S., Drage, D. S., Sharkey, M., & Berresheim, H. (2020). Perfluoroalkyl substances and brominated flame retardants in landfill-related air, soil, and groundwater from Ireland. *Distribution of perfluoroalkyl substances (PFASs) in aquatic plant-based systems: From soil adsorption and plant uptake to effects on microbial community*. Available from https://doi.org/10.1016/j.scitotenv.2019.135834.

Hu, L., Luo, D., Wang, L., Yu, M., Zhao, S., Wang, Y., Mei, S., & Zhang, G. (2021). Levels and profiles of persistent organic pollutants in breast milk in China and their potential health risks to breastfed infants: A review. *Science of the Total Environment*, *753*. Available from https://doi.org/10.1016/j.scitotenv.2020.142028.

Huang, Y., Lu, M., Chen, L., Bai, M., Ouyang, X., & Huang, X. (2020). Development of solid-phase microextraction with multiple interactions-based monolithic fibers for the sensitive determination of perfluoroalkyl phosphonic acids in water and vegetable samples. *Talanta*, 206. Available from https://doi.org/10.1016/j.talanta.2019.120198.

Jeong, T. Y., & Simpson, M. J. (2020). Time-dependent biomolecular responses and bioaccumulation of perfluorooctane sulfonate (PFOS) in *Daphnia magna*. *Comparative Biochemistry and Physiology - Part D: Genomics and Proteomics*, *35*. Available from https://doi.org/10.1016/j.cbd.2020.100701.

Kim, Y. R., White, N., Bräunig, J., Vijayasarathy, S., Mueller, J. F., Knox, C. L., Harden, F. A., Pacella, R., & Toms, L. M. L. (2020). Per- and poly-fluoroalkyl substances (PFASs) in follicular fluid from women experiencing infertility in Australia. *Environmental Research*, *190*. Available from https://doi.org/10.1016/j.envres.2020.109963.

Kleywegt, S., Raby, M., McGill, S., & Helm, P. (2020). The impact of risk management measures on the concentrations of per- and polyfluoroalkyl substances in source and treated drinking waters in Ontario. Canada. *Science of The Total Environment, 748*.

Kovářová, J., & Svobodová, Z. (2008). Perfluorinated compounds: Occurrence and risk profile. *Neuroendocrinology Letters, 29*(5), 599–608.

Kuroda, K., Murakami, M., Oguma, K., Takada, H., & Takizawa, S. (2014). Investigating sources and pathways of perfluoroalkyl acids (PFAAs) in aquifers in Tokyo using multiple tracers. *Science of The Total Environment, 488-489*, 51–60. Available from https://doi.org/10.1016/j.scitotenv.2014.04.066.

Lal, M. S., Megharaj, M., Naidu, R., & Bahar, M. M. (2020). Uptake of perfluorooctane sulfonate (PFOS) by common home-grown vegetable plants and potential risks to human health. *Environmental Technology and Innovation*, 19. Available from https://doi.org/10.1016/j.eti.2020.100863.

Li, B. B., Hu, L. X., Yang, Y. Y., Wang, T. T., Liu, C., & Ying, G. G. (2020). Contamination profiles and health risks of PFASs in groundwater of the Maozhou River basin. *Environmental Pollution, 260*. Available from https://doi.org/10.1016/j.envpol.2020.113996.

Li, P., Oyang, X., Zhao, Y., Tu, T., Tian, X., Li, L., Zhao, Y., Li, J., & Xiao, Z. (2019). Occurrence of perfluorinated compounds in agricultural environment, vegetables, and fruits in regions influenced by a fluorine-chemical industrial park in China. *Chemosphere, 225*, 659–667. Available from https://doi.org/10.1016/j.chemosphere.2019.03.045.

Li, R., Tang, T., Qiao, W., & Huang, J. (2020). Toxic effect of perfluorooctane sulfonate on plants in vertical-flow constructed wetlands. *Journal of Environmental Sciences (China), 92*, 176–186. Available from https://doi.org/10.1016/j.jes.2020.02.018.

Lindstrom, A. B., Strynar, M. J., Delinsky, A. D., Nakayama, S. F., McMillan, L., Libelo, E. L., Neill, M., & Thomas, L. (2011). Application of WWTP biosolids and resulting perfluorinated compound contamination of surface and well water in Decatur, Alabama, USA. *Environmental Science and Technology, 45*(19), 8015–8021. Available from https://doi.org/10.1021/es1039425.

Lindstrom, A. B., Strynar, M. J., & Libelo, E. L. (2011). Polyfluorinated compounds: Past, present, and future. *Environmental Science and Technology, 45*(19), 7954–7961. Available from https://doi.org/10.1021/es2011622.

Liu, H., Pan, Y., Jin, S., Li, Y., Zhao, L., Sun, X., Cui, Q., Zhang, B., Zheng, T., Xia, W., Zhou, A., Campana, A. M., Dai, J., & Xu, S. (2020a). Associations of per-/polyfluoroalkyl substances with glucocorticoids and progestogens in newborns. *Environment International, 140*105636. Available from https://doi.org/10.1016/j.envint.2020.105636.

Liu, L., Liu, Y., Gao, B., Ji, R., Li, C., & Wang, S. (2020b). Removal of perfluorooctanoic acid (PFOA) and perfluorooctane sulfonate (PFOS) from water by carbonaceous nanomaterials: A review. *Critical Reviews in Environmental Science and Technology, 50*(22), 2379–2414. Available from https://doi.org/10.1080/10643389.2019.1700751.

Liu, S., Yang, R., Yin, N., & Faiola, F. (2020c). Effects of per- and poly-fluorinated alkyl substances on pancreatic and endocrine differentiation of human pluripotent stem cells. *Chemosphere, 254*. Available from https://doi.org/10.1016/j.chemosphere.2020.126709.

Liu, Y., Li, A., Buchanan, S., & Liu, W. (2020d). Exposure characteristics for congeners, isomers, and enantiomers of perfluoroalkyl substances in mothers and infants. *Environment International, 144*. Available from https://doi.org/10.1016/j.envint.2020.106012.

Liu, Z., Lu, Y., Shi, Y., Wang, P., Jones, K., Sweetman, A. J., Johnson, A. C., Zhang, M., Zhou, Y., Lu, X., Su, C., Sarvajayakesavaluc, S., & Khan, K. (2017). Crop bioaccumulation and human exposure of perfluoroalkyl acids through multi-media transport from a mega fluorochemical industrial park, China. *Environment International, 106*. Available from https://doi.org/10.1016/j.envint.2017.05.014.

Liu, W., Wu, J., He, W., & Xu, F. (2019a). A review on perfluoroalkyl acids studies: Environmental behaviors, toxic effects, and ecological and health risks. *Ecosystem Health and Sustainability, 5* (1), 1–19. Available from https://doi.org/10.1080/20964129.2018.1558031.

Liu, J., Zhong, G., Li, W., & Mejia Avendaño, S. (2019b). Isomer-specific biotransformation of perfluoroalkyl sulfonamide compounds in aerobic soil. *Science of the Total Environment, 651*, 766–774. Available from https://doi.org/10.1016/j.scitotenv.2018.09.214.

Mejia Avendaño, S., & Liu, J. (2015). Production of PFOS from aerobic soil biotransformation of two perfluoroalkyl sulfonamide derivatives. *Chemosphere, 119*, 1084–1090. Available from https://doi.org/10.1016/j.chemosphere.2014.09.059.

Meng, J., Liu, S., Zhou, Y., & Wang, T. (2019). Are perfluoroalkyl substances in water and fish from drinking water source the major pathways towards human health risk? *Ecotoxicology and Environmental Safety, 181*, 194–201. Available from https://doi.org/10.1016/j.ecoenv.2019.06.010.

Meng, J., Wang, T., Wang, P., Giesy, J. P., & Lu, Y. (2013). Perfluorinated compounds and organochlorine pesticides in soils around Huaihe River: a heavily contaminated watershed in Central China. *Environmental Science and Pollution Research, 20*, 3965–3974. Available from https://doi.org/10.1007/s11356-012-1338-6.

Montagnolli, R. N., Matos Lopes, P. R., Matos Cruz, J., Marina Turini Claro, E., Quiterio, G. M., & Bidoia, E. D. (2017). The effects of fluoride based fire-fighting foams on soil microbiota activity and plant growth during natural attenuation of perfluorinated compounds. *Environmental Toxicology and Pharmacology, 50*, 119–127. Available from https://doi.org/10.1016/j.etap.2017.01.017.

Navarro, I., de la Torre, A., Sanz, P., & de los ÁngelesMartínez, M. (2020). Perfluoroalkyl acids (PFAAs): Distribution, trends and aquatic ecological risk assessment in surface water from Tagus River basin (Spain). *Environmental Pollution, 256*. Available from https://doi.org/10.1016/j.envpol.2019.113511, In this issue.

Pan, Y., Shi, Y., Wang, J., Jin, X., & Cai, Y. (2011). Pilot Investigation of Perfluorinated Compounds in River Water, Sediment, Soil and Fish in Tianjin, China. *Bulletin of Environmental Contamination and Toxicology, 87*, 152–157. Available from https://doi.org/10.1007/s00128-011-0313-0.

Petro, E. M. L., D'Hollander, W., Covaci, A., Bervoets, L., Fransen, E., De Neubourg, D., De Pauw, I., Leroy, J. L. M. R., Jorssen, E. P. A., & Bols, P. E. J. (2014). Perfluoroalkyl acid contamination of follicular fluid and its consequence for in vitro oocyte developmental competence. *Science of the Total Environment, 496*, 282–288. Available from https://doi.org/10.1016/j.scitotenv.2014.07.028.

Princz, J., Jatar, M., Lemieux, H., & Scroggins, R. (2018). Perfluorooctane sulfonate in surface soils: Effects on reproduction in the collembolan, *Folsomia candida*, and the oribatid mite, *Oppia nitens*. *Chemosphere, 208*, 757–763. Available from https://doi.org/10.1016/j.chemosphere.2018.06.020.

Qi, Y., Huo, S., Hu, S., Xi, B., Su, J., & Tang, Z. (2016). Identification, characterization, and human health risk assessment of perfluorinated compounds in groundwater from a suburb of Tianjin, China. *Environmental Earth Sciences, 75*. Available from https://doi.org/10.1007/s12665-016-5415-x.

Qi, X., Zhou, J., Wang, M., Yang, M. R., Tang, X. Y., Mao, X. F., & Wang, T. T. (2019). Perfluorinated compounds in poultry products from the Yangtze River Delta and Pearl River

Delta regions in China. *Science of the Total Environment, 689*, 1079–1086. Available from https://doi.org/10.1016/j.scitotenv.2019.06.258.

Qian, J., Lu, B., Chen, H., Wang, P., Wang, C., Li, K., Tian, X., Jin, W., He, X., & Chen, H. (2019). Phytotoxicity and oxidative stress of perfluorooctanesulfonate to two riparian plants: *Acorus calamus* and *Phragmites communis*. *Ecotoxicology and Environmental Safety, 180*, 215–226. Available from https://doi.org/10.1016/j.ecoenv.2019.04.078.

Qiao, W., Xie, Z., Zhang, Y., Liu, X., Xie, S., Huang, J., & Yu, L. (2018). Perfluoroalkyl substances (PFASs) influence the structure and function of soil bacterial community: Greenhouse experiment. *Science of the Total Environment, 642*, 1118–1126. Available from https://doi.org/10.1016/j.scitotenv.2018.06.113.

Qiao, X., Jiao, L., Zhang, X., Li, X., Hao, S., Kong, M., & Liu, Y. (2020). Contamination profiles and risk assessment of per- and polyfluoroalkyl substances in groundwater in China. *Environmental Monitoring and Assessment, 192*(2). Available from https://doi.org/10.1007/s10661-019-8005-z.

Qu, Y., Jiang, X., Cagnetta, G., Liu, L., Bao, Y., Li, W., Wang, Q., Liang, C., Huang, J., Yang, H., & Yu, G. (2019). Poly- and perfluoroalkyl substances in a drinking water treatment plant in the Yangtze River Delta of China: Temporal trend, removal and human health risk. *Science of the Total Environment*, 696. Available from https://doi.org/10.1016/j.scitotenv.2019.133949.

Rankin, K., Mabury, S. A., Jenkins, T. M., & Washington, J. W. (2016). A North American and global survey of perfluoroalkyl substances in surface soils: Distribution patterns and mode of occurrence. *Chemosphere, 161*. Available from https://doi.org/10.1016/j.chemosphere.2016.06.109.

Ren, Y., Jin, L., Yang, F., Liang, H., Zhang, Z., Du, J., Song, X., Miao, M., & Yuan, W. (2020). Concentrations of perfluoroalkyl and polyfluoroalkyl substances and blood glucose in pregnant women. *Environmental Health: A Global Access Science Source, 19*(1). Available from https://doi.org/10.1186/s12940-020-00640-8.

Sammut, G., Sinagra, E., Sapiano, M., Helmus, R., & de Voogt, P. (2019). Perfluoroalkyl substances in the Maltese environment – (II) sediments, soils and groundwater. *Science of The Total Environment, 682*. Available from https://doi.org/10.1016/j.scitotenv.2019.04.403.

Scher, D. P., Kelly, J. E., Huset, C. A., Barry, K. M., Hoffbeck, R. W., Yingling, V. L., & Messing, R. B. (2018). Occurrence of perfluoroalkyl substances (PFAS) in garden produce at homes with a history of PFAS-contaminated drinking water. *Chemosphere, 196*. Available from https://doi.org/10.1016/j.chemosphere.2017.12.179.

Seo, S. H., Son, M. H., Shin, E. S., Choi, S. D., & Chang, Y. S. (2019). Matrix-specific distribution and compositional profiles of perfluoroalkyl substances (PFASs) in multimedia environments. *Journal of Hazardous Materials, 364*. Available from https://doi.org/10.1016/j.jhazmat.2018.10.012.

Sepulvado, J. G., Blaine, A. C., Hundal, L. S., & Higgins, C. P. (2011). Occurrence and fate of perfluorochemicals in soil following the land application of municipal biosolids. *Environmental Science and Technology, 45*(19), 8106–8112. Available from https://doi.org/10.1021/es103903d.

Seyoum, A., Pradhan, A., Jass, J., & Olsson, P. E. (2020). Perfluorinated alkyl substances impede growth, reproduction, lipid metabolism and lifespan in *Daphnia magna*. *Science of the Total Environment, 737*. Available from https://doi.org/10.1016/j.scitotenv.2020.139682.

Shan, G., Wei, M., Zhu, L., Liu, Z., & Zhang, Y. (2014). Concentration profiles and spatial distribution of perfluoroalkyl substances in an industrial center with condensed fluorochemical facilities. *Science of The Total Environment, 490*. Available from https://doi.org/10.1016/j.scitotenv.2014.05.005.

Shigei, M., Ahrens, L., Hazaymeh, A., & Dalahmeh, S. S. (2020). Per- and polyfluoroalkyl substances in water and soil in wastewater-irrigated farmland in Jordan. *Science of The Total Environment, 716*. Available from https://doi.org/10.1016/j.scitotenv.2020.137057.

Stoiber, T., Evans, S., & Naidenko, O. V. (2020). Disposal of products and materials containing per- and polyfluoroalkyl substances (PFAS): A cyclical problem. *Chemosphere, 260*. Available from https://doi.org/10.1016/j.chemosphere.2020.127659.

Suja, F., Pramanik, B. K., & Zain, S. M. (2009). Contamination, bioaccumulation and toxic effects of perfluorinated chemicals (PFCs) in the water environment: A review paper. *Water Science and Technology, 60*(6), 1533–1554. Available from https://doi.org/10.2166/wst.2009.504.

Sun, R., Wu, M., Tang, L., Li, J., Qian, Z., Han, T., & Xu, G. (2018). Perfluorinated compounds in surface waters of Shanghai, China: Source analysis and risk assessment. *Ecotoxicology and Environmental Safety, 149*, 88–95. Available from https://doi.org/10.1016/j.ecoenv.2017.11.012.

Tan, B., Wang, T., Wang, P., Luo, W., Lu, Y., Romesh, K. Y., & Giesy, J. P. (2014). Perfluoroalkyl substances in soils around the Nepali Koshi River: levels, distribution, and mass balance. *Environmental Science and Pollution Research volume, 21*, 9201–9211. Available from https://doi.org/10.1007/s11356-014-2835-6.

Ullah, S., Alsberg, T., & Berger, U. (2011). Simultaneous determination of perfluoroalkyl phosphonates, carboxylates, and sulfonates in drinking water. *Journal of Chromatography A, 1218*(37), 6388–6395. Available from https://doi.org/10.1016/j.chroma.2011.07.005.

Wang, G., Wang, X., Xing, Z., Lu, J., Chang, Q., & Tong, Y. (2019). Occurrence and distribution of perfluorooctane sulfonate and perfluorooctanoic acid in three major rivers of Xinjiang, China. *Environmental Science and Pollution Research, 26*(27), 28062–28070. Available from https://doi.org/10.1007/s11356-019-05770-2.

Wang, T., Chen, C., Naile, J. E., Khim, J. S., Giesy, J. P., & Lu, Y. (2011). Perfluorinated Compounds in Water, Sediment and Soil from Guanting Reservoir, China. *Bulletin of Environmental Contamination and Toxicology, 87*(6), 74–79. Available from https://doi.org/10.1007/s00128-011-0307-y.

Wang, P., Wang, T., Giesy, J. P., & Lu, Y. (2013). Perfluorinated compounds in soils from Liaodong Bay with concentrated fluorine industry parks in China. *Chemosphere, 91*(6), 751–757. Available from https://doi.org/10.1016/j.chemosphere.2013.02.017.

Wang, T., Wang, P., Meng, J., Liu, S., Lu, Y., Khim, J. S., & Giesy, J. P. (2015). A review of sources, multimedia distribution and health risks of perfluoroalkyl acids (PFAAs) in China. *Chemosphere, 129*, 87–99. Available from https://doi.org/10.1016/j.chemosphere.2014.09.021.

Wang, Q., Zhao, Z., Ruan, Y., Li, J., Sun, H., & Zhang, G. (2018). Occurrence and distribution of perfluorooctanoic acid (PFOA) and perfluorooctanesulfonic acid (PFOS) in natural forest soils: A nationwide study in China. *Science of The Total Environment, 645*, 596–602. Available from https://doi.org/10.1016/j.scitotenv.2018.07.151.

Wei, C., Wang, Q., Song, X., Chen, X., Fan, R., Ding, D., & Liu, Y. (2018). Distribution, source identification and health risk assessment of PFASs and two PFOS alternatives in groundwater from non-industrial areas. *Ecotoxicology and Environmental Safety, 152*, 141–150. Available from https://doi.org/10.1016/j.ecoenv.2018.01.039.

Wen, B., Li, L., Liu, Y., Zhang, H., Hu, X., Shan, X. Q., & Zhang, S. (2013). Mechanistic studies of perfluorooctane sulfonate, perfluorooctanoic acid uptake by maize (*Zea mays* L. cv. TY2). *Plant and Soil, 370*(1–2), 345–354. Available from https://doi.org/10.1007/s11104-013-1637-9.

Wilkinson, J. L., Hooda, P. S., Swinden, J., Barker, J., & Barton, S. (2018). Spatial (bio)accumulation of pharmaceuticals, illicit drugs, plasticisers, perfluorinated compounds and metabolites in river sediment, aquatic plants and benthic organisms. *Environmental Pollution*, *234*, 864–875. Available from https://doi.org/10.1016/j.envpol.2017.11.090.

Wu, J., Junaid, M., Wang, Z., Sun, W., & Xu, N. (2020). Spatiotemporal distribution, sources and ecological risks of perfluorinated compounds (PFCs) in the Guanlan River from the rapidly urbanizing areas of Shenzhen, China. *Chemosphere*, *245*. Available from https://doi.org/10.1016/j.chemosphere.2019.125637.

Xiao, F., Simcik, M. F., Halbach, T. R., & Gulliver, J. S. (2015). Perfluorooctane sulfonate (PFOS) and perfluorooctanoate (PFOA) in soils and groundwater of a U.S. metropolitan area: Migration and implications for human exposure. *Water Research*, *72*, 64–74. Available from https://doi.org/10.1016/j.watres.2014.09.052.

Xu, D., Li, C., Wen, Y., & Liu, W. (2013). Antioxidant defense system responses and DNA damage of earthworms exposed to perfluorooctane sulfonate (PFOS). *Environmental Pollution*, *174*, 121–127. Available from https://doi.org/10.1016/j.envpol.2012.10.030.

Xu, J., Tian, Y. Z., Zhang, Y., Guo, C. S., Shi, G. L., Zhang, C. Y., & Feng, Y. C. (2013). Source apportionment of perfluorinated compounds (PFCs) in sediments: Using three multivariate factor analysis receptor models. *Journal of Hazardous Materials*, *260*, 483–488. Available from https://doi.org/10.1016/j.jhazmat.2013.06.001.

Yang, H. B., Zhao, Y. Z., Tang, Y., Gong, H. Q., Guo, F., Sun, W. H., Liu, S. S., Tan, H., & Chen, F. (2019). Antioxidant defence system is responsible for the toxicological interactions of mixtures: A case study on PFOS and PFOA in *Daphnia magna*. *Science of the Total Environment*, *667*, 435–443. Available from https://doi.org/10.1016/j.scitotenv.2019.02.418.

Zabaleta, I., Bizkarguenaga, E., Iparragirre, A., Navarro, P., Prieto, A., Fernández, L. A., & Zuloaga, O. (2014). Focused ultrasound solid-liquid extraction for the determination of perfluorinated compounds in fish, vegetables and amended soil. *Journal of Chromatography A*, *1331*, 27–37. Available from https://doi.org/10.1016/j.chroma.2014.01.025.

Zhang, G., Pan, Z., Wu, Y., Shang, R., Zhou, X., & Fan, Y. (2019). Distribution of perfluorinated compounds in surface water and soil in partial areas of Shandong Province, China. *Soil and Sediment Contamination*, *28*(5), 502–512. Available from https://doi.org/10.1080/15320383.2019.1635079.

Zhang, D. Q., Wang, M., He, Q., Niu, X., & Liang, Y. (2020a). Distribution of perfluoroalkyl substances (PFASs) in aquatic plant-based systems: From soil adsorption and plant uptake to effects on microbial community. *Environmental Pollution*, *257*. Available from https://doi.org/10.1016/j.envpol.2019.113575.

Zhang, M., Wang, P., Lu, Y., Lu, X., Zhang, A., Liu, Z., Zhang, Y., Khan, K., & Sarvajayakesavalu, S. (2020b). Bioaccumulation and human exposure of perfluoroalkyl acids (PFAAs) in vegetables from the largest vegetable production base of China. *Environment International*, *135*. Available from https://doi.org/10.1016/j.envint.2019.105347.

Zheng, X. M., Liu, H. L., Shi, W., Wei, S., Giesy, J. P., & Yu, H. X. (2012). Effects of perfluorinated compounds on development of zebrafish embryos. *Environmental Science and Pollution Research*, *19*(7), 2498–2505. Available from https://doi.org/10.1007/s11356-012-0977-y.

Zhu, H., & Kannan, K. (2019). Distribution and partitioning of perfluoroalkyl carboxylic acids in surface soil, plants, and earthworms at a contaminated site. *Science of the Total Environment*, *647*, 954–961. Available from https://doi.org/10.1016/j.scitotenv.2018.08.051.

Chapter 5

Fate and transport of perfluorinated compounds in soils and groundwater

Chengliang Li and Longfei Liu
College of Resources and Environment, Shandong Agricultural University, Tai'An, P.R. China

Chapter Outline

5.1 Overview of per-and polyfluoroalkyl substances

Per- and polyfluoroalkyl substances (PFASs) or perfluorinated compounds (PFCs) are a class of man-made emerging persistent organic pollutants, which do not occur naturally but are widespread in the environment. As indicated in the previous chapter, the molecular structure of a PFAS compound usually comprises a hydrophobic perfluoroalkyl or polyfluoroalkyl chain with varying length and a hydrophilic functional group (e.g., sulfonate, carboxylate, sulfonamides, and phosphate). They are often divided into long-chain and short-chain PFASs. The former refers to perfluoroalkyl carboxylic acids (PFCAs) with seven and more fluorinated carbons and perfluoroalkyl

Emerging Contaminants in Soil and Groundwater Systems. DOI: https://doi.org/10.1016/B978-0-12-824088-5.00008-2

sulfonates (PFSAs) with six and more fluorinated carbons, and substances that have the potential to be degraded to long-chain perfluoroalkyl acids (PFAAs) by biological or abiotic processes (Buck et al., 2011). The unique structure of PFASs accounts for their very high oil and water-repellency, thermal stability, and acid-base resistance. PFASs are therefore extensively used in manufacturing and households such as coating, textiles, paper packaging, fire-fighting foams, polishes, and electroplates. However, numerous applications of PFASs have inevitably caused their extensive release into the environment (Chen et al., 2016; ITRC, 2020). Furthermore, due to the hydrophilicity of their functional groups, PFASs are highly water-soluble and easily transported in aquatic media. This facilitates the spread of PFASs into other environmental media by water flow, which in turn enhances the potential eco-toxicity of these compounds. In addition, the higher energy of C − F bond (531.5 KJ/mol) results in the very high stability of PFASs and their resistance to biological and chemical degradation. Consequently, PFASs are ubiquitous and globally detected in soils, water bodies, and organisms, even in the remote areas including the arctic (Chen et al., 2016; ITRC, 2020; USEPA, 2018). They are persistent in the environment and bio-accumulated in high trophic level organisms through bio-magnification, posing huge ecological and health risks. The environmental impacts of the large-scale applications of PFASs have accordingly been the focus of several recent publications (Chen et al., 2016; ITRC, 2020; Sörengård et al., 2019). The details of the chemical properties and environmental impacts of common PFAS compounds can be found in Chapter 4, Occurrences and impacts of perfluorinated compounds in soils and groundwater.

The sources of PFAS emission are predominantly from manufacturing and disposal of PFAS-containing products into the environment (see Chapter 4: Occurrences and Impacts of Perfluorinated Compounds in Soils and Groundwater). Manufacturing processes produce both solid and liquid wastes and these are discharged into aquatic streams with ineffective, little or even no treatment (Chen et al., 2016; ITRC, 2020; Steenland et al., 2020). Similarly, the disposal of spent PFAS-based products into landfills, aquatic, and soil environment can increase the concentrations of these contaminants in the environment. These routes of emission, where PFAS is released through their manufactures or uses, are referred to as "direct" sources of emission. On the other hand, "indirect" sources of emission arise from the formation of PFASs through the transformation/degradation of their precursors in the environment, such as the degradation of fluorotelomer-based products. However, indirect sources may be less significant than direct sources of emission (Prevedouros et al., 2006). To date, perfluorooctanoic acid (PFOA) and perfluorooctanoic sulfonate (PFOS) as the main intermediate products degraded from their precursors are widely distributed in the terrestrial environment. In a survey conducted in Europe, PFOA and PFOS were

TABLE 5.1 Concentrations of PFOA and PFOS in various media.

Sampling site	Environmental medium	Concentration		References
		PFOA	PFOS	
Tennessee River, United States	River water	25–598 ng/L	114 ng/L	Hansen et al. (2002)
Little Hocking, West Virginia, United States	Air	200 ng/m	–	Paustenbach et al. (2007)
	Surface soil	11 μg/kg	–	Paustenbach et al. (2007)
	Drinking water	4 μg/L	–	Paustenbach et al. (2007)
New York, United States	Surface water	10–173 ng/L	0.8–1090 ng/L	Sinclair et al. (2006)
Georgia, United States	Wastewater	1–227 ng/L	1.8–22 ng/L	Loganathan et al. (2007)
Kentucky, United States	Wastewater	22–334 ng/L	7.0–149 ng/L	Loganathan et al. (2007)
California, United States	Wastewater	10–190 ng/L	20–190 ng/L	Plumlee et al. (2008)
Ruhr area, Germany	Surface water	0–3640 ng/L	0–193 ng/L	Skutlarek et al. (2006)
	Drinking water	0–519 ng/L	0–22 ng/L	Skutlarek et al. (2006)
River Elbe, Germany	River water	7.2–9.6 ng/L	0.6–2.9 ng/L	Ahrens et al. (2009)
Hangzhou, China	Tap water	1.1–109 ng/L	1.5–13 ng/L	Fujii et al. (2007)
Pearl River, China	River water	0.85–13 ng/L	0.90–99 ng/L	So et al. (2007)
Yangtze River, China	River water	2.0–260 ng/L	0.01–14 ng/L	So et al. (2007)
Dalian, China	Snow	7.74–56.7 ng/g	37.5–108 ng/g	Li et al. (2010)

(*Continued*)

TABLE 5.1 (Continued)

Sampling site	Environmental medium	Concentration		References
		PFOA	PFOS	
Shanghai, China	Sediment	5.20–203 ng/g	1.57–8.78 ng/g	Enevoldsen and Juhler (2010)
	Soil	3.28–47.5 ng/g	8.58–10.4 ng/g	Li et al. (2021)
	Sludge	9.21–75.5 ng/g	28.1–135 ng/g	Ahrens et al. (2011)
Tangxun Lake, China	Sediment	0.48–6.35 ng/g	10.9–623 ng/g	Zhou et al. (2013)
North China	Soil	0–2.32 ng/g	0–1.88 ng/g	Meng et al. (2015)
Nansi Lake, China	Surface water	34.9–84.6 ng/L	0.49–1.79 ng/L	Cao et al. (2015)
Eastern China	Soil	0.16–62.45 ng/g	0–1.56 ng/g	Chen et al. (2016)
	Surface water	0–223.8 ng/L	0–30.2 ng/L	Chen et al. (2016)
	Groundwater	0–475.0 ng/L	0–94.9 ng/L	Chen et al. (2016)
Lake Chaohu, China	Sediment	0.10 ng/g	0.09 ng/g	Liu et al. (2018)
City of Portsmouth, United Kingdom	Groundwater	0.35 μg/L	2.5 μg/L	ATSDR, 2019)
Po River, Italy	River water	60–1300 ng/L	10 ng/L	Loos et al. (2008)
Tuscany, Italy	Wastewater	0.19–103.9 ng/L	0.088–4.3 ng/L	Ciofi et al. (2018)
	Groundwater	0.14–20.8 ng/L	0.068–2.5 ng/L	Ciofi et al. (2018)
	Surface water	0.10–4.0 ng/L	0.052–0.31 ng/L	Ciofi et al. (2018)
	Drinking water	0.11–12.8 ng/L	0.069–0.16 ng/L	Ciofi et al. (2018)

Yodo River, Japan	River water	4.2–2600 ng/L	0.4–123 ng/L	Lein et al. (2008)
West coast, South Korea	Surface water	0.9–62 ng/L	2.24–651 ng/L	Rostkowski et al. (2006)
West coast, Korea	Surface water	2.95–68.6 ng/g	2.95–68.6 ng/g	Naile et al. (2010)
	Biota	0–1.46 ng/g	0.26–612 ng/g	Naile et al. (2010)
Kakum and Pra Rivers, Ghana	River water	1.78–321 ng/L	77.2–275 ng/L	Essumang et al. (2017)
	Tap water	66–190 ng/L	16.2–168.3 ng/L	Essumang et al. (2017)
Lake Victoria basin, Kenya	Wastewater	1.3–28 ng/L	0.9–18.8 ng/L	Chirikona et al. (2015)
	Sewage sludge	117–673 pg/g	32–286 pg/g	Chirikona et al. (2015)
Wastewater treatment plant, Sweden	Tap water	62–1410 pg/L	0–3.89 ng/L	Filipovic & Berger, (2015)
	Influent	2.9–4.75 ng/L	1.76–6.98 ng/L	Filipovic & Berger, (2015)
	Effluent	3.24–7.82 ng/L	0.822–5.03 ng/L	Filipovic & Berger, (2015)
	Sludge	1.46–3.23 pg/g	2.17–8.82 pg/g	Filipovic & Berger, (2015)

shown to comprise 40% and 60%, respectively, of the PFASs detected in groundwater (Lindstrom et al., 2011).

The concentration of PFOA ranged from 0 ng/L to 4 μg/L in water, and 0 ng/kg to 11 μg/kg in soil (Table 5.1). Besides, PFOA has been found in snow, sediments, drink water, and biotas. The concentration of PFOS ranged from 0 ng/L to 2.5 μg/L in water, and 0 ng/kg to 623 μg/kg in soil (Table 5.1). Tang et al. (2006) reported that the concentration of PFOS reached 1650 mg/L in the wastewater in the United States. Similarly, PFOS has also been detected in these environmental media. The toxicity of PFOA and PFOS is well-known. Due to their persistence and toxicity, the production and discharge of PFASs including PFOA and PFOS are now restricted or prohibited in many countries and regions.

5.2 Interactions with soils

5.2.1 Sorption of per- and polyfluoroalkyl substances

Soil is a kind of special environmental medium and is composed of various granular minerals, organic matter (OM), moisture, air, microorganisms, etc. Compared with water and atmosphere, soil medium has poor fluidity and is easily affected by other environmental media. Therefore the types and physical and chemical properties of soil can strongly affect the transport of PFASs. Atmospheric deposition, direct discharge of landfill leachate, and media containing PFAS can lead to the emergence and increase of PFASs in soils. PFAS can enter the groundwater through infiltration through soils. The atmospheric transport and deposition of PFASs into soil surface occur on large regional and global range. Therefore PFASs can be observed across a wide range of locations, but it is not the main source of PFASs in soils. Disposal of PFAS-containing wastes including industrial wastewater and waste products may be the main sources of PFASs in soils.

Rankin et al. determined the distribution of 32 PFASs in surface soils at 62 locations representing all continents (North America $n = 33$, Europe $n = 10$, Asia $n = 6$, Africa $n = 5$, Australia $n = 4$, South America $n = 3$, and Antarctica $n = 1$) (Rankin et al., 2016). Quantifiable levels of perfluoroalkyl carboxylates (PFCAs: PFHxA-PFTeDA) were observed in all samples with total concentrations ranging from 29 to 14,300 pg/g (dry weight); while perfluoroalkane sulfonates (PFSAs: PFHxS, PFOS, and PFDS) were detected in all samples but one, ranging from <LOQ-3270 pg/g, confirming the global distribution of PFASs in terrestrial settings. The geometric mean PFCA and PFSA concentrations were observed to be higher in the northern hemisphere (930 and 170 pg/g) compared to the southern hemisphere (190 and 33 pg/g). PFOA and PFOS were the most commonly detected analytes at concentrations up to 2670 and 3100 pg/g, respectively. Given the soils were collected from locations absent of direct human activity, these results suggest that the

atmospheric long-range transport of neutral PFASs followed by oxidation and deposition is a significant source of PFCAs and PFSAs to soils.

Sorption of PFASs onto soils is very important, involves complex behaviors, and is one of the primary processes that influence their fate and transport in soil and groundwater systems. Enevoldsen and Juhler studied the sorption characteristics of six PFASs including perfluoroheptanoic acid (PFHpA), PFOA, perfluorononanoic acid (PFNA), perfluorodecanoic acid (PFDA), perfluorobutane sulfonate (PFBS), and PFOS in two soils (sandy and clayey soil), and described the isotherms with the Freundlich model (Enevoldsen & Juhler, 2010). The results indicate that the adsorption characteristics are influenced by the soil type and PFAS property. The authors also found a strong correlation between the organic carbon normalized sorption coefficient (K_{oc}) and molecular weight of PFAS. The adsorption and desorption coefficient (K_d) values of PFASs on the two soils are $0.1 - 33$ and $0.3 - 65$ L/kg. The results suggest that soil has high adsorption capacity and low desorption of PFASs. It should be noted that soil can play an important role in retaining PFASs and thus protecting groundwater from PFAS pollution.

The sorption behaviors of PFOS, PFOA, and PFBS in six soils with contrasting characteristics were studied by Milinovic et al. (2015). Both the Freundlich and linear models were appropriate to describe the sorption behaviors of PFASs in soils, and enabled the authors to derive solid–liquid distribution coefficients (K_d) for each compound in each soil. They found that K_d values increased from 19 to 295 mL/g for PFOS, from 2.2 to 38 mL/g for PFOA, and from 0.4 to 6.8 mL/g for PFBS, and were positively correlated with the organic carbon content of the soils. K_{OC} values obtained from the correlations were 710, 96, and 17 mL/g for PFOS, PFOA, and PFBS, respectively. Whereas K_d values decreased in the sequence PFOS > PFOA > PFBS, desorption yields were lower than 13% for PFOS, from 24% to 58% for PFOA, and from 32% to 60% for PFBS. This shows that the physicochemical characteristics of PFASs, basically their hydrophobicity, controlled their sorption behavior in soils, with PFOS being the most irreversibly sorbed PFAS.

Lee and Mabury investigated the sorption of perfluoroalkyl phosphonates (PFPAs) and perfluoroalkyl phosphinates (PFPiAs), homologs of PFAAs with their perfluorocarbon tails attached through a carbon–phosphorus (C–P) bond to either a phosphonate or phosphinate headgroup, in seven soils of varying geochemical parameters [four loam soils (A – D) were collected around Southern Ontario, Canada, while another loam soil (E) was collected in Athens, Georgia, United States] (Lee & Mabury, 2017). The Elliott silt loam (F) and Florida Pahokee peat (G) were obtained, air-dried, and presieved, from the International Humic Substances Society (Golden, Colorado), and used as received (Fig. 5.1). The sorption equilibrium was typically reached within 6–24 h, and analysis of the Masurf-spiked $CaCl_2$

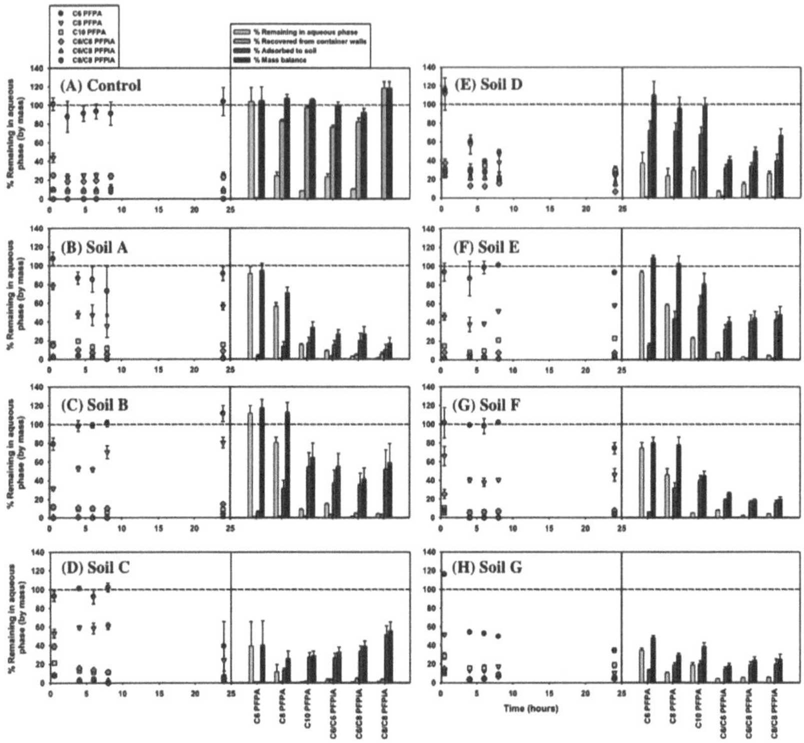

FIGURE 5.1 Sorption kinetics of perfluoroalkyl phosphonates (PFPAs) and perfluoroalkyl phosphinates (PFPiAs) in seven different soils. (A) Control samples with no soils were used to test for the potential of abiotic degradation and adsorption to the container walls. (B–H) Sorption kinetics of PFPAs and PFPiAs in seven soils of varying geochemical parameters (soil A: Sandy loam, soil B: Clay loam, soil C: Loam, soil D: Sandy loam, soil E: Sandy clay loam, soil F:Silt loam and soil G: Peat). Percent of PFPAs and PFPiAs (by mass) remaining in the aqueous phase upon equilibration with each soil type over time (*left*). Mass balances of each PFPA and PFPiA congener, as determined from their percent distribution in the aqueous, soil, and container surface phases upon reaching equilibrium (*right*). Each data point represents the arithmetic mean percent of the triplicate ($n = 3$) samples. The *error bar* represents the standard error. *Reprinted with permission from Lee, H., & Mabury, S. (c. 2017). Sorption of perfluoroalkyl phosphonates and perfluoroalkyl phosphinates in soils.* Environmental Science and Technology, 51*(6), 3197–3205. https://doi.org/10.1021/acs.est.6b04395. Copyright (2017) American Chemical Society.*

controls revealed significant losses (> 75%) of the C8 and C10 PFPAs and all three PFPiAs to the container walls (Fig. 5.1A), but similar declines in aqueous concentrations were also observed in Masurf-spiked water only. The fact that C6 PFPA remained mostly in the aqueous phase of both sets of controls and also in the aqueous phases equilibrated with the majority of the soils suggests complexation with Ca^{2+} and Hg^{2+} from the biocide may only minimally impact PFPA and PFPiA sorption and this behavior is expected to persist for the longer chain lengths as this interaction should be predominantly driven by the anionic headgroup of the PFPAs and PFPiAs. As expected, the presence of soils greatly reduced this surface adsorption to mixtures that was analyzed for mass balance distribution among the aqueous, soil, and container phases (Fig. 5.1B − H).

The majority of the sorption isotherms measured for the PFPAs and PFPiAs in the seven soils were nonlinear, with the majority of the *n* values ranging between 0.7 and 1, although there were some exceptional cases of lower *n* values (Fig. 5.2). Nonlinearity in Freundlich isotherms is characteristic of saturation of the soil surface sites available for the sorption by high analyte concentrations, nonuniform interactions within the soil, and competitive sorption in a multisolute system. Sediment sorption of PFCAs and PFSAs, present as a mixture or in the presence of other surfactants at high concentrations (μg/L − mg/L), has been shown to exhibit nonlinear isotherms, while sediments spiked individually with PFOA and PFOS at much lower concentrations (ng/L) exhibited predominantly linear isotherms.

Wei et al. investigated the sorption and desorption of PFOS onto six soils with different physicochemical properties (Wei et al., 2017). The sorption kinetics of PFOS on the six soils demonstrates that PFOS sorption requires less than 48 h to reach equilibrium. The kinetic curves are well fitted by the pseudo-second-order kinetic model, suggesting that chemisorption is important to PFOS sorption on soils. PFOS sorption isotherms can be described by the Freundlich model well for all six soils ($R^2 = 0.979–0.999$). The correlation analysis between K_F of PFOS and the physicochemical properties of the soils show a positive correlation between K_F and Al_2O_3, soil organic carbon

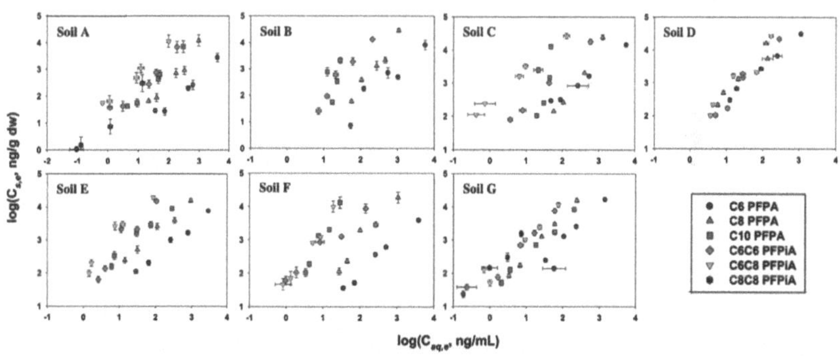

FIGURE 5.2 Sorption isotherms of perfluoroalkyl phosphonates and perfluoroalkyl phosphinates in seven different soils. Equilibrium concentrations measured in the soil and aqueous phases are denoted as $C_{s,e}$ [ng/g dry weight (dw)] and $C_{aq,e}$ (ng/mL), respectively. Each data point represents the arithmetic mean concentration of the triplicate ($n = 3$) samples. The *error bar* represents the standard error. *Rearrange the order of the graphs with permission from Lee, H., & Mabury, S. (c. 2017). Sorption of perfluoroalkyl phosphonates and perfluoroalkyl phosphinates in soils.* Environmental Science and Technology, 51*(6), 3197–3205. https://doi.org/10.1021/acs.est.6b04395. Copyright (2017) American Chemical Society.*

and Fe_2O_3. The authors concluded that hydrophobic interaction, ion exchange, surface complexing, and hydrogen-bonding may all play important roles in PFOS sorption onto the soil samples.

Johnson et al. studied the adsorption of PFOS on different soil and mineral surfaces (Johnson et al., 2007). They found that PFOS adsorption capacities of the soils and minerals increase with the increase of equilibrium concentrations. Isotherms and calculated solid/solution distribution ratio experiments indicate that PFOS adsorption is significant but smaller than hydrocarbon analogs or organic compounds of similar molecular weight.

Clay and oxide minerals in soils may show different sorption ability to PFASs. PFOA is adsorbed more strongly on goethite than on δ-Al_2O_3, but not at all on illite (Fig. 5.3A) (Li et al., 2012). This difference is likely due to the different electrostatic interactions. The pH values in supernatants ranged from 7.63 to 7.76 for δ-Al_2O_3, from 6.92 to 6.99 for goethite and from 6.75 to 6.97 for illite. The points of zero charge (PZC) are 8.1 for δ-Al_2O_3 (Tombácz & Szekeres, 2001), 9.3 for goethite (Antelo et al., 2005), and 3.5 for illite (Lan et al., 2007). Under the experimental conditions, goethite should carry more positive charges than δ-Al_2O_3; while illite should be negatively charged. In this pH range, PFOA is predominantly negatively charged because its pK_a is 3.8 (Goss, 2008). Therefore it can be concluded that electrostatic interaction causes various PFAS sorption affinities among goethite, δ-Al_2O_3 and illite. Other studies have shown that surface area normalized adsorption of PFASs on clay and oxide minerals in soils follows the order of Goethite < Kaolinite < High iron sand < Ottawa sand standard (Johnson et al., 2007). The results indicate that mineral composition can strongly affect sorption of PFASs in soils, which can play an important roles

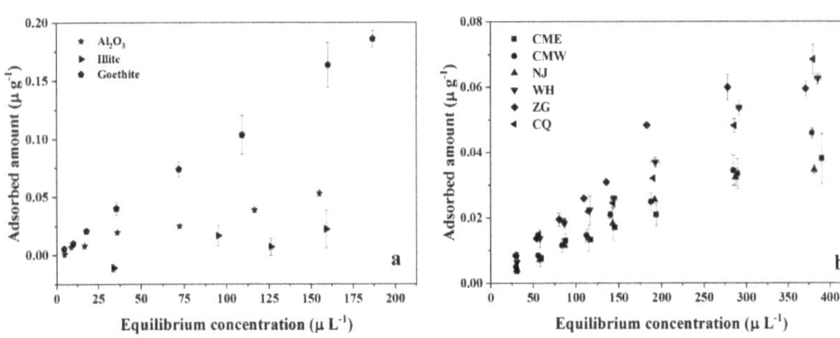

FIGURE 5.3 Adsorption isotherms of perfluorooctanoate on (A) sediment model components and (B) Yangtze River sediments (*CME*, Chongming East; *CMW*, Chongming West; *CQ*, Chongqing; *NJ*, Nanjing; *WH*, Wuhan; *ZG*, Zigui). *Reproduced from Li, C., Ji, R., Schäffer, A., Sequaris, J., Amelung, W., Vereecken, H., & Klumpp, E. (c. 2012). Sorption of a branched nonylphenol and perfluorooctanoic acid on Yangtze River sediments and their model components.* Journal of Environmental Monitoring, 14*(10), 2653–2658. https://doi.org/10.1039/c2em30394a. With permission from the Royal Society of Chemistry.*

controlling the interactions and behaviors of PFASs in soil and groundwater systems.

Zhao et al. studied the sorption of four PFAAs (PFOS, PFHxS, PFOA, and PFHxA) on three typical minerals [montmorillonite (MM), kaolinite (KL), and hematite (HM)] (Fig. 5.4) (Zhao et al., 2014). The sorption of PFOS and PFHxS was much stronger than that of PFOA and PFHxA. The sorption of each PFAA on the soil minerals followed an order of HM > KL > MM, even though MM was positively while KL and HM were negatively charged. This implies that the sorption is driven by some other interactions besides electrostatic attraction. Surface complexing and hydrogen-bonding could make great contributions to the sorption of PFOS on the minerals. The results are important for understanding the transport and fate of PFAAs in soils and groundwater.

Several studies have examined the adsorption of PFASs onto other sorbents that have properties and compositions similar to soils including activated sludge and sediments. The obtained knowledge is valuable and can be used to improve the understanding of the sorption behaviors of PFASs in soils and groundwater.

Zhou et al. studied the adsorption behaviors of PFOS and PFOA on different activated sludge from sewage treatment plants (Zhou et al., 2010). The adsorption kinetics indicate that the adsorption equilibrium can be reached within 11 h. The influences of pH and zeta potential imply that electrostatic repulsion hinders the adsorption, but the compositions of polysaccharide and protein on the bacterial surface of activated sludge are favorable for the adsorption of PFOS and PFOA. In addition, the living microorganisms, which are abundant in soils, can also promote the adsorption of PFOA and PFOS. Compared with the adsorption of other PFASs with different length

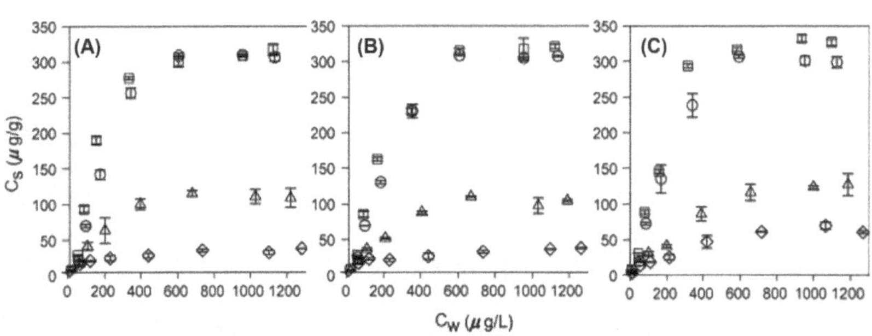

FIGURE 5.4 Sorption isotherms of perfluoroalkyl acids on three soil minerals: ○PFOS, □PFHxS, △PFOA, and ◇PFHxA on (A) MM, (B) KL, and (C) HM. *Rearrange the order of the graphs according to Zhao, L., Bian, J., Zhang, Y., Zhu, L., & Liu, Z. (c. 2014). Comparison of the sorption behaviors and mechanisms of perfluorosulfonates and perfluorocarboxylic acids on three kinds of clay minerals.* Chemosphere, 114, *51–58. https://doi.org/10.1016/j.chemosphere.2014.03.098. Copyright (2014), with permission from Elsevier.*

of C−F chain and functional groups, hydrophobic interaction plays an important role in the adsorption process of PFOS and PFOA on activated sludge. The calculated distribution coefficient indicates that PFOS has a higher adsorption tendency to activated sludge than PFOA.

Adsorption isotherms of PFOA on the Yangtze River sediments are presented in Fig. 5.3B. The isotherms are slightly nonlinear and can be properly simulated by the Freundlich model (Li et al., 2012). The K_{OC} (L/kg) values of PFOA were calculated based on the organic carbon content and the distribution coefficient (K_d) simulated in 0−200 μg/L equilibrium concentration range by the linear model. The K_{OC} values for PFOA on sediments range from 18.32 to 40.53 L/kg. A higher K_{OC} value of 129 L/kg on freshwater sediment can be explained by the bridging effect of Ca^{2+} (Higgins & Luthy, 2006), which could not be important to its K_{OC} in soils with high calcium levels. In fact, bivalent cations like Ca^{2+} can act as a bridge between the anionic sorbate and the negatively charged surface leading to an enhanced adsorption of PFASs in soils. Higgins and Luthy systematically studied the adsorption behavior of anionic PFCAs and PFSAs with varying chain lengths to natural sediments of varying iron oxide and organic carbon contents (Higgins & Luthy, 2006). They found that organic carbon, rather than iron oxide content, is the dominant parameter affecting PFAS adsorption. This indicates the importance of hydrophobic interactions between PFASs and OM in soils and groundwater. However, PFOA did not show any sorption on Suwannee River natural OM in the presence of Na^+. Under neutral and alkaline conditions, natural OM is negatively charged and the anionic form of PFOA is prevailing. Therefore the electrostatic repulsion between PFASs and natural OM in soils should also be considered as a significant factor to their sorption and behaviors in the subsurface.

Ololade et al. investigated the distinguishing sorption behaviors of PFOS on various residual sediments under oxic and anoxic condition (Ololade et al., 2016). They found that sediment components and redox properties change with oxic/anoxic condition had an opposite influence on the sorption of PFOS on the various sediments containing different components (Fig. 5.5 and Table 5.2). The oxic ambient promoted the sorption of PFOS on sediment after extracting iron and manganese ($S\text{-}_{FeMn}$). However, the sorption capability of PFOS under anoxic condition was higher than that under oxic condition on sediment after extracting manganese ($S\text{-}_{Mn}$). In addition, the isotherm of $S\text{-}_{Mn}$, sediment removed OM by 0.1 M NaOCl ($S\text{-}_{OM1}$) and 30% H_2O_2 ($S\text{-}_{OM2}$) and original sediment (UNTD) under oxic condition, was not from the original point except $S\text{-}_{FeMn}$. They concluded that the adsorption was mainly due to the native bound adsorbate. The fact that adsorption isotherm could get back to the original point only after the iron composition was removed indicates that the native bonded PFOS on iron composition may be responsible for the cross-over type isotherm. In contrast to the results obtained from S-FeMn and S-Mn, the alteration of PFOS concentration

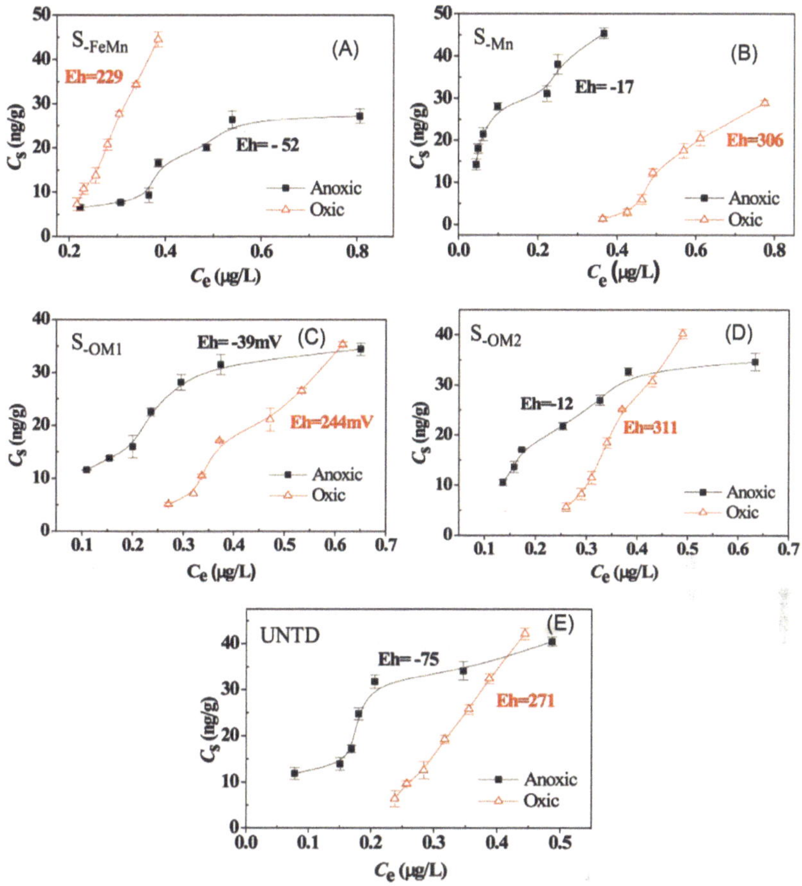

FIGURE 5.5 Comparative sorption of perfluorooctanesulfonate under oxic and anoxic conditions on different types of sediments (A) S-FeMn, (B) S-Mn, (C) S-OM1, (D) S-OM2, (E) UNTD. *Reprinted from Ololade, I., Zhou, Q., & Pan, G. (c. 2016). Influence of oxic/anoxic condition on sorption behavior of PFOS in sediment.* Chemosphere, 150, *798–803. https://doi.org/10.1016/j.chemosphere.2015.08.068. Copyright (2016), with permission from Elsevier.*

seemed to affect the sorption trend on S_{-OM1}, S_{-OM2}, and UNTD. Anoxic condition was beneficial to sorption of PFOS at the range of low concentration. However, oxic ambient promoted the sorption of PFOS with higher concentration on these three sediments.

5.2.2 Important factors

5.2.2.1 *Alkyl chain length*

Ahrens et al. studied the partitioning behavior of different PFASs between pore water and soils (Ahrens et al., 2009). Short-chain PFASs are mainly

TABLE 5.2 Original sorption data of PFOS under oxic and anoxic conditions in different types of sediments including S-FeMn, S-Mn, S-OM1, S-OM2, and UNTD.

Sample	Anoxic	C_e (μg/L)	0.223	0.3069	0.366	0.3856	0.4851	0.5395	0.8058
S-FeMn		Cs (ng/g)	6.57	7.725	9.37	16.575	20.095	26.419	27.265
		Error bar	0.6364	0.2192	1.68291	0.86974	0.74246	2.05202	1.63342
	Oxic	C_e (μg/L)	0.2168	0.2303	0.2548	0.2791	0.3035	0.3388	0.3848
		C_s (ng/g)	7.328	10.7855	13.8095	20.8335	27.8585	34.449	44.606
		Error bar	1.41421	1.26077	1.72039	1.26077	0.61306	0.45962	1.7211
S_{-Mn}	Anoxic	C_e (μg/L)	0.0441	0.049	0.0616	0.0979	0.2224	0.249	0.3658
		C_s (ng/g)	14.235	18.04	21.455	28.08	31.105	38.04	45.37
		Error bar	1.32229	1.06066	1.52028	0.77782	1.84555	2.37588	1.24451
	Oxic	C_e (μg/L)	0.3632	0.4255	0.4634	0.4905	0.5691	0.6108	0.775
		C_s (ng/g)	1.4739	2.98	5.9621	12.3783	17.6044	20.4648	28.9955
		Error bar	0.58987	0.7665	1.16658	0.95417	1.74175	1.72082	0.61306
S_{-OM1}	Anoxic	C_e (μg/L)	0.1094	0.1542	0.2003	0.2361	0.2954	0.3735	0.6507
		C_s (ng/g)	11.62	13.83	15.985	22.555	28.185	31.555	34.47
		Error bar	0.09899	0.29698	2.11425	0.78489	1.46371	1.87383	1.18794
	Oxic	C_e (μg/L)	0.271	0.3198	0.3361	0.3713	0.4716	0.5339	0.615
		C_s (ng/g)	5.1592	7.208	10.5575	17.148	21.137	26.644	35.3915
		Error bar	0.3066	0	0.30618	0.15274	2.14678	0.45962	0.46033

S_{-OM2}	Anoxic	C_e (μg/L)	0.1359	0.1584	0.1733	0.2535	0.3262	0.3823	0.6336
		C_s (ng/g)	10.56	13.66	17.07	21.725	26.955	32.71	34.65
		Error bar	0.62225	1.0748	0.18385	0.74246	1.03945	0.74953	1.7112
	Oxic	C_e (μg/L)	0.26	0.29	0.31	0.34	0.37	0.43	0.49
		C_s (ng/g)	5.697	8.292	11.5325	18.483	25.148	30.763	40.2695
		Error bar	0.87823	1.10733	1.26077	1.02672	0.15274	1.0253	0.91995
UNTD	Anoxic	C_e (μg/L)	0.0778	0.1509	0.1691	0.1806	0.2059	0.3462	0.487
		C_s (ng/g)	11.885	13.962	17.236	24.775	31.76	34.15	40.495
		Error bar	1.26714	1.31522	0.87116	1.32229	1.51321	1.99404	0.96874
	Oxic	C_e (μg/L)	0.238	0.257	0.284	0.317	0.355	0.387	0.444
		C_s (ng/g)	6.4605	9.7005	12.625	19.316	25.7985	32.4975	42.221
		Error bar	1.72039	0.57771	1.86252	0.88388	1.02601	1.07268	1.22612

Reprinted from Ololade, I. A., Zhou, Q., & Pan, G. (2016). Influence of oxic/anoxic condition on sorption behavior of PFOS in sediment. *Chemosphere, 150,* 798–803. https://doi.org/10.1016/j.chemosphere.2015.08.068.

distributed in the water phase, while long-chain PFASs are mainly distributed in the solid phase. Alkyl chain length of PFAS compounds is considered to be the dominant factor affecting their migration and distribution. Short-chain PFCAs ($C \leq 7$) were found exclusively in pore water, while long-chain PFCAs ($C \geq 11$) were found only in the solid phase. In addition, PFSAs bind more strongly to the solid than PFCAs. Based on these results, it can conclude that perfluorocarbon chain length and functional group are the dominating parameters influencing the partitioning behavior of PFASs in soils (Ahrens et al., 2009). Furthermore, the length of alkyl chain is the dominant structural feature influencing PFAS sorption, with each CF_2 moiety contributing $0.50 - 0.83$ log units to the measured distribution coefficients (Higgins & Luthy, 2006).

Increasing binding strength between PFASs and soils with increasing alkyl chain length was observed for both PFSAs (C_4, C_6, and C_8) and PFCAs (C_3-C_{10}) (Fig. 5.6). For the PFSAs and PFCAs, the increase in log K_{OC} per additional CF_2 moiety was 0.83 and 0.60 log units, respectively. Short-chained PFASs such as PFBA, PFPeA, and PFBS were weakly sorbed (less than 10% on average), whereas long-chained PFASs (PFOS, FOSA, C_9-C_{11}, and C_{13} PFCAs) were sorbed strongly (on average, 99%–100%). The increased sorption with increased PFCA chain length is in agreement with previous results of Ahrens et al. (2010) and Higgins and Luthy (2006), who found suspended particulate matter and sediment partitioning coefficients to increase with 0.50–0.75 log units per CF_2 moiety. For the PFSAs, however, previous studies found an increase of 0.40–0.60 log units per CF_2 moiety, which is lower as compared to this study (Higgins & Luthy, 2006; Milinovic et al., 2015). The increased sorption upon increasing chain length is attributed to an increase in PFAS hydrophobicity with each CF2 moiety. For the most long-chained PFASs (i.e., PFDoDA and PFTeDA), however, no increased sorption was observed upon increasing perfluorocarbon chain length. This observation might be explained by possible strong binding of the most long-chained PFCAs to dissolved organic carbon, which would decrease their observed K_{OC} values. Alternatively, the more or less equal average sorption among the C9–C11 and C13 PFCAs could be attributed to steric hindrances originating from the increased rigidity reported for PFCA alkyl chains longer than C10.

Wang et al. studied the adsorption behavior of PFOS and PFBS on humic acid (HA) modified boehmite by free energy (Eq. (1)) (Wang, Shih, & Leckie, 2015; Wang, Shih, Ma, & Li, 2015). PFOS has eight perfluorinated carbons and PFBS only has four carbons. A previous study showed that the free energy of hydrophobic interaction is usually related to the chain length of PFASs (Higgins & Luthy, 2006). Therefore $\Delta G_{Hydrophobic\ Interaction,\ HA-PFOS}$ is much higher than $\Delta G_{Hydrophobic\ Interaction,\ HA\text{-}PFBS}$. In addition, if hydrophobic interaction is dominant in the PFOS and PFBS sorption on the HA-saturated surface, the sulfonate groups of PFOS and PFBS should be

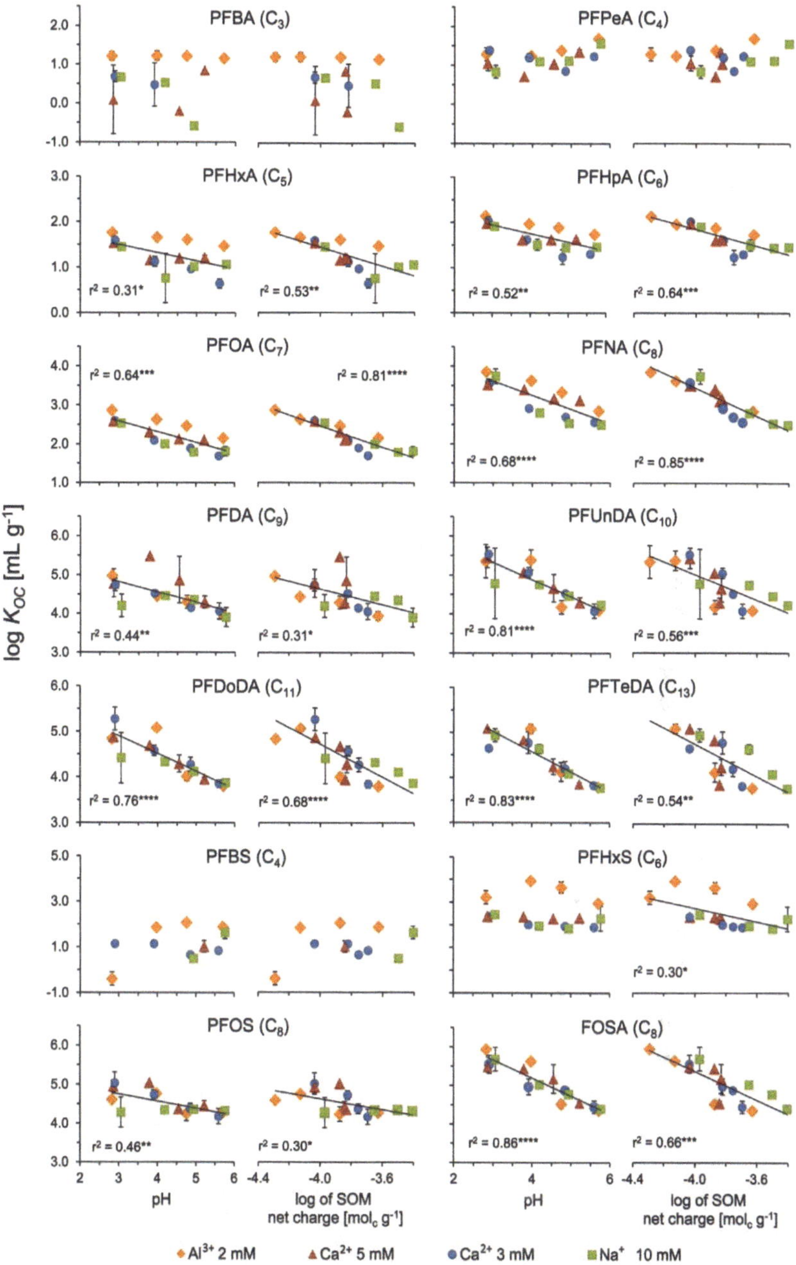

FIGURE 5.6 Effect of pH and simulated soil organic matter (SOM) bulk net charge on log K_{OC} of 14 target PFASs with the treatments of Al3 + (2 mM), Ca2 + (5 mM), Ca2 + (3 mM), and Na + (10 mM) in the soil sorption experiment. Each data point represents the average of duplicates. log K_{OC} values of 0, 1, 2, 3, 4, and 5 correspond to sorbed fractions of 0.5%, 4.5%, 33.1%, 83.6%, 98.1%, and 99.8%, respectively. $^{*}P \leq 0.5$, $^{**}P \leq 0.01$, $^{***}P \leq 0.001$, $^{****}P \leq 0.0001$. *Reprinted from Pereira, H., Ullberg, M., Dan, B., Gustafsson, J., & Ahrens, L. (c. 2018). Sorption of perfluoroalkyl substances (PFASs) to an organic soil horizon—Effect of cation composition and pH.* Chemosphere, 207, *183–191. https://doi.org/10.1016/j.chemosphere.2018.05.012. Copyright (2018), with permission from Elsevier.*

oriented further away from the negative charge of HA to minimize the electrostatic repulsion. Because the chain length of PFOS is around double than that of PFBS, the sulfonate group of PFOS is further way from the negatively charged HA than that of PFBS. $\Delta G_{Electrostatic\ Interaction,\ HA-PFOS}$ is much lower than $\Delta G_{Electrostatic\ Interaction,\ HA-PFBS}$, and, therefore, PFOS may partition more than PFBS into the sorbed HA on the boehmite surface, due to its relatively longer carbon–fluorine chain (Fig. 5.7).

$$\Delta G_{Sorption} = \Delta G_{Hydrophobic\quad Interaction,\&\#x000A0;HA-PFASs} + \Delta G_{Electrostatic\quad Interaction,\&\#x000A0;HA-PFASs} \tag{1}$$

$\Delta G_{Sorption}$ is the total free energy associated with the PFAS sorption on the HA-saturated boehmite. $\Delta G_{Hydrophobic\ Interaction,\ HA-PFASs}$ is the free energy initiated by the hydrophobic interaction between the carbon–fluorine chain of PFAS and the hydrophobic fraction of HA. $\Delta G_{Electrostatic\ Interaction,\ HA-PFASs}$ is the free energy produced by the electrostatic repulsion between the sulfonate group of PFAS and the negative charge of HA.

5.2.2.2 pH

Most PFASs exist in the form of anions. Therefore, for anionic PFASs, solution pH may strongly affect their interactions with soils. The effects of electrolyte pH on PFAS adsorption and desorption have been investigated in several studies (Liu et al., 2020; Tang et al., 2010). In general, when the pH value of electrolyte increases, PFAS adsorption capacity of soils also increases. When pH reduces, proton in electrolyte can neutralize the negative

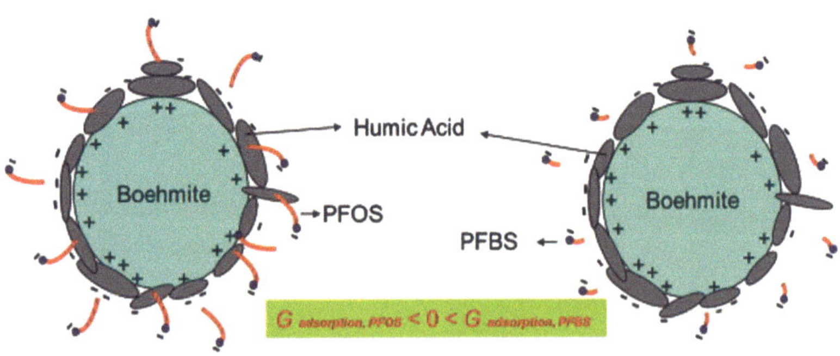

FIGURE 5.7 Conceptual model for the different sorption behavior of perfluorooctanoic sulfonate (PFOS) and perfluorobutane sulphonat (PFBS) on the HA-modified boehmite surface. PFOS can partition into the sorbed HA on the boehmite surface because it has a relatively longer carbon–fluorine chain length than PFBS. *Reprinted from Wang, F., Shih, K., & Leckie, J. (c. 2015). Effect of humic acid on the sorption of perfluorooctane sulfonate (PFOS) and perfluorobutane sulfonate (PFBS) on boehmite.* Chemosphere, 118*(1), 213–218. https://doi.org/10.1016/j.chemosphere.2014.08.080. Copyright (2015), with permission from Elsevier.*

charge on the soil surface to reduce the electrostatic repulsion between the negative charged soil surface and the anionic PFASs. Several studies have demonstrated that the enrichment of PFASs in soils increases with the increase of pH (Ahrens & Bundschuh, 2014; Li et al., 2012; Tang et al., 2010).

Zhao et al. investigated the effect of solution pH (2.27−11.16) on the sorption of four PFAAs (PFOS, PFHxS, PFOA, and PFHxA) to soil mineral surfaces [montmorillonite (MM), kaolinite (KL), and hematite (HM)] (Fig. 5.8) (Zhao et al., 2014). The four PFAAs are mainly deprotonated within the studied pH range and present as anions. The pH_{pzc} values of the three minerals are 7.2 for MM, 3.6 for KL, and 5.9 for HM. The minerals are positively charged due to the protonation of the surface when the solution pH is lower than their pH_{pzc} values, otherwise they are negatively charged. At lower pH, a strong electrostatic attraction exists between the positively charged mineral surface and the negatively charged PFAAs, resulting in a relatively high sorption affinity. As the solution pH increases gradually, the mineral surface becomes from positive to neutral, and even negative gradually. Thus the electrostatic attraction vanishes and other interactions (including electrostatic repulsion) predominate, leading to weak sorption of PFAAs on the minerals.

5.2.2.3 Cationic and anionic species

PFAS sorption on soils also increases with increasing solution Ca^{2+} concentration, suggesting that electrostatic interactions may play an important role in this process. The increased concentration of Ca^{2+} can neutralize the negative charge on the soil surface to reduce the electrostatic repulsion between

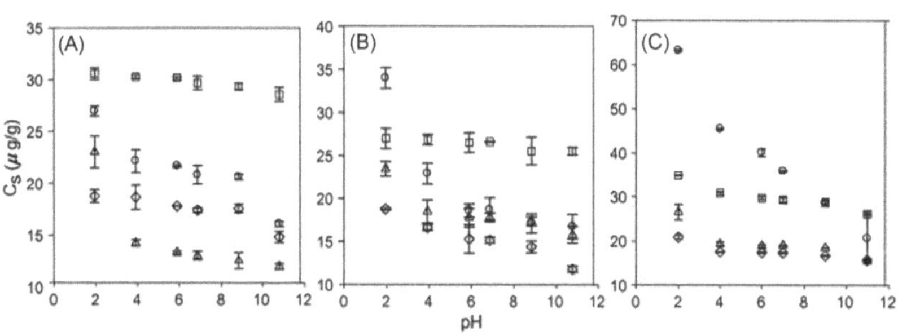

FIGURE 5.8 Impact of pH on the adsorption of perfluoroalkyl acids (○PFOS, □PFHxS, △PFOA, ◇PFHxA) on soil minerals: (A) MM, (B) KL, and (C) HM. *Rearrange the order of the graphs according to Zhao, L., Bian, J., Zhang, Y., Zhu, L., & Liu, Z. (c. 2014). Comparison of the sorption behaviors and mechanisms of perfluorosulfonates and perfluorocarboxylic acids on three kinds of clay minerals.* Chemosphere, 114, *51–58. https://doi.org/10.1016/j.chemosphere.2014.03.098. Copyright (2014), with permission from Elsevier.*

the negative charged soil surface and the anionic PFASs. Meanwhile, the alteration on pH value can affect the sorption of Ca^{2+} onto the soil surface to some extent, which may influence the sorption of PFASs on soils through electrostatic effect and Ca-bridging mechanism. Tang et al. investigated the adsorption of PFOS on goethite and silica (Tang et al., 2010). They found that the adsorption of PFOS on the surface of silica was only slighted affected by solution pH, ionic strength, and Ca^{2+} concentration. In contrast, the adsorption of PFOS on goethite increased significantly at low pH and high Ca^{2+} concentrations, which may be attributed to the electrostatic and Ca-bridging mechanisms that promote the sorption of negatively charged PFOS molecules onto the positively charged goethite surface (Tang et al., 2010).

Pereira et al. found that, for PFASs with intermediate chain length (i.e., C_5-C_8 PFCAs and PFHxS), the log K_{OC} increased with the addition of cations to the soil in the following order: Al^{3+} (2 mM) > Ca^{2+} (5 mM) > Ca^{2+} (3 mM) = Na^+ (10 mM) (Fig. 5.6) (Pereira et al., 2018). Moreover, for the C_5-C_8 PFCAs, the effects on sorption due to different cation treatments were more pronounced at higher pH as compared to lower pH values. These effects are consistent with the observations of Higgins and Luthy (2006) and Zhang et al. (2013) (effect of $Ca^{2+} > Na^+$), and Wang, Shih, and Leckie (2015) ($Al^{3+} > Ca^{2+} > Na^+$). Except for stronger sorption in the Al^{3+} treatment, there was no clear effect of either the pH value or added cations for two of the three shortest PFASs, that is, PFBA (C_3) and PFBS (C_4). By contrast, for the most long-chained PFASs (i.e., C_9-C_{11} and C_{13} PFCAs, PFOS, and FOSA), there were few statistically significant effects of the added cations on the observed sorption (Table 5.3). In other words, the strong effects of added cations that were seen for the PFASs of intermediate length but were not seen for the substances with the longest chain length.

Fig. 5.9 illustrates the sorption of four PFAAs (PFOS, PFHxS, PFOA, and PFHxA) on three soil minerals [montmorillonite (MM), kaolinite (KL), and hematite (HM)] as a function of the concentration of monovalent cation Na^+ or divalent cation Ca^{2+} (Zhao et al., 2014). The results reveal that the sorption of PFOA and PFHxA increased slightly with an increase in ionic strength, suggesting that electrostatic interaction was weak and hydrophobic interaction may play an important role in the sorption of PFCAs on minerals surface. The sorption of PFOS increased significantly with an increase in ionic strength for both Na^+ and Ca^{2+}, while it was much less significant for PFHxS. At the pH in this study (about 6.9), the surface of MM was positively charged, while KL and HM were negatively charged. All the PFAAs were in anionic forms. With increasing the concentrations of Na^+/Ca^{2+}, the electrostatic repulsion between the negatively charged KL and HM surface and negatively charged PFOS molecules was reduced, which promotes the sorption of PFOS at higher concentrations of Na^+/Ca^{2+}. For the positively charged MM, the electrostatic attraction between PFOS anions and mineral

TABLE 5.3 Paired Student's *t*-test (double-sided, type 1) testing significance ($P \leq 0.05$) for differences in log K_{OC} averages between the four cation treatments in the sorption experiment.

Tested hypothesis	log K_{OC} [Al^{3+} (2 mM) > Ca^{2+} (5 mM)]	log K_{OC} [Al^{3+} (2 mM) > Ca^{2+} (3 mM)]	log K_{OC} [Al^{3+} (2 mM) > Na^{+} (10 mM)]	log K_{OC} [Ca^{2+} (5 mM) > Ca^{2+} (3 mM)]	log K_{OC} [Ca^{2+} (5 mM) > Na^{+} (10 mM)]	log K_{OC} [Ca^{2+} (3 mM) > Na^{+} (10 mM)]
PFBA		S				
PFPeA	S		S			
PFHxA	S	S	S			
PFHpA	S	S	S		S	
PFOA	S	S	S	S	S	
PFNA		S	S	S	S	
PFDA					S	
PFUnDA						
PFDoDA					S	
PFTeDA					S	
FOSA						
PFBS						
PFHxS	S	S	S	S		

Notes: S corresponds to a significant difference in average log K_{OC} and blanks cells correspond to a nonsignificant difference. The log K_{OC} values of the study roughly followed log-normal distributions. The sign of any significant difference between two cation treatments was determined by comparison of the respective log K_{OC} means. The results should be viewed as indicative as paired samples from different cation treatments have similar, but not equal, pH values.

Source: Reprinted from Pereira, H. C., Ullberg, M., Dan, B. K., Gustafsson, J. P., & Ahrens, L. (2018). Sorption of perfluoroalkyl substances (PFASs) to an organic soil horizon–Effect of cation composition and pH. *Chemosphere, 207*, 183–191. https://doi.org/10.1016/j.chemosphere.2018.05.012. Copyright (2018), with permission from Elsevier.

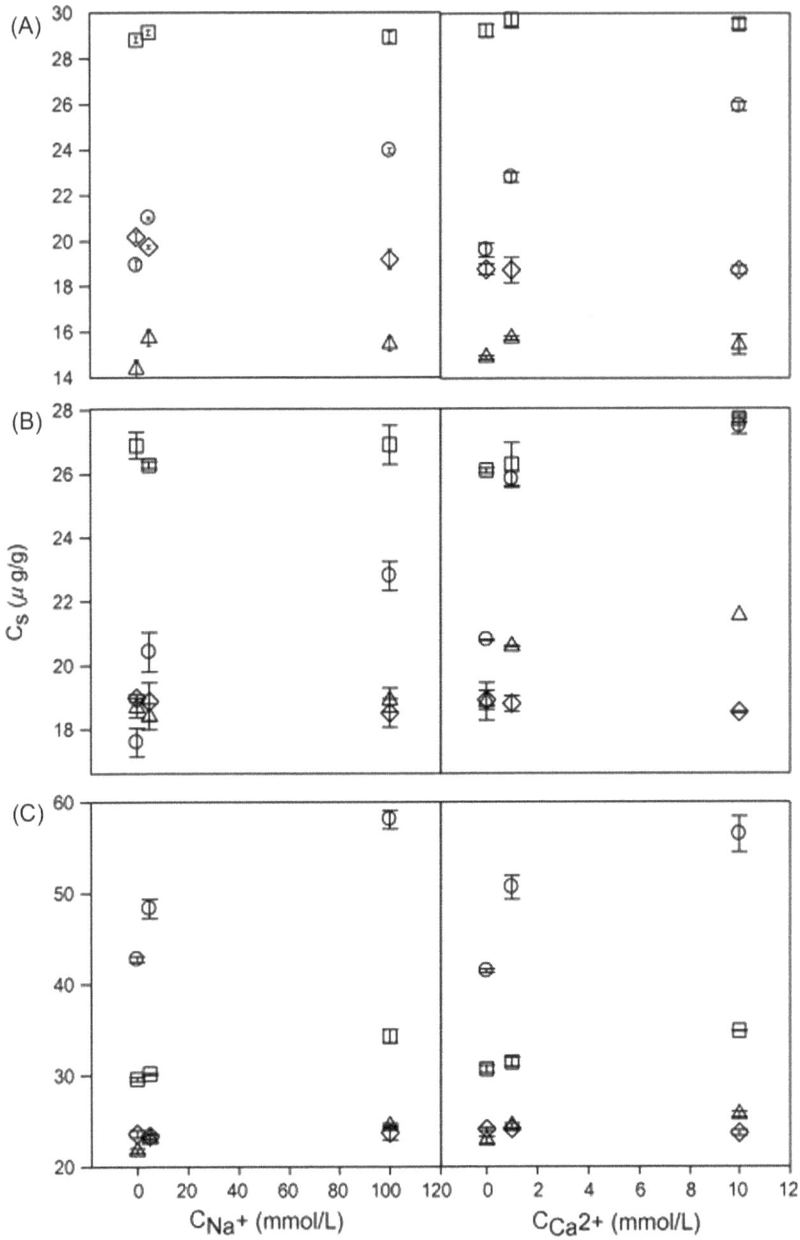

FIGURE 5.9 Effects of Na + and Ca2 + on the sorption of PFAAs (○PFOS, □PFHxS, △PFOA, ◇PFHxA) on soil minerals: (A) MM, (B) KL, and (C) HM. *Rearrange the order of the graphs according to Zhao, L., Bian, J., Zhang, Y., Zhu, L., & Liu, Z. (c. 2014). Comparison of the sorption behaviors and mechanisms of perfluorosulfonates and perfluorocarboxylic acids on three kinds of clay minerals.* Chemosphere, 114, *51–58. https://doi.org/10.1016/j.chemosphere.2014.03.098. Copyright (2014), with permission from Elsevier.*

surfaces could be weakened at higher concentration of Na^+/Ca^{2+} due to the electrical double layer compression effect (Tang et al., 2010). While at increasing ionic strength, the electrostatic repulsive force that repels two adjacent PFOS anions was greatly suppressed. This suppression on PFOS–PFOS electrostatic repulsive force was stronger than PFOS-surface electrostatic interaction at higher concentrations of Na^+ and Ca^{2+}, giving a rise in the sorption of PFOS on MM. But for PFHxS, the smaller molecular size compared to PFOS makes it easier to enter the micropores of the minerals and hardly affected by the solution chemistry.

Phosphate (P) is an important soil nutrient; however, it is also a well-known pollutant and a primary factor of eutrophication. P can convey negative charges to the soil particle surfaces, thereby changing the adsorption of pollutants including PFASs in soils. The effects of soil characteristics and phosphate competition on the adsorption of PFOS on soils were determined in a previous study (Qian et al., 2017). Three types of soils, including original soil, the soil from which OM was removed, and the soil from which both OM and ferric oxides were removed, were used as adsorbents for PFOS sorption in solutions with or without P (Fig. 5.10). The addition of P in the soil solutions reduced the K_f of PFOS on three tested soils by approximately 25%, 50%, and 15%, respectively. For the binary system of PFOS and P, soil with higher ferric oxide content showed greater K_f reduction after P addition; whereas soil with higher OM content showed less K_f reduction.

5.2.2.4 Organic matter

OM can play an important role in affecting the interaction between PFASs and soils (Johnson et al., 2007). The enrichment of different PFASs in soils may increase with the increase of OM content due to the strong hydrophobic interactions between PFASs and OM in soils. However, previous studies

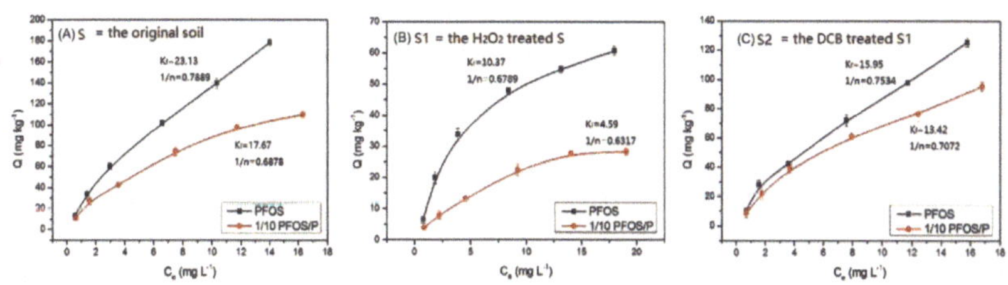

FIGURE 5.10 Effect of phosphate (P) on perfluorooctane sulfonate sorption by three soil samples [(A) original soil (S); (B) after removing SOM from S (S1); (C) after removing ferric oxides from S1 (S2)] and the best-fit Freundlich isotherm model results. *Symbols* and *lines* are experimental data and model results, respectively. *Rearrange the order of the graphs according to Qian, J., Shen, M., Wang, P., Wang, C., Hou, J., Ao, Y., Liu, J., & Li, K. (c. 2017). Adsorption of perfluorooctane sulfonate on soils: Effects of soil characteristics and phosphate competition.* Chemosphere, 168, *1383–1388. https://doi.org/10.1016/j.chemosphere.2016.11.114. Copyright (2017), with permission from Elsevier.*

have shown no significant sorption of PFOA on HA due to strong electrostatic repulsion (Koopal et al., 2004). The authors also observed no significant binding between the anionic surfactant (sodium dodecyl sulfate) and purified Aldrich HA. On the other hand, there were strong sorption of cationic surfactants such as dodecyl- and cetylpyridinium chloride on the HA, which can be attributed to both electrostatic attraction and hydrophobic interaction (Koopal et al., 2004). These results suggest the effect OM in soils on the sorption of anionic PFOAs is depending the properties of OM as well as the soils. In fact, the presence of OM such as HA in soils can change their surface properties and thus influence the sorption of PFASs by altering the electrostatic and hydrophobic interactions.

Zhang et al. studied the effects of OM in the fate and transport of PFOS in the environment (Zhang et al., 2013). They isolated three major components in soil/sediment: humin/kerogen, humic/fulvic acid (HA/FA), and inorganic component after removing organics, and then explored their contributions to PFOS adsorption using batch adsorption experiments and molecular dynamic simulations (Fig. 5.11 and Table 5.4). The adsorption capacity of PFOS reduced with the decrease in TOC of treated soil/sediment samples, suggesting that organic substances play a key role in PFOS adsorption. Furthermore, humin/kerogen is the dominant organic substance for PFOS adsorption on soil/sediment. The results suggest that the humin/kerogen component dominated the PFOS adsorption due to its aliphatic features where hydrophobic effect and phase transfer are the primary adsorption mechanism. Compared with the humin/kerogen, the HA/FA component contributed less to the PFOS adsorption because of its hydrophilic and polar characteristics. The electrostatic repulsion between the polar groups of HA/FA and PFOS anions was attributable to the reduced PFOS adsorption. When the soil OM was extracted, the inorganic component also played a nonnegligible role because PFOS molecules might form surface complexes on SiO_2 surface.

5.3 Retention, transport, and release

5.3.1 Fate and transport of per- and polyfluoroalkyl substances in soils and groundwater

Risk assessment must be conducted as accurately as possible to support effective decision-making because of the high potential health risks associated with PFAS-contaminated sites. Accurate risk assessment requires a comprehensive and definitive understanding of the retention, transport, and release behaviors of PFASs in subsurface, including the accurate determination of their transport potential from sources to groundwater and soils.

The environmental fate of PFASs is controlled by the transport, distribution, and transformation processes after they are released into the

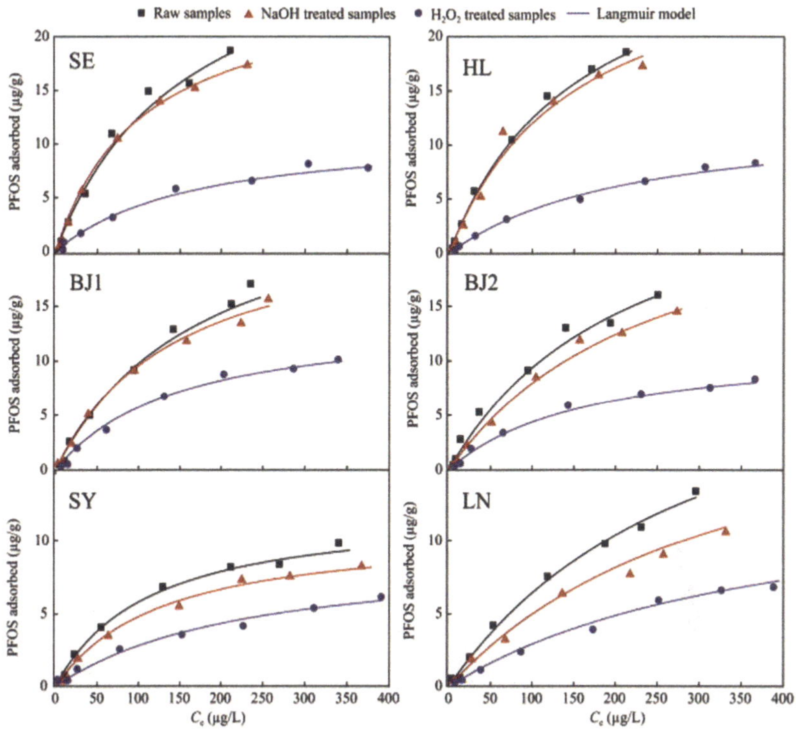

FIGURE 5.11 Adsorption isotherms of perfluorooctanoic sulfonate (PFOS) on soil/sediment samples at pH = 7, PFOS = 10–500 μg/L, adsorbents = 15 g/L. Sediment (SE) and soils (HL, BJ, SY, LN) were treated by NaOH and H_2O_2 to remove HA/FA and organic substances including HA/FA and humin/kerogen, respectively. *Reprinted from Zhang, R., Yan, W., & Jing, C. (c. 2015). Experimental and molecular dynamic simulation study of perfluorooctane sulfonate adsorption on soil and sediment components.* Journal of Environmental Sciences (China), 29, *131–138. https://doi.org/10.1016/j.jes.2014.11.001. Copyright (2015), with permission from Elsevier.*

environment (Fig. 5.12). PFASs can be released into the environment from both point and nonpoint sources during producing and after using products containing PFASs. Some precursor compounds are eventually transformed into stable PFASs by biological and physical-chemical processes. These stable PFASs can be distributed in the atmosphere emissions, wastewater discharges, and solid waste disposals, and find their ways into the surface water, soils and groundwater, oceans, and even polar glaciers (Ahrens & Bundschuh, 2014; Concawe, 2016). For the volatile PFASs, evaporation and diffusion may be the main transport path (Ahrens & Bundschuh, 2014; Concawe, 2016). However, water flow can serve as a significant path for the transport of PFASs in soil and groundwater systems.

TABLE 5.4 Properties of collected soil/sediment samples.

Samples	pH	ς Potential (mV)	TOC (%)	
			Raw samples	NaOH treated samples
LN	7.3	−12.3	1.00 ± 0.15	0.59 ± 0.09
BJ1	7.7	−10.4	1.62 ± 0.19	1.05 ± 0.12
SY	6.7	−13.0	2.08 ± 0.40	0.91 ± 0.11
BJ2	7.4	−8.6	1.87 ± 0.30	0.45 ± 0.02
HL	7.0	−11.8	6.62 ± 0.10	4.21 ± 0.30
SE	8.5	−14.6	8.28 ± 0.31	4.94 ± 0.23

Source: Reprinted from Zhang, R., Yan, W., & Jing, C. (2015). Experimental and molecular dynamic simulation study of perfluorooctane sulfonate adsorption on soil and sediment components. *Journal of Environmental Sciences (China), 29,* 131–138. https://doi.org/10.1016/j.jes.2014.11.001. Copyright (2015), with permission from Elsevier.

It is clear that short-range atmospheric transport and deposition can cause PFAS contamination in terrestrial and aquatic systems near the points of significant emissions, which can introduce PFASs into soils and groundwater. PFASs released from industrial emissions are likely related to particulate matters and have high possibility to settle on the ground in dry weather. A previous study suggests that ammonium perfluorooctanoate air emissions from a fluoropolymer manufacturing facility can be transported to the well field, deposited onto the soil, and then migrate downward with precipitation into the underlying aquifer (Davis et al., 2007).

5.3.2 Impact factors

Many factors can influence the fate and transport of PFASs in soil and groundwater systems and they can be roughly putted in two categories: environmental conditions and PFAS properties. Environmental conditions usually affect the interactions between PFASs and soils (Li et al., 2012). These conditions mainly include soil properties [e.g., natural organic carbon (NOC) content, permeability, surface charge, exchange capacity, clay minerals, water content, etc.], oxidation-reduction conditions, pore water chemistry (e.g., pH, ionic strength, ion types, NOC content), and the presence of cocontaminants. On the other hand, the inherent properties of PFASs, such as the length of alkyl chain, the ionic state, the type and amount of functional groups, and the degree of fluorination can strongly influence the partitioning, sorption, and transformation behaviors of PFASs to control their retention, transport, and release in soils and groundwater. Generally speaking, the

FIGURE 5.12 Pathways of per- and polyfluoroalkyl substances (PFASs) into the environment and their fate. (A) Agricultural sources such as biosolids application and irrigation with reclaimed wastewater. (B) Emissions from manufacturing and subsequent atmospheric transport. (C) Interactions with plants in soils that lead to accumulation and potentially transformation. (D) Plant uptake and potential phytovolatilization. (E) Interactions between water, soil gas, nonaqueous-phase liquids, and soil where microorganisms play a governing role in PFAS transformation. *Reprinted from Sharifan, H., Bagheri, M., Wang, D., Burken, J., Higgins, C., Liang, Y., Liu, J., Schaefer, C., & Blotevogel, J. (c. 2021). Fate and transport of per- and polyfluoroalkyl substances (PFASs) in the vadose zone.* Science of the Total Environment, 771. *https://doi.org/10.1016/j.scitotenv.2021.145427. Copyright (2021), with permission from Elsevier.*

transport of PFASs in soils and groundwater is strongly influenced by these factors.

5.3.2.1 Natural organic carbon

PFASs may be attracted to NOC in soils through hydrophobic interactions, which can enhance their retention and reduce their mobility in soils and groundwater. This retention process is similar to the filtration of PFASs by granular activated carbon in water treatment systems. The interaction between PFASs and NOC is related to the specific organic-carbon partition coefficient (K_{oc}), the PFAS-specific octanol-water partition coefficient (K_{ow}), and the PFAS- and soil-specific distribution coefficient (K_d). Among them, K_{oc} may be the dominant indicator. Short-chain PFASs are mostly hydrophilic and are often more mobile in soils and groundwater than long-chain ones, which have a higher hydrophobicity and hence tend to bind to NOC in soils through hydrophobic interactions. As results, the retention of PFASs in soils increases with increasing chain length as well as with increasing solid phase fraction of organic carbon in soils.

Gelrich et al. studied the effect of soil NOC content on the leaching behavior of PFASs in soil (Fig. 5.13) (Gellrich et al., 2012). The authors changed the NOC content of the soil by adding different amounts of sewage sludge (the NOC content of sewage sludge 1 and sewage sludge 2 were 7% and 14%, respectively). They found that PFBA, PFBS, PFPeA, and PFHxA were not retained and eluted together with the tracer (Fig. 5.13E). PFHxS and PFHpA, which have the same fluorinated carbon chain length, showed similar breakthrough curves (BTCs) over a period of more than 70 weeks. For PFHpA, a fast first desorption was observed followed by a slower second desorption.

After adding sewage sludge 1 (Fig. 5.13A), the breakthrough of PFBA and PFBS appeared to be complete after 1 year. Thus complete elution took twice the time taken for the other tubes with the same flow velocity. The PFASs showed a stronger adsorption to the sewage sludge, probably due to a higher content of organic carbon, which was a factor for the degree of

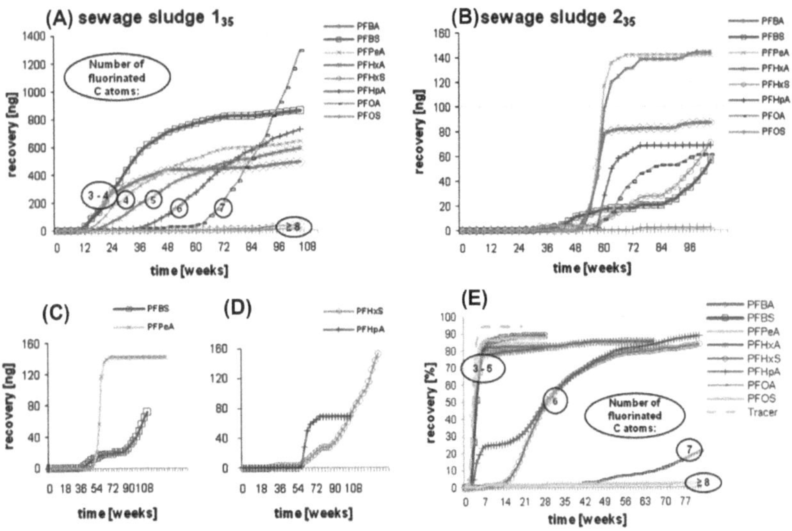

FIGURE 5.13 Breakthrough curves of per- and polyfluoroalkyl substances (PFASs) in soil with (A) sewage sludge 1, (B–D) sewage sludge 2, (C and D) excerpts of (B) showing only PFASs with the same fluorinated carbon chain length, (E) no sewage sludge. *Rearrange the order of the graphs according to Gellrich, V., Stahl, T., & Knepper, T. (c. 2012). Behavior of perfluorinated compounds in soils during leaching experiments.* Chemosphere, 87*(9), 1052–1056. https://doi.org/10.1016/j.chemosphere.2012.02.011. Copyright (2012), with permission from Elsevier.*

adsorption to soils or sediment. This sludge had a content of organic carbon of 7%, whereas the soil only had 2%.

In the percolated water from the sewage sludge 2 (Fig. 5.13B), no PFASs could be detected in the first 50 weeks. This can be attributed to the high organic carbon content of 14%. PFBA, PFPeA, and PFHxA eluted with sharp narrow BTCs from week 50 to week 70. The elution of PFHpA and PFOA, again, started 10 weeks later. PFOS and other PFASs with longer carbon chains could not be detected in the percolating water after 120 weeks. The concentrations measured were only about a fifth of those measured from the sewage sludge 1. For this treatment, the PFASs did not only elute according to the chain length as it did for the other sewage sludge, but there are clear differences concerning the functional groups. As illustrated in Fig. 5.13C,D, the sulfonic acids show a different leaching behavior than the carboxylic acids of the same fluorinated chain length. The desorption started later for all of them, but then the PFCAs eluted faster than from sewage sludge 1, whereas the PFSAs eluted only slowly and erratically.

Lyu et al. (2019) studied the transport and retention of PFOA in three soils with different NOC contents (Fig. 5.14). They found that the first breakthrough of PFOA in the desert, black, and red soils was at ~1.93, 2.86, and 3.08 PVs, respectively, under 1.0 mM $CaCl_2$ condition; and postponed to 2.48, 4.85, and 3.78 PVs, respectively, under 10.0 mM $CaCl_2$ condition. The BTCs of PFOA in the desert soil were approximately symmetric in shape, while the BTCs in black and red soils exhibited asymmetrical shape and high degrees of tailing. The retention of PFOA under both solution ionic strength conditions generally followed the order of red soil > black soil > desert soil. Fig. 5.14A,B shows a higher transport of PFOA in desert soil than that in black soil under both solution ionic strength conditions. The zeta potentials of the desert soil (1 mM: -15.70 ± 0.10 mV, 10 mM: -9.11 ± 0.10 mV) were very close to those of the black soil (1 mM: -13.02 ± 0.10 mV, 10 mM: -8.53 ± 0.97 mV) under both ionic strength conditions. So the NOC content of the black soil (2.57%) was much higher than that of the desert soil (0.05%), which may be the key factor affecting PFOA retention and transport in the two soils. This confirms the key role of soil NOC in PFOA mobility in soils and suggests that hydrophobic interaction is an important mechanism controlling the retention and transport of PFOA in soils.

5.3.2.2 *Solution chemistry*

Xing et al. used a two-site nonequilibrium transport model to describe the transport behaviors and mechanisms of PFOA and PFOS in saturated sand columns with consideration of the coupled effect of solution cation type and pH (Fig. 5.15) (Xing et al., 2021). In the study, PFOA showed higher mobility than PFOS, and divalent cation inhibited their transport by increasing the nonequilibrium interactions between them and the sand. When there was

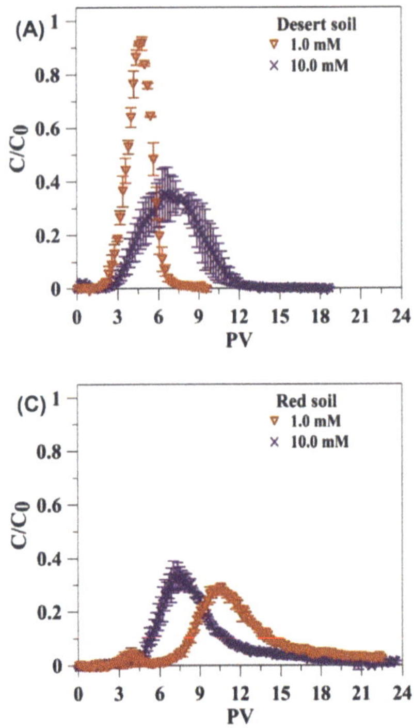

FIGURE 5.14 Breakthrough curves of PFOA in (A) desert soil, (B) black soil, and (C) red soil columns under 1.0 and 10.0 mM $CaCl_2$ solutions at pH5.6. *Error bars* represent standard deviations from replicate experiments ($n = 2$). *Reprinted from Lyu, X., Liu, X., Sun, Y., Ji, R., Gao, B., & Wu, J. (c. 2019). Transport and retention of perfluorooctanoic acid (PFOA)in natural soils: Importance of soil organic matter and mineral contents, and solution ionic strength.* Journal of Contaminant Hydrology, 225. *https://doi.org/10.1016/j.jconhyd.2019.03.009. Copyright (2019), with permission from Elsevier.*

only monovalent cation in the solution, pH had little influence on the mobility of PFOA because it had limited affinity with the sand. However, polyvalent cations provided additional adsorption sites for PFOA through cation bridging and thus escalated the effect of pH. Differently, decreasing pH inhibited the transport of PFOS more significantly, and the effect was stronger than that of changing cation type.

Lyu et al. (2019) found that solution ionic strength strongly affected the transport of PFOA in three soils (Fig. 5.14). For the negatively charged desert and black soils (Fig. 5.14A,B), increasing ionic strength (1.0–10.0 mM) strongly retarded PFOA BTCs (desert soil: 1.93–2.48 PV; black soil: 2.86–4.85 PV) and reduced its transport (desert soil: 96.13 ± 1.15% to 94.71 ± 1.96%; black soil: 92.52 ± 2.32% to 83.52 ± 0.65%). The desert soil (1 mM: −15.70 ± 0.10 mV, 10 mM: −9.11 ± 0.10 mV) and black soil (1 mM: −13.02 ± 0.10 mV, 10 mM: −8.53 ± 0.97 mV) became less negatively

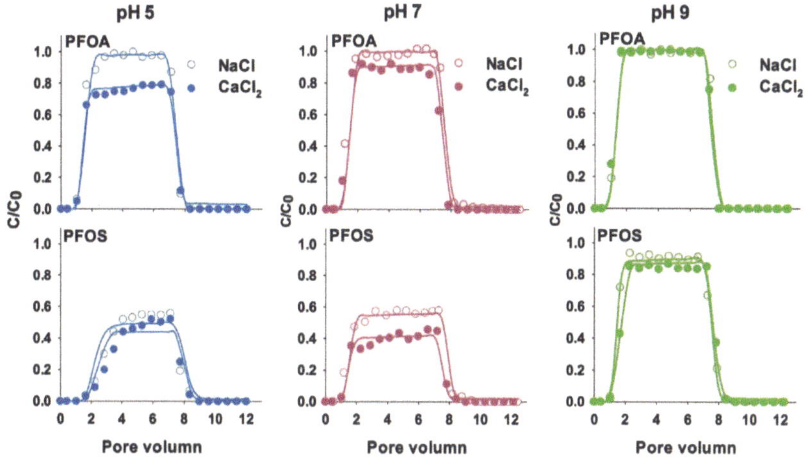

FIGURE 5.15 Breakthrough curves of PFOA and perfluorooctanoic sulfonate at different pH in NaCl and $CaCl_2$ solutions (10 mM ionic strength). *Symbols* and *lines* represent measurement and model simulation, respectively. *Reprinted from Xing, Y., Li, Q., Chen, X., Fu, X., Ji, L., Wang, J., Li, T., & Zhang, Q. (c. 2021). Different transport behaviors and mechanisms of perfluorooctanoate (PFOA) and perfluorooctane sulfonate (PFOS) in saturated porous media.* Journal of Hazardous Materials, 402. *https://doi.org/10.1016/j.jhazmat.2020.123435. Copyright (2021), with permission from Elsevier.*

charged at a higher Ca^{2+} concentration. The electrostatic repulsion between the two negatively charged soil surface and anion PFOA molecules was thus weakened with increasing concentrations of Ca^{2+}. In contrast, the effect of Ca^{2+} on PFOA transport in positively charged red soil was completely different from that in the two negatively charged soils. Fig. 5.14C shows that increasing concentrations of Ca^{2+} led to higher mobility of PFOA in the red soil ($80.27 \pm 2.33\%$ to $89.08 \pm 2.26\%$). The peak breakthrough concentrations under two ionic strengths were 0.28 (1.0 mM) and 0.33 (10.0 mM). So increasing Ca^{2+} concentration may have a negative effect on PFOA adsorption onto positively charged minerals (e.g., alumina and boehmite), with potential mechanisms of Ca^{2+} bridging of carboxyl groups of PFASs compounds. This may be responsible for the higher transport of PFOA at higher Ca^{2+}.

5.3.2.3 Moisture content

As surfactants, PFASs tend to accumulate at air–water interfaces and may stay in the vadose zone for a long time before contaminating groundwater. However, the extent of retention and transport processes of PFASs in the vadose zone are still not well understood. Guo et al. applied a mathematical model to simulate PFOS transport in the vadose zone at a fire-training area site impacted by aqueous film-forming foam (Guo et al., 2020). They found that air–water interfacial adsorption has significant impact on PFOS retention in soils, amplified by low water content due to gravity drainage. The total retardation factors of PFOS range from 233 to 1355 in sand and 146 to 792 in soil used in their study. The simulations also show that it may take several decades or even longer for PFOS to reach groundwater. Counterintuitively, lower water content in the sand—due to stronger drainage and weaker capillary retention—led to greater retardation factors than those of the soil. Also, most PFOS is adsorbed at air–water interfaces with only 1%–2% in the aqueous phase.

In a recent study, miscible-displacement experiments were conducted to investigate the impact of adsorption at the air–water and decane–water interfaces on PFAS retention and transport in porous media (Brusseau et al., 2019). Two types of media were used, a quartz sand and a soil, and PFOS was used as the model PFAS. The BTCs of PFOS transport in the media under unsaturated conditions shifted noticeably rightward (delayed arrival) compared to those under saturated conditions (Fig. 5.16A), indicating greater

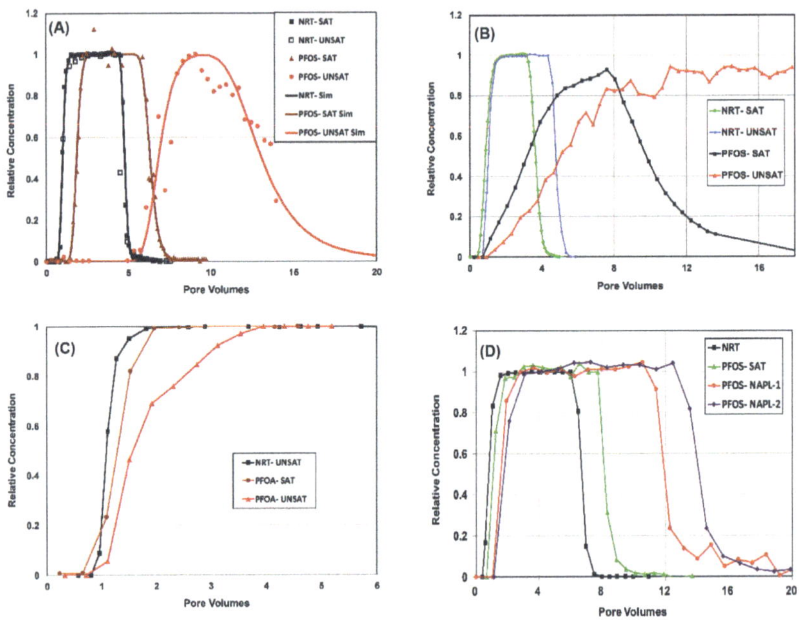

FIGURE 5.16 Breakthrough curves for transport of (A) a nonreactive tracer (NRT) and perfluorooctanoic sulfonate (PFOS) under saturated (SAT) and unsaturated (UNSAT, Sw = 0.66) conditions in the sand; (B) a nonreactive tracer (NRT) and PFOS under saturated (SAT) and unsaturated (UNSAT, Sw = 0.66) conditions in the soil; (C) a nonreactive tracer (NRT) and perfluorooctanoic acid under saturated (SAT) and unsaturated (UNSAT, Sw = 0.78) conditions in the sand; (D) PFOS in the absence (SAT) and presence of decane NAPL (NAPL, Sw = 0.62) in the sand. *Reprinted from Brusseau, M., Yan, N., Van Glubt, S., Wang, Y., Chen, W., Lyu, Y., Dungan, B., Carroll, K., & Holguin, F. (c. 2019). Comprehensive retention model for PFAS transport in subsurface systems.* Water Research, 148, *41–50. https://doi.org/10.1016/j.watres.2018.10.035. Copyright (2019), with permission from Elsevier.*

PFOS retained in the media due to adsorption at the air–water or decane–water interface. PFOS retardation factors in the soil were 7.3 and 3.6 in unsaturated and saturated soils, respectively (Fig. 5.16B). The retardation factor of PFOS transport in the sand was 7 in the unsaturated system, much higher than that (1.8) in the saturated system (Fig. 5.16C). Air–water interfacial adsorption is thus a significant sink for PFOS in the two media, contributing more than 80% and 32% of the total retention for the sand and the soil, respectively. For the experiments conducted with decane residual emplaced within the sand, adsorption at the decane–water interface contributed more than 70% of the total retention for PFOS in the media (Fig. 5.16D).

Most soils contain both fixed- and variably charged surfaces, such that the net charge on the soils, as well as the charge of functional groups of individual PFAS, can be strongly influenced by solution pH. For example, the net negative charge on most clay minerals can result in electrostatic interactions with cationic functional groups that are present on some PFASs. However, many PFASs, such as PFOA and PFOS, are anionic at environmentally relevant pH, such that electrostatic repulsion in soils dominated by negatively charged minerals would enhance the mobility of anionic and zwitter ionic PFASs. Changes in pH potentially impact these electrostatic processes by altering surface charges (or possibly the ionic nature of PFASs). As a result, decreases in pH have been shown to enhance sorption by anionic PFASs such as PFOS and other PFAS.

5.3.2.4 Medium property

Natural media with different surface property may have distinct influence on the transport of PFASs in soils (Lyu, Liu, Sun, et al., 2020; Lyu, Liu, Wu, et al., 2020). Compared with uncoated sand, Al/Fe oxyhydroxide-coated sand

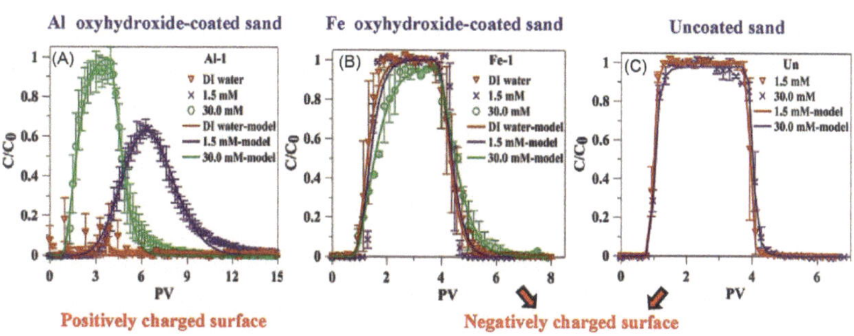

FIGURE 5.17 Breakthrough curves for perfluorooctanoic acid in (A) Al oxyhydroxide-coated sand, (B) Fe oxyhydroxide-coated sand, and (C) uncoated sand under three NaCl concentrations (0, 1.5, and 30.0). *Reprinted from Lyu, X., Liu, X., Wu, X., Sun, Y., Gao, B., & Wu, J. (c. 2020). Importance of Al/Fe oxyhydroxide coating and ionic strength in perfluorooctanoic acid (PFOA) transport in saturated porous media.* Water Research, 175. *https://doi.org/10.1016/j.watres.2020.115685. Copyright (2020), with permission from Elsevier.*

has more abundant functional groups. Thus a clear increase in PFOA retardation in Al/Fe oxyhydroxide-coated sand (retardation factor: Al: 1.87–5.58, Fe: 1.28–4.05), higher than those in uncoated sand (1.00–1.05), was observed (Fig. 5.17), which can be attributed to stronger electrostatic attraction between anionic PFOA and Al/Fe oxyhydroxide-coated sand surface. Higher ionic strength can significantly inhabit PFOA retention and retardation in positively charged sand. The considerable retention of PFOA ($\sim$90%) in deionized water than those in 1.5 mM and 30.0 mM NaCl ($<$10%) clearly proves the role of competitive adsorption of Cl^- on PFOA transport in positively charged sand. In contrast, higher ionic strength (0–30 mM NaCl) can slightly increase PFOA retardation in negatively charged sand, confirming the dominance of electrostatic interaction.

Soil surface roughness can also play important roles in affecting the retention and transport of PFASs in soils and groundwater. Lyu, Liu, Wu, et al. (2020) examined the influence of surface roughness on PFOA transport in unsaturated porous media including glass beads, quartz sand, and limestone (Fig. 5.18). They found that decreasing moisture content significantly increased the air–water interfacial adsorption of PFOA and thus led to its greater retardation in all three types of porous media. The authors emphasize that rougher surface (limestone > quartz sand > glass beads) and smaller grain size (i.e., large solid specific surface area) can significantly enhance PFOA retardation in soils and groundwater under unsaturated conditions. In addition, the greater retardation of PFOA in unsaturated porous media can be attributed to the larger air–water interfacial areas associated with rougher surface and smaller grain size.

5.4 Conceptual models of per- and polyfluoroalkyl substance retention

So far, the research on modeling the fate and transport of PFASs in soils and groundwater is still far from complete. Nevertheless, the main processes governing the retention of PFASs in soils and groundwater can be generally summarized as the conceptual models (Fig. 5.19) in following sections.

5.4.1 Solid-phase adsorption

In the past decade, the adsorption of PFASs on solids has become the focus of research. At the same time, the effect of solid-phase adsorption on PFAS transport in subsurface porous media (soils) has also been investigated (Chen et al., 2016; ITRC, 2020). Due to the nature of the molecular structure of PFASs, the adsorption of PFASs on porous media is very complicated. As mentioned earlier, cationic and anionic PFASs may have different electrostatic interactions with soil particles including soils colloids, leading to differences in the adsorption potential of different types of PFASs on solid

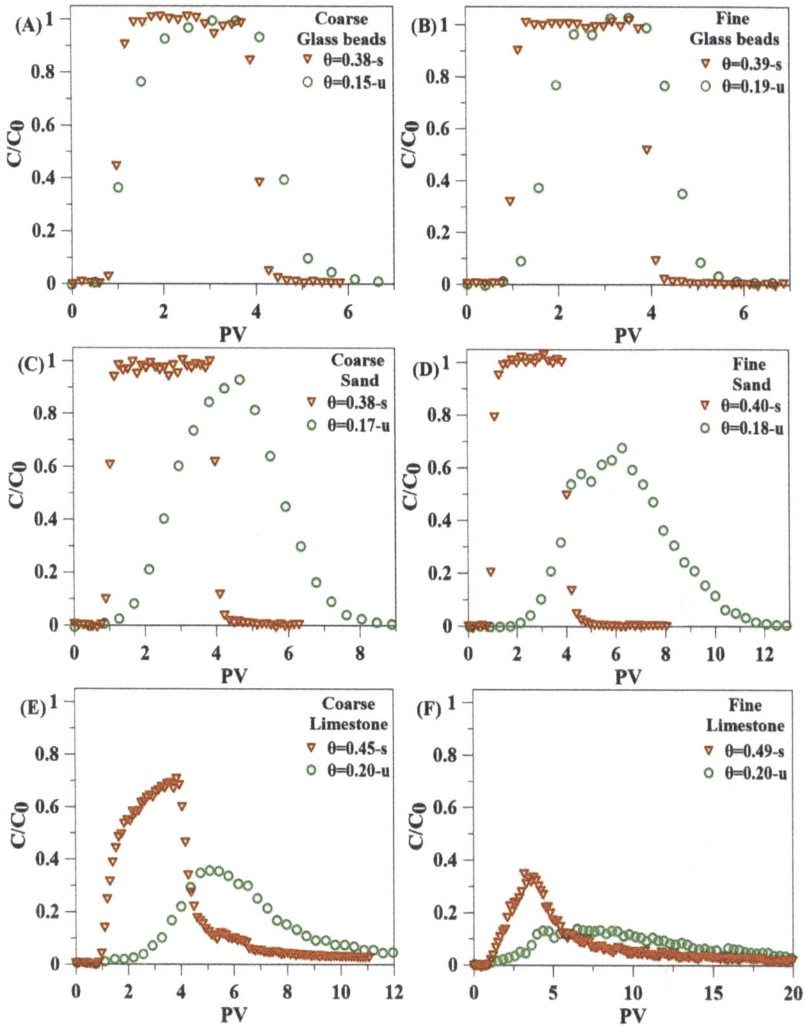

FIGURE 5.18 Breakthrough curves of PFOA in columns packed with two-sized (A, B) glass beads, (C, D) quartz sand, (E, F) limestone under different moisture content conditions, s represents saturated conditions, u represents unsaturated conditions, and θ is moisture content. *Reprinted from Lyu, X., Liu, X., Sun, Y., Gao, B., Ji, R., Wu, J., & Xue, Y. (c. 2020). Importance of surface roughness on perfluorooctanoic acid (PFOA) transport in unsaturated porous media.* Environmental Pollution, 266. *https://doi.org/10.1016/j.envpol.2020.115343. Copyright (2020), with permission from Elsevier.*

phase in porous media. The geochemical properties of soils, such as magnitude and nature of NOC, and magnitude and type of clay and metal minerals, play a key role in the retention and transport of PFASs in subsurface. In addition, the solubility of PFASs in water can also affect their solid phase

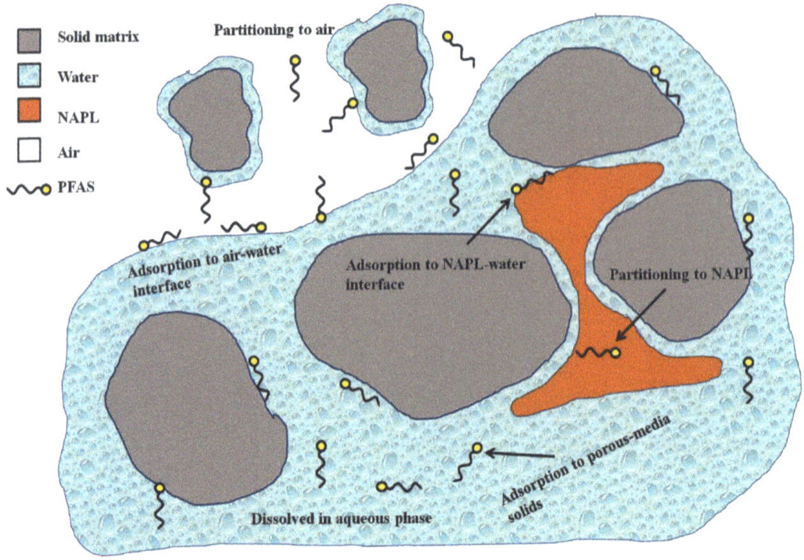

FIGURE 5.19 Transport and fate of per- and polyfluoroalkyl substance in source zones. *Redrawn according to Brusseau, M. (c. 2018). Assessing the potential contributions of additional retention processes to PFAS retardation in the subsurface.* Science of the Total Environment, 613–614, *176–185. Copyright (2018), with permission from Elsevier. https://doi.org/10.1016/j.scitotenv.2017.09.065.*

adsorption. For example, PFOS and PFOA have relatively high water solubility and generally have low adsorption potential on soils. Furthermore, it is well-known that the adsorption of solid porous media is sensitive to water chemistry. This has been demonstrated for PFAS that factors such as ionic strength, salt composition, pH, and presence of cocontaminants affect solid-phase adsorption.

5.4.2 Partitioning to air

Partitioning to the soil atmosphere can serve as a way of retention during aqueous-phase transport of volatile PFASs in the vadose zone. As gas-phase diffusion coefficients are several orders of magnitude greater than aqueous values, gas-phase diffusive may be a significant transport mechanism of PFASs in soils. Air–water partition coefficients vary greatly among the different PFASs. Measured and calculated air–water partition coefficients for PFOA and PFOS are in the 3 – 10 range (Ding & Peijnenburg, 2013). Given their small coefficients, partitioning of PFOA and PFOS to the soil atmosphere in the vadose zone is unlikely to be a significant retention process. On the contrary, the coefficients of some other PFASs, such as FTOHs and fluorotelomer olefins are quite large, which are about 0.05 − 4 and 100 − 10,000, respectively (Ding & Peijnenburg, 2013). Thus air–water partitioning is anticipated to be a significant process influencing the transport behavior of these compounds. Indeed, atmospheric transport of the fluorotelomers has been identified as the probable cause of the observed widespread distributions of PFASs in the environment.

5.4.3 Adsorption to air–water interface

PFAS contamination is likely to be present in the vadose zone at many sites as a result of waste disposal, spillage, and specified applications. PFOS, PFOA, and other related PFASs of concern are surfactants, and by their nature will accumulate at air–water interfaces. Thus it is anticipated that adsorption on air–water interfaces may be a significant retention and attenuation mechanism for these PFAS in the unsaturated soils.

Fluorinated surfactants are typically more surface active than the analogous standard hydrocarbon surfactants. This is related to the unique nature of the "tail" portion of the compound. The tails of fluorinated surfactants are hydrophobic due to their nonpolar, fully saturated nature, just as are the tails of standard hydrocarbon surfactants. However, unlike hydrocarbon surfactants, the tails of fluorinated surfactants are also lipophobic due to the low polarizability of the C–F bond. This unique property results in greater reductions in surface tension caused by fluorinated surfactants. So, PFAS can preferentially form films at the air–water interface, with the hydrophobic C–F tail oriented toward the air and the hydrophilic headgroup dissolved in the water. This behavior influences aerosol-based transport and deposition, suggesting that accumulation of PFASs at water surfaces will occur. In addition, this preference for the air–water interface may have important implications for vadose zone transport, where unsaturated conditions provide significant air–water interfacial area. This includes the potential for enhanced retention at the water table and the capillary fringe, which is the subject of significant ongoing research. For example, Brusseau showed that adsorption of PFOS and PFOA at the air–water interface can increase the retardation factor for aqueous-phase transport (Brusseau, 2018); this interfacial process accounted for approximately 50% of the total retention in a model system with 20% air saturation.

5.4.4 Nonaqueous-phase liquids-water partitioning

Another potential source of retardation of organic compounds is retention by organic liquids in nonaqueous-phase liquids (NAPL) source zones in soils and groundwater. This is of potential significance for PFASs given that they often occur in the presence of NAPL contamination at mixed-waste sites. For example, a primary source of PFAS contamination is the prior use of aqueous film-forming foams for fire suppression. In some cases, sites that served as designated fire-training areas were also used as disposal sites for solvent and oil wastes. In these settings, the released petroleum hydrocarbon fuel forms a NAPL zone into which the PFASs may partition and accumulate along the NAPL–water interface (Brusseau, 2018). These processes may result in increased PFAS mass retained in the NAPL source zones, increased PFAS sorption and resulting retardation, and greater persistence of dissolved PFAS.

5.4.5 Adsorption at the nonaqueous-phase liquids–water interface

The adsorption of constituents at the NAPL–water interface is similar to that at the air–water interface. PFASs, especially surfactant-based ones, can accumulate at fluid–fluid interfaces, including NAPL–water interfaces. Thus NAPL–water interface adsorption may be important to the retention and transport of PFASs in the source zones of NAPL-contaminated soils and groundwater. It should also be noted that contaminants in the gas phase can also be adsorbed at the air–NAPL interfaces in soil and groundwater systems.

5.5 Conclusion and perspectives

PFASs have been banned by many countries because of their persistent harm to the environment and human health. Because of their unique properties, they are allowed to be used in some applications, such that the release of these chemicals into soils and groundwater continues. In this chapter, we summarize the fate and transport of PFAS in soils and groundwater, which are influences by various factors including their classification and basic physical-chemical properties, sources and applications, interaction with soils, and retention and release in soils. At the same time, the distribution and partition model of PFASs in soils are analyzed. According to the results of the current studies, the main source of PFASs in the environment is the processing and use of PFAS raw materials and the large-scale application of PFAS-containing aqueous fire-fighting foam. The contribution of leachate from landfills and sewage treatment plants is relatively small. Aqueous fire-fighting foam, PFAS-containing products, and sewage entering groundwater and soils can cause serious damage to the subsurface ecosystem. Due to the nondegradability of PFASs, they can last for a long time. PFASs are retained in soils mainly in two ways: (1) hydrophobic adsorption to naturally occurring solid organic particles and (2) surface adsorption to charged mineral surfaces. Both of them are closely related to the physical and chemical properties of soils and PFASs. Additional investigations, including long-term field monitoring and numerical modeling, are needed to provide insights in order to better understand the retention and transport of PFASs in soils and groundwater.

PFAS release has resulted in historical and continuing human and ecological exposures to these toxic chemicals. None of the current remedial actions is sufficient to prevent the ecological exposures, nor reliably prevent human exposures. As the fate, transport, bioaccumulation potential, and toxicology of PFAS are gained broader understanding and recognizing, the remediation of PFAS-contaminated soils and groundwater is also facing challenges. The substrates that need to be remediated include both soil and water. It is necessary to carefully consider the remediation goals of each medium and their specific characteristics, such as types of soil minerals, flow rates of groundwater, coexisting pollutants, natural OM, and various concentrations and types of PFASs that need to be treated. However, the extremely stable molecular structures and physical-chemical properties of PFASs make them persistent and difficult to be removed from soils and groundwater. The current methods to repair PFAS-contaminated soils/ groundwater mainly include sorption and stabilization (e.g., silica-based adsorbent, granular activated carbon, injectable particulate carbon, ion-exchange resins, activated carbon fibers, etc.), destruction technologies (e.g., thermal treatment, chemical oxidation, biological methods/bioremediation, electron beam, etc.), and separation technologies (e.g., soil washing, lique-fractionation, etc.). Each method has its advantages as well as shortcomings and limitations. Among them, the sorption and stabilization is regarded as one of the promising remediation method for PFASs in subsurface. However, this process requires a large amount of adsorbent as well as substantial installation efforts, which increase the cost of restoration and may change the land use. Several other methods, alone or in combination, generally are only tested at the laboratory scale. Their rationality and feasibility of field applications still need further research and optimization.

References

Ahrens, L., & Bundschuh, M. (2014). Fate and effects of poly- and perfluoroalkyl substances in the aquatic environment: A review. *Environmental Toxicology and Chemistry*, *33*(9), 1921–1929. Available from https://doi.org/10.1002/etc.2663.

Ahrens, L., Shoeib, M., Harner, T., Lane, D. A., Guo, R., & Reiner, E. J. (2011). Comparison of annular diffusion denuder and high volume air samplers for measuring per- and polyfluoroalkyl substances in the atmosphere. *Analytical Chemistry*, *83*(24), 9622–9628. Available from https://doi.org/10.1021/ac202414w.

Ahrens, L., Taniyasu, S., Yeung, L. W. Y., Yamashita, N., Lam, P. K. S., & Ebinghaus, R. (2010). Distribution of polyfluoroalkyl compounds in water, suspended particulate matter and sediment from Tokyo Bay, Japan. *Chemosphere*, *79*(3), 266–272. Available from https://doi.org/10.1016/j.chemosphere.2010.01.045.

Ahrens, L., Yamashita, N., Yeung, L. W. Y., Taniyasu, S., Horii, Y., Lam, P. K. S., & Ebinghaus, R. (2009). Partitioning behavior of per- and polyfluoroalkyl compounds between pore water and sediment in two sediment cores from Tokyo Bay, Japan. *Environmental Science and Technology*, *43*(18), 6969–6975. Available from https://doi.org/10.1021/es901213s.

Ahrens, L., Yamashita, N., Yeung, L. W. Y., Taniyasu, S., Horii, Y., Lam, P. K. S., & Ebinghaus, R. (2009). Partitioning behavior of per- and polyfluoroalkyl compounds between pore water and sediment in two sediment cores from Tokyo Bay, Japan. *Environmental Science and Technology*, *43*(18), 6969–6975. Available from https://doi.org/10.1021/es901213s.

Antelo, J., Avena, M., Fiol, S., López, R., & Arce, F. (2005). Effects of pH and ionic strength on the adsorption of phosphate and arsenate at the goethite-water interface. *Journal of Colloid and Interface Science*, *285*(2), 476–486. Available from https://doi.org/10.1016/j.jcis.2004.12.032.

ATSDR. (2019). Per and polyfluoroalkyl substances (PFAS) in the Pease Tradeport Public Water System. *Health Consultation*. Available from https://www.atsdr.cdc.gov/HAC/pha/pease/Pease-Tradeport-Public-Water-PFAS-HC-508.pdf.

Brusseau, M. L. (2018). Assessing the potential contributions of additional retention processes to PFAS retardation in the subsurface. *Science of the Total Environment*, *613–614*, 176–185. Available from https://doi.org/10.1016/j.scitotenv.2017.09.065.

Brusseau, M. L., Yan, N., Van Glubt, S., Wang, Y., Chen, W., Lyu, Y., Dungan, B., Carroll, K. C., & Holguin, F. O. (2019). Comprehensive retention model for PFAS transport in subsurface systems. *Water Research*, *148*, 41–50. Available from https://doi.org/10.1016/j.watres.2018.10.035.

Buck, R. C., Franklin, J., Berger, U., Conder, J. M., Cousins, I. T., Voogt, P. D., Jensen, A. A., Kannan, K., Mabury, S. A., & van Leeuwen, S. P. J. (2011). Perfluoroalkyl and polyfluoroalkyl substances in the environment: Terminology, classification, and origins. *Integrated Environmental Assessment and Management*, *7*(4), 513–541. Available from https://doi.org/10.1002/ieam.258.

Cao, Y., Cao, X., Wang, H., Wan, Y., & Wang, S. (2015). Assessment on the distribution and partitioning of perfluorinated compounds in the water and sediment of Nansi Lake, China. *Environmental Monitoring and Assessment*, *187*(10), 611. Available from https://doi.org/10.1007/s10661-015-4831-9.

Chen, S., Jiao, X. C., Gai, N., Li, X. J., Wang, X. C., Lu, G. H., Piao, H. T., Rao, Z., & Yang, Y. L. (2016). Perfluorinated compounds in soil, surface water, and groundwater from rural areas in eastern China. *Environmental Pollution*, *211*, 124–131. Available from https://doi.org/10.1016/j.envpol.2015.12.024.

Concawe. (2016). Environmental fate and effects of poly- and perfluoroalkyl substances (PFAS). https://www.concawe.eu/publication/environmental-fate-and-effects-of-poly-and-perfluoroalkyl-substances-pfas-report-no-816.

Chirikona, F., Filipovic, M., Ooko, S., & Orata, F. (2015). Perfluoroalkyl acids in selected wastewater treatment plants and their discharge load within the Lake Victoria basin in Kenya. *Environmental Monitoring and Assessment*, *187*(5), 238. Available from https://doi.org/10.1007/s10661-015-4425-6.

Ciofi, L., Renai, L., Rossini, D., Ancillotti, C., Falai, A., Fibbi, D., Bruzzoniti, M. C., Santana-Rodriguez, J. J., Orlandini, S., & Del Bubba, M. (2018). Applicability of the direct injection liquid chromatographic tandem mass spectrometric analytical approach to the sub-ngL^{-1} determination of perfluoro-alkyl acids in waste, surface, ground and drinking water samples. *Talanta*, *176*, 412–421. Available from https://doi.org/10.1016/j.talanta.2017.08.052.

Davis, K. L., Aucoin, M. D., Larsen, B. S., Kaiser, M. A., & Hartten, A. S. (2007). Transport of ammonium perfluorooctanoate in environmental media near a fluoropolymer manufacturing facility. *Chemosphere*, *67*(10), 2011–2019. Available from https://doi.org/10.1016/j.chemosphere.2006.11.049.

Ding, G., & Peijnenburg, W. J. G. M. (2013). Physicochemical properties and aquatic toxicity of poly- and perfluorinated compounds. *Critical Reviews in Environmental Science and Technology*, *43*(6), 598–678. Available from https://doi.org/10.1080/10643389.2011.627016.

Enevoldsen, R., & Juhler, R. K. (2010). Perfluorinated compounds (PFCs) in groundwater and aqueous soil extracts: Using inline SPE-LC-MS/MS for screening and sorption characterisation of perfluorooctane sulphonate and related compounds. *Analytical and bioanalytical chemistry*, *398*(3), 1161–1172. Available from https://doi.org/10.1007/s00216-010-4066-0.

Essumang, D. K., Eshun, A., Hogarh, J. N., Bentum, J. K., Adjei, J. K., Negishi, J., Nakamichi, S., Habibullah-Al-Mamun, M., & Masunaga, S. (2017). Perfluoroalkyl acids (PFAAs) in the Pra and Kakum River basins and associated tap water in Ghana. *Science of the Total Environment*, *579*, 729–735. Available from https://doi.org/10.1016/j.scitotenv.2016.11.035.

Filipovic, M., & Berger, U. (2015). Are perfluoroalkyl acids in waste water treatment plant effluents the result of primary emissions from the technosphere or of environmental recirculation? *Chemosphere*, *129*, 74–80. Available from https://doi.org/10.1016/j.chemosphere.2014.07.082.

Fujii, S., Polprasert, C., Tanaka, S., Lien, N. P. H., & Qiu, Y. (2007). New POPs in the water environment: Distribution, bioaccumulation and treatment of perfluorinated compounds–A review paper. *Journal of Water Supply Research and Technology-Aqua*, *56*, 313–326. Available from https://doi.org/10.2166/aqua.2007.005.

Gellrich, V., Stahl, T., & Knepper, T. P. (2012). Behavior of perfluorinated compounds in soils during leaching experiments. *Chemosphere*, *87*(9), 1052–1056. Available from https://doi.org/10.1016/j.chemosphere.2012.02.011.

Goss, K. U. (2008). The pKa values of PFOA and other highly fluorinated carboxylic acids. *Environmental Science and Technology*, *42*(2), 456–458. Available from https://doi.org/10.1021/es702192c.

Guo, B., Zeng, J., & Brusseau, M. L. (2020). A mathematical model for the release, transport, and retention of per- and polyfluoroalkyl substances (PFAS) in the Vadose Zone. *Water Resources Research*, *56*(2). Available from https://doi.org/10.1029/2019WR026667.

Hansen, K. J., Johnson, H. O., Eldridge, J. S., Butenhoff, J. L., & Dick, L. A. (2002). Quantitative characterization of trace levels of PFOS and PFOA in the Tennessee River. *Environmental Science and Technology*, *36*(8), 1681–1685. Available from https://doi.org/10.1021/es010780r.

Higgins, C. P., & Luthy, R. G. (2006). Sorption of perfluorinated surfactants on sediments. *Environmental Science and Technology*, *40*(23), 7251–7256. Available from https://doi.org/10.1021/es061000n.

ITRC. (2020). PFAS—Per- and polyfluoroalkyl substances. PFAS Fact Sheets. https://pfas-1.itrc-web.org.

Johnson, R. L., Anschutz, A. J., Smolen, J. M., Simcik, M. F., & Lee Penn, R. (2007). The adsorption of perfluorooctane sulfonate onto sand, clay, and iron oxide surfaces. *Journal of Chemical and Engineering Data*, *52*(4), 1165–1170. Available from https://doi.org/10.1021/je060285g.

Koopal, L. K., Goloub, T. P., & Davis, T. A. (2004). Binding of ionic surfactants to purified humic acid. *Journal of Colloid and Interface Science, 275*(2), 360–367. Available from https://doi.org/10.1016/j.jcis.2004.02.061.

Lan, Y., Deng, B., Kim, C., & Thornton, E. C. (2007). Influence of soil minerals on chromium (VI) reduction by sulfide under anoxic conditions. *Geochemical Transactions, 8*. Available from https://doi.org/10.1186/1467-4866-8-4.

Lee, H., & Mabury, S. A. (2017). Sorption of perfluoroalkyl phosphonates and perfluoroalkyl phosphinates in soils. *Environmental Science and Technology, 51*(6), 3197–3205. Available from https://doi.org/10.1021/acs.est.6b04395.

Lein, N. P. H., Fujii, S., Tanaka, S., Nozoe, M., & Tanaka, H. (2008). Contamination of perfluorooctane sulfonate (PFOS) and perfluorooctanoate (PFOA) in surface water of the Yodo River basin (Japan). *Desalination, 226*, 338–347. Available from https://doi.org/10.1016/j.desal.2007.01.247.

Li, C., Ji, R., Schäffer, A., Sequaris, J. M., Amelung, W., Vereecken, H., & Klumpp, E. (2012). Sorption of a branched nonylphenol and perfluorooctanoic acid on Yangtze River sediments and their model components. *Journal of Environmental Monitoring, 14*(10), 2653–2658. Available from https://doi.org/10.1039/c2em30394a.

Li, Z., Lyu, X., Gao, B., Xu, H., Wu, J., & Sun, Y. (2021). Effects of ionic strength and cation type on the transport of perfluorooctanoic acid (PFOA) in unsaturated sand porous media. *Journal of Hazardous Materials, 403*. Available from https://doi.org/10.1016/j.jhazmat.2020.123688.

Li, F., Zhang, C., Qu, Y., Chen, J., Chen, L., Liu, Y., & Zhou, Q. (2010). Quantitative characterization of short- and long-chain perfluorinated acids in solid matrices in Shanghai, China. *Science of the Total Environment, 408*(3), 617–623. Available from https://doi.org/10.1016/j.scitotenv.2009.10.032.

Lindstrom, A. B., Strynar, M. J., & Libelo, E. L. (2011). Polyfluorinated compounds: Past, present, and future. *Environmental Science and Technology, 45*(19), 7954–7961. Available from https://doi.org/10.1021/es2011622.

Liu, W., He, W., Wu, J., Qin, N., He, Q., & Xu, F. (2018). Residues, bioaccumulations and biomagnification of perfluoroalkyl acids (PFAAs) in aquatic animals from Lake Chaohu, China. *Environmental Pollution, 240*, 607–614. Available from https://doi.org/10.1016/j.envpol.2018.05.001.

Liu, L., Liu, Y., Gao, B., Ji, R., Li, C., & Wang, S. (2020). Removal of perfluorooctanoic acid (PFOA) and perfluorooctane sulfonate (PFOS) from water by carbonaceous nanomaterials: A review. *Critical Reviews in Environmental Science and Technology, 50*(22), 2379–2414. Available from https://doi.org/10.1080/10643389.2019.1700751.

Loganathan, B. G., Sajwan, K. S., Sinclair, E., Senthil, K. K., & Kannan, K. (2007). Perfluoroalkyl sulfonates and perfluorocarboxylates in two wastewater treatment facilities in Kentucky and Georgia. *Water Reseach, 41*(20), 4611–4620. Available from https://doi.org/10.1016/j.watres.2007.06.045.

Loos, R., Locoro, G., Huber, T., Wollgast, J., Christoph, E. H., de Jager, A., Manfred Gawlik, B., Hanke, G., Umlauf, G., & Zaldívar, J.-M. (2008). Analysis of perfluorooctanoate (PFOA) and other perfluorinated compounds (PFCs) in the River Po watershed in N-Italy. *Chemosphere, 71*(2), 306–313. Available from https://doi.org/10.1016/j.chemosphere.2007.09.022.

Lyu, X., Liu, X., Sun, Y., Gao, B., Ji, R., Wu, J., & Xue, Y. (2020). Importance of surface roughness on perfluorooctanoic acid (PFOA) transport in unsaturated porous media. *Environmental Pollution, 266*. Available from https://doi.org/10.1016/j.envpol.2020.115343.

Lyu, X., Liu, X., Sun, Y., Ji, R., Gao, B., & Wu, J. (2019). Transport and retention of perfluorooctanoic acid (PFOA)in natural soils: Importance of soil organic matter and mineral contents, and solution ionic strength. *Journal of Contaminant Hydrology, 225*. Available from https://doi.org/10.1016/j.jconhyd.2019.03.009.

Lyu, X., Liu, X., Wu, X., Sun, Y., Gao, B., & Wu, J. (2020). Importance of Al/Fe oxyhydroxide coating and ionic strength in perfluorooctanoic acid (PFOA) transport in saturated porous media. *Water Research, 175*. Available from https://doi.org/10.1016/j.watres.2020.115685.

Meng, J., Wang, T., Wang, P., Zhang, Y., Li, Q., Lu, Y., & Giesy, J. P. (2015). Are levels of perfluoroalkyl substances in soil related to urbanization in rapidly developing coastal areas in North China? *Environmental Pollution, 199*, 102–109. Available from https://doi.org/10.1016/j.envpol.2015.01.022.

Milinovic, J., Lacorte, S., Vidal, M., & Rigol, A. (2015). Sorption behaviour of perfluoroalkyl substances in soils. *Science of the Total Environment, 511*, 63–71. Available from https://doi.org/10.1016/j.scitotenv.2014.12.017.

Naile, J. E., Khim, J. S., Wang, T., Chen, C., Luo, W., Kwon, B.-O., Park, J., Koh, C.-H., Jones, P. D., Lu, Y., & Giesy, J. P. (2010). Perfluorinated compounds in water, sediment, soil and biota from estuarine and coastal areas of Korea. *Environmental Pollution, 158*(5), 1237–1244. Available from https://doi.org/10.1016/j.envpol.2010.01.023.

Ololade, I. A., Zhou, Q., & Pan, G. (2016). Influence of oxic/anoxic condition on sorption behavior of PFOS in sediment. *Chemosphere, 150*, 798–803. Available from https://doi.org/10.1016/j.chemosphere.2015.08.068.

Paustenbach, D. J., Panko, J. M., Scott, P. K., & Unice, K. M. (2007). A methodology for estimating human exposure to perfluorooctanoic acid (PFOA): A retrospective exposure assessment of a community (1951-2003). *Journal of Toxicology and Environmental Health-Part A, 70*(1), 28–57. Available from https://doi.org/10.1080/15287390600748815.

Pereira, H. C., Ullberg, M., Dan, B. K., Gustafsson, J. P., & Ahrens, L. (2018). Sorption of perfluoroalkyl substances (PFASs) to an organic soil horizon–Effect of cation composition and pH. *Chemosphere, 207*, 183–191. Available from https://doi.org/10.1016/j.chemosphere.2018.05.012.

Plumlee, M. H., Larabee, J., & Reinhard, M. (2008). Perfluorochemicals in water reuse. *Chemosphere, 72*(10), 1541–1547. Available from https://doi.org/10.1016/j.chemosphere.2008.04.057.

Prevedouros, K., Cousins, I. T., Buck, R. C., & Korzeniowski, S. H. (2006). Sources, fate and transport of perfluorocarboxylates. *Environmental Science and Technology, 40*(1), 32–44. Available from https://doi.org/10.1021/es0512475.

Qian, J., Shen, M., Wang, P., Wang, C., Hou, J., Ao, Y., Liu, J., & Li, K. (2017). Adsorption of perfluorooctane sulfonate on soils: Effects of soil characteristics and phosphate competition. *Chemosphere, 168*, 1383–1388. Available from https://doi.org/10.1016/j.chemosphere.2016.11.114.

Rankin, K., Mabury, S. A., Jenkins, T. M., & Washington, J. W. (2016). A North American and global survey of perfluoroalkyl substances in surface soils: Distribution patterns and mode of occurrence. *Chemosphere, 161*, 333–341. Available from https://doi.org/10.1016/j.chemosphere.2016.06.109.

Rostkowski, P., Yamashita, N., So, I. M. K., Taniyasu, S., Lam, P. K. S., Falandysz, J., Lee, K. T., Kim, S. K., Khim, J. S., Im, S. H., Newsted, J. L., Jones, P. D., Kannan, K., & Giesy, J. P. (2006). Perfluorinated compounds in streams of the Shihwa Industrial Zone and Lake Shihwa, South Korea. *Environmental Toxicology and Chemistry, 25*(9), 2374–2380. Available from https://doi.org/10.1897/05-627r.1.

Sinclair, E., Mayack, D. T., Roblee, K., Yamashita, N., & Kannan, K. (2006). Occurrence of perfluoroalkyl surfactants in water, fish, and birds from New York State. *Archives of Environmental Contamination and Toxicology*, *50*(3), 398–410. Available from https://doi.org/10.1007/s00244-005-1188-z.

Skutlarek, D., Exner, M., & Färber, H. (2006). Perfluorinated surfactants in surface and drinking waters. *Environmental Science and Pollution Research*, *13*(5), 299–307. Available from https://doi.org/10.1065/espr2006.07.326.

So, M. K., Miyake, Y., Yeung, W. Y., Ho, Y. M., Taniyasu, S., Rostkowski, P., Yamashita, N., Zhou, B. S., Shi, X. J., Wang, J. X., Giesy, J. P., Yu, H., & Lam, P. K. S. (2007). Perfluorinated compounds in the Pearl River and Yangtze River of China. *Chemosphere*, *68* (11), 2085–2095. Available from https://doi.org/10.1016/j.chemosphere.2007.02.008.

Sörengård, M., Kleja, D. B., & Ahrens, L. (2019). Stabilization and solidification remediation of soil contaminated with poly- and perfluoroalkyl substances (PFASs). *Journal of Hazardous Materials*, *367*, 639–646. Available from https://doi.org/10.1016/j.jhazmat.2019.01.005.

Steenland, K., Fletcher, T., Stein, C. R., Bartell, S. M., Darrow, L., Lopez-Espinosa, M. J., Barry Ryan, P., & Savitz, D. A. (2020). Review: Evolution of evidence on PFOA and health following the assessments of the C8 Science Panel. *Environment International*, *145*. Available from https://doi.org/10.1016/j.envint.2020.106125.

Tang, C. Y., Fu, Q. S., Robertson, A. P., Criddle, C. S., & Leckie, J. O. (2006). Use of reverse osmosis membranes to remove perfluorooctane sulfonate (PFOS) from semiconductor wastewater. *Environmental Science and Technology*, *40*(23), 7343–7349. Available from https://doi.org/10.1021/es060831q.

Tang, C. Y., Shiang Fu, Q., Gao, D., Criddle, C. S., & Leckie, J. O. (2010). Effect of solution chemistry on the adsorption of perfluorooctane sulfonate onto mineral surfaces. *Water Research*, *44*(8), 2654–2662. Available from https://doi.org/10.1016/j.watres.2010.01.038.

Tombácz, E., & Szekeres, M. (2001). Effects of impurity and solid-phase dissolution on surface charge titration of aluminium oxide. *Progress in Polymer Science*, *117*, 18–26. Available from https://doi.org/10.1007/3-540-45405-5_4.

USEPA. (2018). PFAS national leadership summit materials. https://www.epa.gov/pfas/pfas-national-leadership-summitmaterials.

Wang, F., Shih, K., & Leckie, J. O. (2015). Effect of humic acid on the sorption of perfluorooctane sulfonate (PFOS) and perfluorobutane sulfonate (PFBS) on boehmite. *Chemosphere*, *118*(1), 213–218. Available from https://doi.org/10.1016/j.chemosphere.2014.08.080.

Wang, F., Shih, K., Ma, R., & Li, X. Y. (2015). Influence of cations on the partition behavior of perfluoroheptanoate (PFHpA) and perfluorohexanesulfonate (PFHxS) on wastewater sludge. *Chemosphere*, *131*, 178–183. Available from https://doi.org/10.1016/j.chemosphere.2015.03.024.

Wei, C., Song, X., Wang, Q., & Hu, Z. (2017). Sorption kinetics, isotherms and mechanisms of PFOS on soils with different physicochemical properties. *Ecotoxicology and Environmental Safety*, *142*, 40–50. Available from https://doi.org/10.1016/j.ecoenv.2017.03.040.

Xing, Y., Li, Q., Chen, X., Fu, X., Ji, L., Wang, J., Li, T., & Zhang, Q. (2021). Different transport behaviors and mechanisms of perfluorooctanoate (PFOA) and perfluorooctane sulfonate (PFOS) in saturated porous media. *Journal of Hazardous Materials*, *402*. Available from https://doi.org/10.1016/j.jhazmat.2020.123435.

Zhang, C., Yan, H., Li, F., Hu, X., & Zhou, Q. (2013). Sorption of short- and long-chain perfluoroalkyl surfactants on sewage sludges. *Journal of Hazardous Materials*, *260*, 689–699. Available from https://doi.org/10.1016/j.jhazmat.2013.06.022.

Zhang, R., Yan, W., & Jing, C. (2015). Experimental and molecular dynamic simulation study of perfluorooctane sulfonate adsorption on soil and sediment components. *Journal of Environmental Sciences (China)*, *29*, 131–138. Available from https://doi.org/10.1016/j.jes.2014.11.001.

Zhao, L., Bian, J., Zhang, Y., Zhu, L., & Liu, Z. (2014). Comparison of the sorption behaviors and mechanisms of perfluorosulfonates and perfluorocarboxylic acids on three kinds of clay minerals. *Chemosphere*, *114*, 51–58. Available from https://doi.org/10.1016/j.chemosphere.2014.03.098.

Zhou, Q., Deng, S., Zhang, Q., Fan, Q., Huang, J., & Yu, G. (2010). Sorption of perfluorooctane sulfonate and perfluorooctanoate on activated sludge. *Chemosphere*, *81*(4), 453–458. Available from https://doi.org/10.1016/j.chemosphere.2010.08.009.

Zhou, Z., Liang, Y., Shi, Y., Xu, L., & Cai, Y. (2013). Occurrence and transport of perfluoroalkyl acids (PFAAs), including short-chain PFAAs in Tangxun Lake, China. *Environmental Science and Technology*, *47*(16), 9249–9257. Available from https://doi.org/10.1021/es402120y.

Chapter 6

Occurrences and impacts of engineered nanoparticles in soils and groundwater

Yulin Zheng[1], Xian Chen[2], Nedgine D. Joseph[3], Yue Zhang[1], Hao Chen[4] and Bin Gao[1]

[1]Department of Agricultural and Biological Engineering, University of Florida, Gainesville, FL, United States, [2]School of Chemistry and Environmental Engineering, Jiangsu University of Technology, Changzhou, P.R. China, [3]Department of Chemical Engineering, University of Florida, Gainesville, FL, United States, [4]Department of Agriculture, University of Arkansas at Pine Bluff, Pine Bluff, AR, United States

Chapter Outline

6.1 Overview

Engineered nanoparticles (ENPs) are nanomaterials conceived, designed, and intentionally produced by human with high homogeneity (Hochella et al., 2019). Owning to their unique properties, ENPs have been extensively researched and applied in a variety of areas such as electronic, biomedical, pharmaceutical, cosmetic, and functional material applications to improve human living conditions. ENPs have also been applied in several green

Emerging Contaminants in Soil and Groundwater Systems. DOI: https://doi.org/10.1016/B978-0-12-824088-5.00005-7

technologies including environmental remediation, renewable energy, and agricultural nutrition to benefit the natural environment and ecosystems. One estimated annual global production rate of ENPs reached 3.6 million metric tons in 2020, compared with 0.9 million metric tons in 2010 (Janković & Plata, 2019). The global nanotechnology market is expected to exceed $125 billion USD in the year of 2024 (Bernd et al., 2018). Rapid production bloom and massive commercial applications of ENPs have inevitably caused their release into the environment as an emerging contaminant and raised many concerns over their potential risks. Much research effort has been devoted to exploring the environment occurrence of ENPs and their potential impacts on the ecosystems, particularly with respect to soil and groundwater systems. This chapter thus mainly focuses on the environmental occurrences, sources, and impacts of ENPs in soils and groundwater.

6.2 Classification of nanoparticles

6.2.1 Definition

Nanometer (nm) is a length unit and 1 nm equals 10^{-9} m. Nanomaterials (NMs) or nanoparticles (NPs) are commonly referred as materials with at least one dimension lying in the range of 1–100 nm (USEPA, 2007). Although there is no official definition of NPs due to different scientific and policy views, small size, high-specific surface area, and enhanced reactivity over bulk materials are universally accepted as their main characteristics. The fundamental properties of NPs, such as size, shape, chemical composition, and atomic structure in nanoscale range, make them distinct from bulk materials at macroscale level. Fig. 6.1 shows the classification (A) and size range (B) of NPs.

6.2.2 Natural nanoparticles

In the last 30 years, many studies reported that a large fraction of natural nanoparticles (NNPs) exist at the nanoscale range in the environment. NNPs

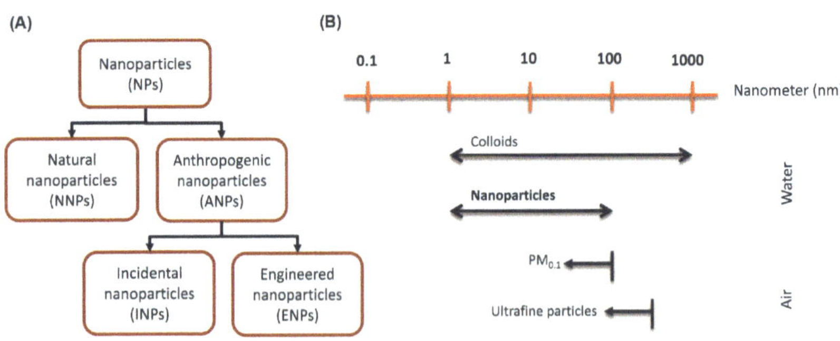

FIGURE 6.1 The classification (A) and size range (B) of NPs.

naturally occur within the global biogeochemical cycle or mechanical processes of earth and can be found almost everywhere in the natural environmental media (Hochella et al., 2019). In fact, NPs have been presented ubiquitously since the earth's origin at an estimated order of thousands of terra-grams (Tg, 10^{12} g). The natural weathering and mineral formation processes in soils are the main force driving NNP production on earth. In addition, the mechanical grinding at faults during earthquakes can be the origin of some NNPs in the environment (Bakshi et al., 2015). Natural clays, produced from those processes, are by far the most abundant inorganic NNPs, accounting for the majority of the terrestrial 10^7 to 10^8 Tg NNPs reservoir (Hochella et al., 2019). Groundwater, rivers, glaciers, wind, and ocean currents are the primary movers of NNPs. NNPs are abundant in groundwater from both geochemical and biotic processes with typical concentration in the lower mg/L level (Baumann, 2010). Airborne NNPs were found with concentration in range from 10^6 to 10^8 NPs per liter of air depending on the air conditions (Bakshi et al., 2015). Mineral dust storm and volcanoes emission are the two largest drivers of airborne NNPs, estimated together at 342 Tg/year (Hochella et al., 2019). Rivers, and to a much lesser extent glaciers, deliver approximately 10^3 to 10^4 Tg of NNPs to soils and sediments each year (Hochella et al., 2019).

6.2.3 Engineered nanoparticles

Incidental nanoparticles (INPs) and ENPs are produced by indirect and direct human influences and activities, and thus are often called anthropogenic nanoparticles (ANPs, Fig. 6.1A). INPs have been produced with the development of humanity but unintentionally. For example, they can be created mechanically by striking two rock together in the Stone Age. Natural or anthropogenic combustion processes generate nanosized soot and carbon black, which are INP by-products from incomplete combustion and recondensation of fossil fuel consumption and biomass burning (Nowack & Bucheli, 2007). From wood fire BBQ to industrial power plant emission, carbon black emitted into the atmosphere as an INP is considered as a critical global air pollution source (Fawole et al., 2016).

In contrast to INPs, ENPs are conceived, designed, and intentionally produced by human with high homogeneity (Hochella et al., 2019). With tunable properties, ENPs have been extensively researched and applied in a variety of areas such as electronic, biomedical, pharmaceutical, cosmetic, and functional materials to improve human life. In addition, ENPs have been used in many green technologies such as environmental remediation, renewable energy, and sustainable agricultural nutrition. The estimated annual global production rate of ENPs was 3.6 million metric tons in 2020, compared with 0.9 million metric tons in 2010 (Janković & Plata, 2019). The global nanotechnology market is expected to exceed $125 billion USD in the year of 2024 (Bernd et al., 2018). Scientists estimate that there are thousands of terra-grams (Tg, 10^{12} g) of NNPs, 1–10 Tg of

INPs, and less than 1 Tg of ENPs moving annually among different compartments of earth by different natural and anthropogenic forces. Despite their small proportion of the total amount of existing NPs, the quantity of ENPs in the environment continues to increase with the prosperity of nanotechnology since the beginning of 21st century.

6.2.3.1 Classification of ENPs

The design and synthesis of novel ENPs and NMs with various properties have contributed to development of many advanced technologies in the past decades. In general, ENPs can be classified into different groups based on their dimensionality, morphology, state, and chemical composition, as shown in Fig. 6.2. Nanoscale (<100 nm) is the critical criteria for the dimensionality of ENPs. Zero-dimensional (0D) ENPs have all three dimensions less than 100 nm and can be of irregular or regular (e.g., cube, sphere, hollow sphere, etc.) shape. One-dimensional (1D) ENPs have two dimensions in nanoscale, including nanorods, nanowires, nanofibers, carbon nanotubes (CNTs), etc. Two-dimensional (2D) ENPs contain only one nanoscale dimension and mainly include single- and multilayered sheets such as graphene and graphene oxide (GO), thin films, nanoplates, and nanocoating. Materials with nanocomposition, nanostructures, or nano-pore-networks are three-dimensional (3D) ENMs, which have outside dimensions beyond 100 nm (Saleh, 2020). Here, only 0–2D ENPs are considered as emerging contaminants in soil and groundwater systems and discussed in detail in this and the following chapters.

Flatness, sphericity, and aspect ratio are important factors controlling the morphology of ENPs. The high and low aspect ratios of ENPs usually depend on their shapes, which can be characterized by high-resolution microscopic visualization techniques such as scanning electron microscopy (SEM) and transmission electron microscopy (TEM). In term of state and uniformity, ENPs can be classified as isometric, inhomogeneous, dispersed, suspended, and aggregated. The aggregation behaviors of ENPs depend on their surface electromagnetic properties.

Chemical composition is also one of the most important criteria to characterize and classify ENPs, especially in the content of their environmental occurrences and impacts. In this chapter, ENPs specifically refer to intentionally produced nanoscale particles that are homogeneous in size, shape, and structure. Fig. 6.3 presents the semiquantitative data of current and developing ENPs and nanotechnologies (A) and the global production volumes of abundant ENPs (B). ENPs are diverse in production level, amount of application, technology readiness level (applied research & development, demonstration, and commercial). According to previous studies, the global production and abundance of ENPs, listed in order of approximate mass (metric ton) from the highest to the lowest, are SiO_2, TiO_2, nano-clay composites, ZnO, iron (Fe)-based ENPs, ZrO_2, Al_2O_3, CeO_2, cellulose, CNTs and graphene (and derivatives), Cu, Ag, quantum

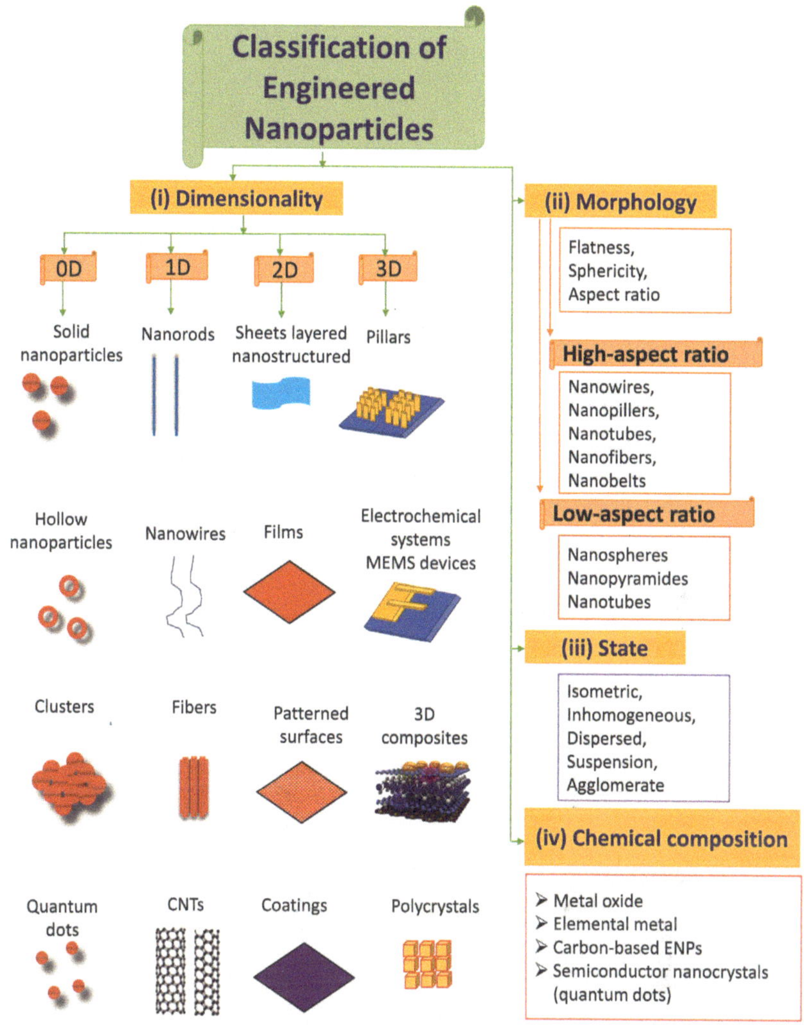

FIGURE 6.2 Schematic illustration of the classification of ENPs based on different criteria. *Modified from Saleh, T. (2020). Nanomaterials: Classification, properties, and environmental toxicities.* Environmental Technology and Innovation, 20. *https://doi.org/10.1016/j.eti.2020.101067.*

dots (QDs), and other nanocomposites (Bernd et al., 2018; Janković & Plata, 2019). ENPs have market demands from various sources and are applied in many commercial products (Fig. 6.4). Up to 0.3 Tg of these ENPs are released into landfills, soil, surface water, groundwater, and air annually (Bernd et al., 2018). These globally abundant ENPs can be classified based on their chemical compositions into four types: (1) metal oxide ENPs such as TiO_2, SiO_2, ZnO, Al_2O_3, CeO_2, and other metal (multimetallic) oxides; (2) metal ENPs such as

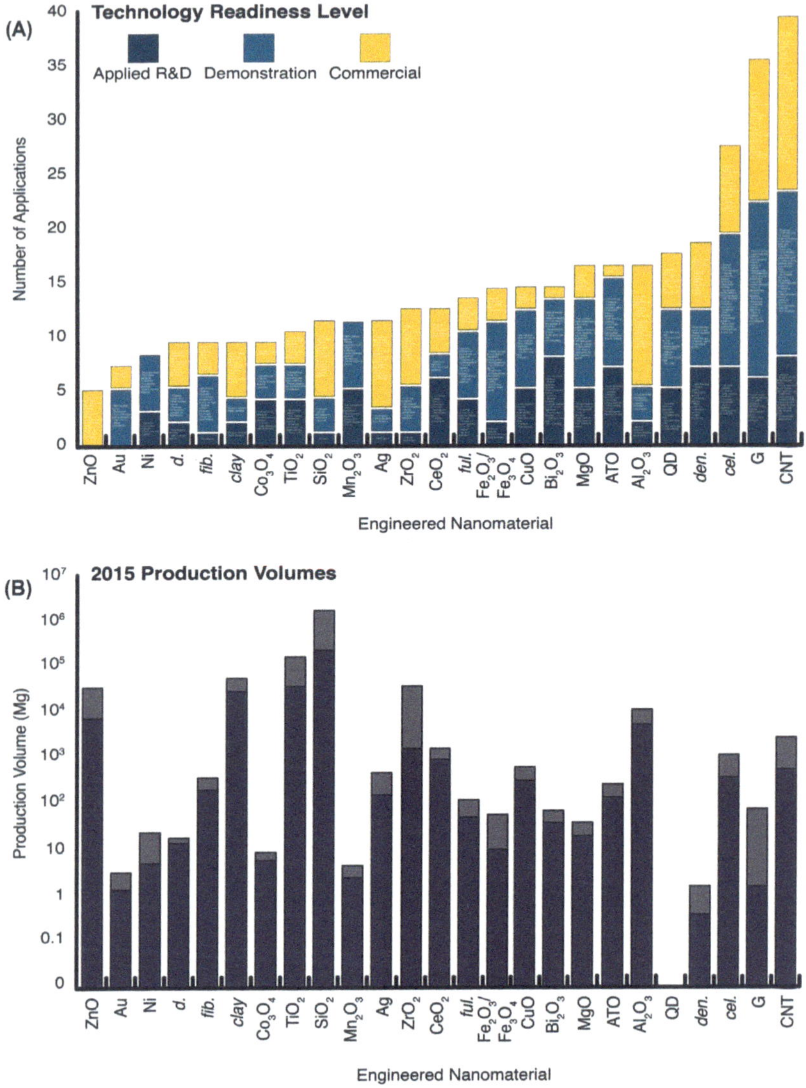

FIGURE 6.3 (A) Presents the semiquantitative data of current and developing ENPs and nanotechnologies and (B) the global production volumes of abundant ENPs. *Modified from Janković, N., & Plata, D. (2019). Engineered nanomaterials in the context of global element cycles.* Environmental Science: Nano, *6(9), 2697–2711. https://doi.org/10.1039/c9en00322c.*

zero-valent Fe, Cu, Ag, and Au (gold); (C) carbon ENPs such as fullerene, CNTs, graphene, and its derivatives [e.g., GO and reduced graphene oxide (rGO)], carbon dots, and carbon nanofiber; and (4) semiconductor nanocrystals (QDs) such as zinc sulfide (ZnS) and cadmium sulfide (CdS).

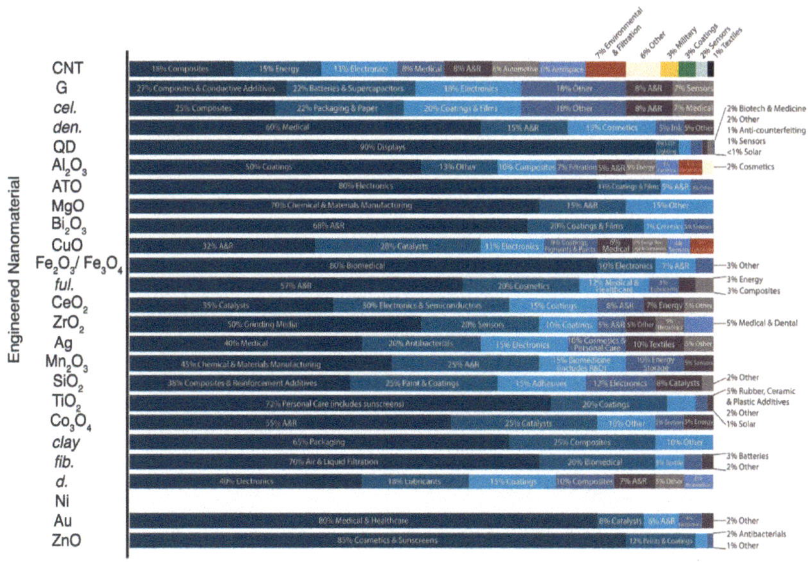

FIGURE 6.4 ENPs market demand by application category. *A and R*, Academia and research. *Modified from Janković, N., & Plata, D. (2019). Engineered nanomaterials in the context of global element cycles.* Environmental Science: Nano, *6(9), 2697–2711. https://doi.org/10.1039/c9en00322c.*

6.2.3.2 Metal oxide ENPs

Metal oxide ENPs are the most abundant ones in global production and massively used in material, chemical, electronic, and personal care industries. Bulk forms of metal oxides have been extensively produced for many years. With the boom of the nanotechnology, nano forms of those metal oxides have been also massively produced and applied in commercial products.

Silica (SiO_2) ENPs: Silicon dioxide also known as silica. In many parts of the world, SiO_2 is commonly found in the nature as quartz, which is the major constituent of sand. Nano-SiO_2 is the most produced ENP by mass, reaching around 3.15 million metric ton in 2020 with a 9%–10% projected growth rate. It is also evenly distributed in various of applications, including composites and reinforcements (38%), paints and coatings (25%), adhesives (15%), electronics (12%), and catalysts (8%) (Janković & Plata, 2019). Owing to the abundance and unique properties, nano-SiO_2 is one of the most promising ENPs and functionalized building blocks for both commercial and research uses (Jarvie et al., 2009).

Titanium dioxide (TiO_2) ENPs: Titanium dioxide also known as titania is one of the excellent band-gap semiconductors and has a large energy gap of 3.2 eV. As a nano-photocatalyst, TiO_2 ENPs can be photoactivated by near UV light. They are mainly used in personal care products (PCPs; such as sunscreens) and commercial coatings, accounting 72% and 20% of the total usage, respectively. TiO_2 ENPs are excellent additives to these products and coatings because they not only provide effective UV radiation filtering but also have the benefit of being invisible. Other applications of TiO_2 ENPs include rubber, ceramic, and plastic additives (5%), solar energy (1%), and others (2%). TiO_2 ENPs have an estimated global production of 40,000–245,000 metric tons in 2020 with anticipated growth rate of 4%–11% (Janković & Plata, 2019). Although TiO_2 ENPs present large specific surface area benefitting from small particle size, their commercial applications have faced the challenge with their tendency to aggregate very easily forming larger particles, which is undesirable for their catalytic efficiency (Weir et al., 2012).

Zinc oxide (ZnO) ENPs: ZnO is a direct band-gap semiconductor with band-gap energy of 3.36 eV at room temperature, high exciton binding energy of 60 meV, and high dielectric constant (Singh et al., 2007). The remarkable properties of ZnO NPs have attracted the many research interests on ultraviolet light-emitting devices, chemical sensors, and solar cells in the past few years (Picca et al., 2015). Similar to nano-TiO_2, ZnO ENPs are mainly used in cosmetics (85%), followed by coatings (12%). In fact, paints, coatings, and cosmetics are the major consumers of metal oxide ENPs. These applications have high anticipated or inherent release rates and/or distinct exposure routes to the environment.

6.2.3.3 Metal ENPs

Silver (Ag) and gold (Au) ENPs are typical elemental state metal ENPs, which have substantial research and applications because of their promising potential in the fields of medical and health care, cosmetics, electronics, and catalysis. The unique surface and optical properties of Au NPs have widely increased their uses as a diagnosis tool for disease, drug delivery, and biosensor applications (Sengani et al., 2017). Ag NPs are used for a wide range of commercial applications to restrict microbial growth in PCPs, textile, etc. A detailed case study on the environmental occurrences and impacts of Ag ENPs is presented at the end of this chapter. Iron ENPs, better known as nano zero-valent iron (nZVI), are commonly used as a reducing agent and highly effective sorbent for various inorganic and organic contaminants in the environment, which has been widely applied in many environmentally friendly techniques, especially soil and groundwater remediation (Wang et al., 2019; Wang, Gao, et al., 2016).

6.2.3.4 Carbon ENPs

Fullerenes, CNTs, and graphene are the main classes of carbon ENPs, which are presented in Fig. 6.5. The fullerenes may be C60 or C70 with carbon atoms in the form of pentagonal and hexagonal shape by sp^2 hybridization. CNTs are tubular elongated structure having 1–2 nm diameter (Liné et al., 2017). The structure of CNTs resembles with graphene sheets rolling upon themselves, which can be categorized into single walled (SWNTs), double walled (DWNTs), and multiwalled

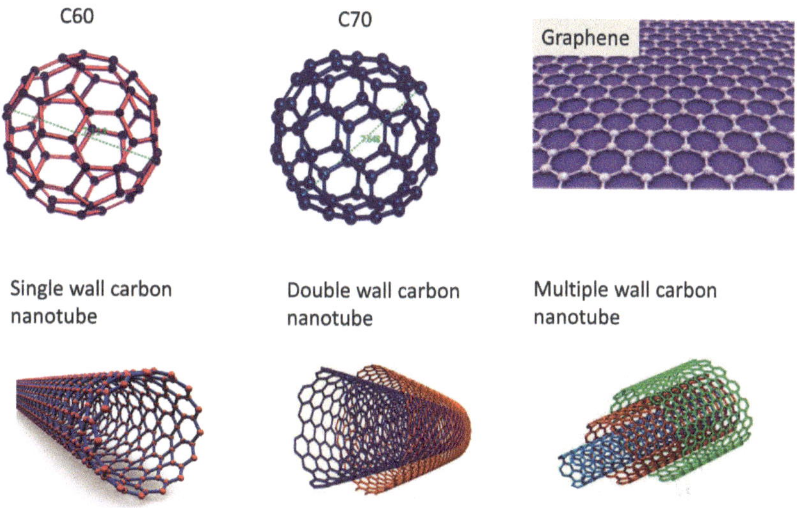

FIGURE 6.5 Different types of carbon ENPs including fullerenes (C60 and C70), graphene and its derivatives, and carbon nanotubes (CNTs). *Modified from Ijaz, I., Gilani, E., Nazir, A., & Bukhari, A. (2020). Detail review on chemical, physical and green synthesis, classification, characterizations and applications of nanoparticles.* Green Chemistry Letters and Reviews, *13(3), 59–81. https://doi.org/10.1080/17518253.2020.1802517.*

(MWNTs) (Liné et al., 2017). Graphene is an allotropic form of carbon. It is a hexagonal network of honeycomb carbon atom lattice having 2D planar surface with 1 nm in thickness (Ijaz et al., 2020; Yang et al., 2019). GO is similarly composed of a single sheet of graphenic carbon that contains areas of disrupted aromaticity where carbon atoms are oxidized. rGO is GO in a form of reduced oxygen functional groups and a greater proportion of graphenic carbon. The discovery and invention of carbon ENPs won several Noble Prizes and have fueled the development of nanotechnology in the 21st century. Nowadays, the applications of carbon ENPs have been extended to a wide range of commercial products and cutting-edge technologies, including energy, electronics, biomedicine, automotive, aerospace, environmental remediation, and others (Janković & Plata, 2019).

6.2.3.5 Quantum dots

Semiconductor QDs are size-dependent tunable fluorescent nanocrystals that have unique optical and electrical properties (Li et al., 2015). QDs commonly consist of a metalloid crystalline core (e.g., cadmium selenide) surrounded by a shell (e.g., zinc sulfide). They are increasingly being used for optical displays and imaging with exceptional advantages in fluorescent and brightness. QDs have a synthetic history of approximately 20 years, and are predicted with rapid increase on production volume, 14–50 metric tons per year in 2020 (Janković & Plata, 2019).

6.3 Sources, occurrences, and pathways of ENPs

6.3.1 Pathways

The general pathways of NPs in the environment are presented in Fig. 6.6. NNPs include naturally occurred NPs from volcanic particle emissions, dust storm, forest fires, etc. Both INPs and ENPs are considered as ANPs, which can enter the environment through different pathways. INPs may come from nonpoint sources and often incidentally produced and released with other contaminants. ENPs in the environment can be traced from the life-cycle orientated release, where 98% – 99.9% of ENPs produced are expected to flow into their designed applications (Keller & Lazareva, 2013), and 0.1%–2% of the ENPs are estimated to be released. ENP release is closely related with the life cycle of their containing products. Most of them end up at production facilities, the uses of the product, and disposal facilities, such as landfills and wastewater treatment plants (WWTPs). Accidental release of ENPs is possible, such as emission during production or transport (Nowack & Bucheli, 2007). Intentional injection of ENPs into the environment is also possible, where iron oxide NPs and nZVI are applied to the natural environment as agricultural fertilization and groundwater remediation (Baumann, 2010). Accidental release, intended release, and waste disposal are critical pathways

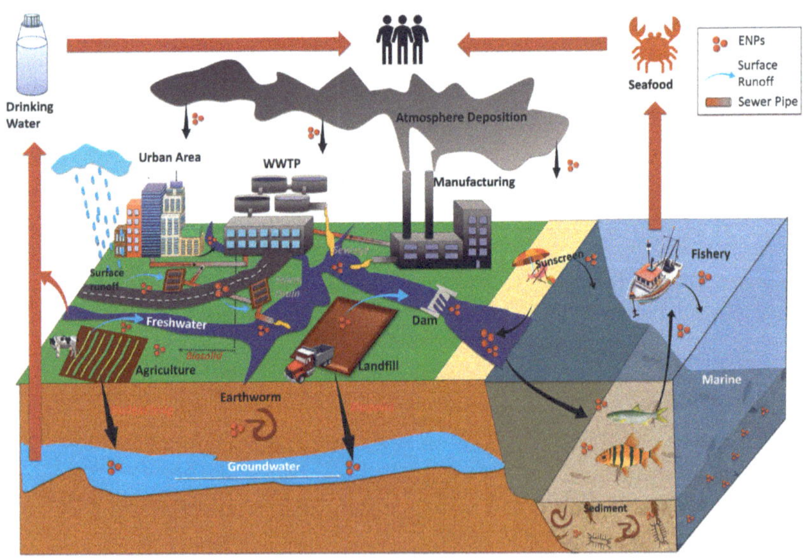

FIGURE 6.6 Pathways of ENPs entering the environment. *Modified from Wu, P., Huang, J., Zheng, Y., Yang, Y., Zhang, Y., He, F., Chen, H., Quan, G., Yan, J., Li, T., & Gao, B. (2019). Environmental occurrences, fate, and impacts of microplastics.* Ecotoxicology and Environmental Safety, 184. *https://doi.org/10.1016/j.ecoenv.2019.109612.*

of ENPs into air, soil, groundwater, and surface water to affect the ecosystems at varying degrees.

6.3.2 Environmental concentrations of ENPs

The rise of ENP applications has brought considerable concerns from scientists and the public over their environmental occurrences and potential risks. The current levels of ENPs in the environment are critical. However, quantification techniques of ENPs in environmental samples are still limited, especially in solid samples, such as soils, sediments, and biological tissues. ENPs in the environment are almost undetectable due to their extremely low concentrations, comparing with background NNPs. Even if they are detectable, it is still difficult to differentiate ENPs from NNPs (Sun et al., 2016). Currently, most of the information on the concentrations of ENPs in the environment is from model simulations. Table 6.1 presents the predicted environment concentrations of selected ENPs in water and soil.

Different modeling methods have been applied to simulate ENP release and their concentrations in the environment (Gottschalk et al., 2013, 2015). The conceptualization, parametrization, and validation of the models are crucial and should cover the whole life cycle of ENPs and their containing products. Material flow analysis methods have commonly applied in previous studies mainly with two approaches: top-down and bottom-up (Gottschalk et al., 2013). The top-down method starts with the market fraction of ENP containing products and their predicted usages and possible release pathways to estimate environmental concentrations of ENPs. In contrast, the bottom-up method includes calculations of the whole picture of commercial products

TABLE 6.1 Environmental concentrations of selected ENPs.

ENP	Surface water (ng/L)	Soil (μg/kg)	Air (ng/m^3)	References
Ag	0.015	0.074	0.007	Gottschalk et al. (2015)
CeO_2	0.004	0.546	0.1	Gottschalk et al. (2015)
SiO_2	5.34	62.82	13.92	Giese et al. (2018)
TiO_2	3.27	5.50	0.8	Gottschalk et al. (2015)
ZnO	0.45	0.372	0.04	Gottschalk et al. (2015)
CNT	0.001	0.248	0.042	Gottschalk et al. (2015)

Modified from Gottschalk, F., Lassen, C., Kjoelholt, J., Christensen, F., & Nowack, B. (2015). Modeling flows and concentrations of nine engineered nanomaterials in the Danish environment. *International Journal of Environmental Research and Public Health, 12*(5), 5581–5602. https://doi.org/10.3390/ijerph120505581.

and possible applications. Next, all products were analyzed based on the release of ENPs during the manufacturing processes, the use of the products, and the recycling and disposal processes. In both methods, ENP release is considered from their production to disposal including their possible usages and applications. Both point source (e.g., release from production plants and waste disposal sites) and nonpoint sources that unintentional release from products (e.g., PCPs and sunscreen) are included in the analysis.

Contaminant fate modeling is widely applied to predict the fate, transport, and distribution of ENPs upon their release into the environment (Praetorius et al., 2012). The fate and transport models are necessary to predict where the ENPs end up in the environmental compartments, using existing knowledge of particle and colloid science, such as classical colloid filtration model and Derjaguin–Landau–Verwey–Overbeek (DLVO) theory (Tian et al., 2010). However, due to the natural complexity of ENPs (aggregation, sedimentation/deposition, dissolution, and transformation), the fate and transport models of ENPs are still in limited uses to predict their environmental concentration.

Keller and Lazareva first estimated the production and emission of ENPs at the global scale based on probabilistic material flow analysis (Keller & Lazareva, 2013). Fig. 6.7A presents the global quantities, applications, and the end-life disposal of 10 major ENPs that exceeded the production of 100 metric ton per year in 2010. Based on the major applications of ENPs, such as automobiles, electronics, and solar energy, 60% – 86% of ENP containing products are expected to end up in landfills. Some commercial products, such as PCPs (including sunscreens and cosmetics), coatings, paints, and pigments, may end up in wastewater and be discharged into air, water, and soil systems. WWTP, waste incineration plant (WIP), and landfill are the major waste disposal facilities but may also be sources of ENPs. Improper management of biosolids generated from WWTP, ashes from WIP, and leachates of landfill can deliver ENPs into the environment. The global distribution of estimated environmental release of selected ENPs is also displayed on the world map in Fig. 6.7B. Asia shows about 50% of the total ENP release owning to its 52% world population and large production of ENP containing products. Europe is in the second place in term of ENP release, followed by North America. Differences in population density, technology level, and waste management strategy may cause the large regional variations in the release of ENPs into air, water, and soil. The following sections mainly focus on the sources and occurrences of ENPs in soils and groundwater.

6.3.3 ENPs in soils and groundwater

6.3.3.1 *Air deposition*

In the natural environment, one of the major sources of NPs in soils is from the direct air deposition. Airborne NPs may deposit directly or be carried by

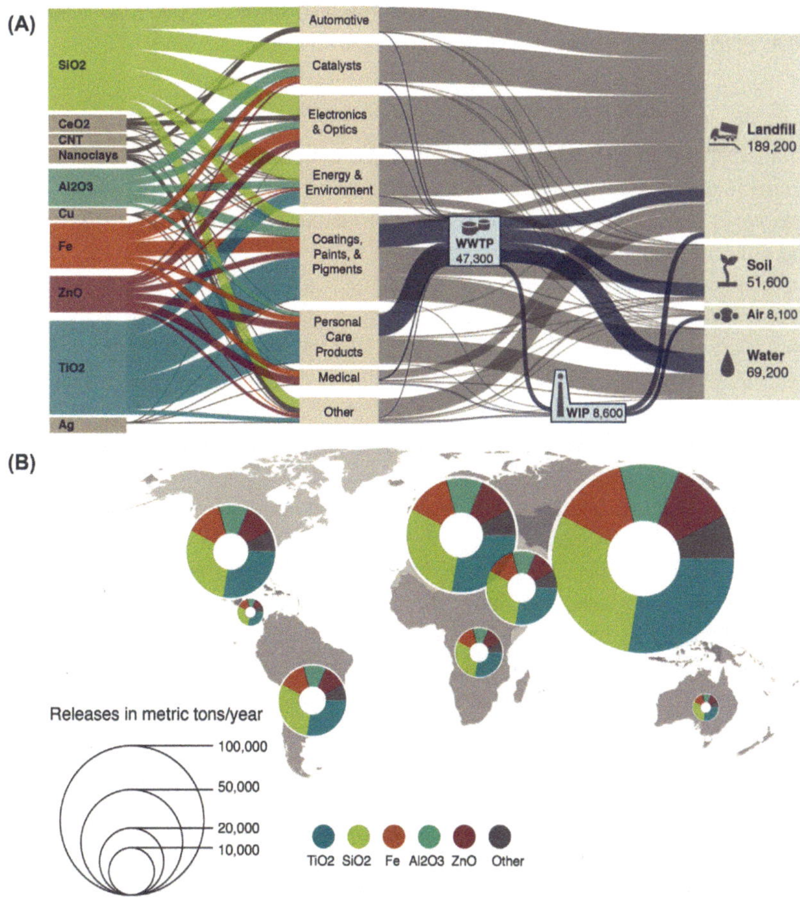

FIGURE 6.7 (A) Estimated global mass flow of ENMs (in metric tons per year) in quantities, applications, and the end-life disposal, considering high production and release estimates as of 2010 and (B) global distribution of estimated ENP release to the environment. *Modified from Keller, A. A., & Lazareva, A. (2013). Predicted releases of engineered nanomaterials: From global to regional to local.* Environmental Science & Technology Letters, *65–70. https://doi.org/10.1021/ez400106t.*

precipitation onto soil surface. Nevertheless, the deposition of ENPs from air to soil is not as much as that of NNPs. ENPs may be emitted into the ambient air from the manufacture, handling, and applications of ENPs. CeO_2 ENPs as a diesel additive are emitted into air during combustion. In fact, vehicle emissions are one the major sources of the ENPs in air, most of which are deposited to soil. With different traffic intensities, locations, and distances, soils near the highways may accumulate ENPs deposited from air (Park et al., 2008). Air monitoring can present strong visual and supportive evidences on the emissions of ENPs during their manufacturing and use (Methner et al., 2012). Because of their small sizes, ENPs can escape from filters from both point and nonpoint sources to the air and eventually deposit to soil.

6.3.3.2 Landfill

Municipal solid waste (MSW) landfill is the major sink for many commercial products that contain ENPs and is identified to be one of the predominant compartments of ENPs by the material flow modeling (Dulger et al., 2016). Since the concern over ENP release is still in its early stage, there are no regulations on monitoring and controlling ENPs in landfill. Many studies have conducted laboratory scale leaching tests of ENPs such as TiO_2 and ZnO from MSW samples (Part et al., 2018, 2020). About 3%–19% of nano TiO_2 and 7%–20% of nano ZnO were leached out from the solid matrix and presented in landfill leachate under conditions of different pH, ionic strength, and natural organic matter (NOM) concentrations (Dulger et al., 2016; Sakallioglu et al., 2016). These facts suggest that MSW landfill can release ENPs into soil and groundwater systems. Based on the predictions of several material flow models, up to 190,000 metric tons of the 10 major ENPs will ultimately end up in landfills at the global scale. Some investigations have confirmed the occurrences and abundance of NPs in landfill leachate and soils; however, it should be noted that these detected NPs may come from natural and anthropogenic sources (Mitrano et al., 2016, 2017). Mitrano et al. observed the existence of ENPs, especially metal-based ENPs, in landfill leachate of Swiss (Mehrabi et al., 2017; Mitrano et al., 2016; Part et al., 2018). Leaching of ENPs from landfills to soils is caused by weathering and degradation of ENP-containing products through both aerobic and anaerobic processes.

The fate, mobility, and transformation of ENPs in a landfill depend on many factors, such as how ENPs are embedded within the waste products, ENP properties, and landfill characteristics based on laboratory investigations. To date, no in situ studies have been published aiming at the release and transformation of ENPs in landfills (Part et al., 2018). The complexity of landfill matrix makes their characteristics change significantly over time, where pH, ionic strength, organics, and temperature may affect the leaching of ENPs into the soil below. Semiconductor QDs used in diverse optoelectronics have been investigated by scientists on their leaching in Austrian landfill sites (Part et al., 2020). CdSe and CdTe QDs with different stabilization techniques may undergo sedimentation due to the loss of surface surfactants and the degradation of shells in landfill leachate (Part et al., 2018). The changed QDs with short linkers may be highly mobile and thus should be closely monitored in landfill soils.

6.3.3.3 Biosolid and compost

WWTP and WIP are two important technological compartments for the disposal of ENP-containing products (Part et al., 2018). Moreover, compost process is often used to treat organic enriched waste (food waste) and convert them into beneficial products for gardening and organic farming. The final processed products from WWTP, WIP, and composting are called sludge/biosolid, ash, and compost, respectively. Once ENP containing products are processed, large fraction of ENPs can be released and eventually end up in the sludge residues, biosolids, ash, and compost. In some countries, compost and biosolids are land applied. About 63% and 55% of biosolids are applied to soils as fertilizer, landfill cover, or agricultural amendment in the United States and EU, respectively (Li et al., 2015). Long-term land application of biosolids has raised much concern over the potential accumulation and ecotoxicity of ENPs in the amended soils. Land application of biosolids is currently subject to several toxic metal concentration limits according to the United States EPA Part 503 Biosolids Rule; however, little information exists on unregulated metals and ENPs (Yang et al., 2019). Gottschalk et al. evaluated the flows and concentrations of ENPs in the Danish environment (Gottschalk et al., 2015). Their results (Table 6.2) show high concentrations of ENPs in disposal facilities such as WWTP and WIP. Moreover, ENPs in soil have higher levels than in other environmental compartments because sludge or biosolid treated soils are the primary sinks for ENPs (Gottschalk et al., 2015).

Stamou and Antizar-Ladislao (2016) conducted life cycle modeling and leaching test on compost containing commercial Ag and TiO_2 ENPs and found that 5% and 15% of these ENPs could be released from the compost. These studies suggest that ENPs from food, packaging, and PCPs would be accumulated on top soil layers if compost is land applied. Yang et al. (2014) inspected multiple United States EPA biosolid land applied soil sites with long-term applications of 6.6–74 dry tons biosolid per hectare per year over 5–25 years. They found the accumulation of Ti (1428 mg/kg), Ag (16 mg/kg), Ce (45 mg/kg), and Zn (634 mg/kg) were significantly higher than that of the control soil samples. It should be noted that most of the Ag ENPs were found in surface soil and the measured concentration decreased with soil depth. In contrast, Ti and Ce had similar distributions in soils from 0 to 100 cm depth. Ti ENP, approximately 50 nm in diameter, was also identified by TEM in the biosolid amended soil (Kim et al., 2012). These results indicate that metal ENPs may behave differently in soils depending on their properties and the environmental conditions.

Although landfilling is still a common practice for waste disposal, thermal waste treatment or incineration is becoming an alternative for both MSW and biosolids. For example, China is estimated to expand its WIP capacity from 3% in 2011 to 30% by 2020 (Walser et al., 2012). Some countries, like Japan and Singapore, have use WIP as their primary waste disposal practice due to limited land resources (Walser et al., 2012). During the thermal treatment in WIP, ash is generated in the combustion chamber that consists bottom ash (residues in the chamber) and fly ash (catch by the air

TABLE 6.2 Concentrations of ENPs in natural and technical compartments of Danish environment.

Compartment	Unit	Photostable TiO_2		Photocatalytic TiO_2		ZnO		Ag		CNT		CeO_2	
		Mode	Range	Mode	Range	Mode	Range	Mode	Range	Mode	Range	Mode	Range
Sewage treatment effluent	ng/L	13,000	3400–92,000	1.6	0.4–14	NA	NA	0.5	0.012–59	0.3	0.1–3.5	9.3	1.1–60
Sewage treatment sludge	μg/kg	770,000	69,000–1,500,000	85,000	9300–230,000	NA	NA	82	4.2–250	7.6	2.7–62	350	44–2300
Waste mass incinerated	μg/kg	15,000	1400–32,000	2800	300–6800	300	40–1500	15sd	10–23	800	440–1300	180	21–930
Bottom ash	μg/kg	33,000	340–88,000	6000	700–18,000	700	100–3900	35	21–66	76	27–710	360	50–2500
Fly ash	μg/kg	170,000	17,000–430,000	30,000	3300–90,000	3600	500–19,000	170	100–330	330	88–4800	2200	240–12,000
Surface water (fresh water)	ng/L	3	0.6–100	0.27	0.05–7	0.45	0.09–13	0.015	0–0.044	0.001	0.0002–0.015	0.004	0.0006–0.1
Sea water	ng/L	0.30	0.04–1	0.02	0.004–0.099	0.04	0.006–0.4	Below pg/L	Below pg/L	Below pg/L	Below pg/L	Below pg/L	Below pg/L
Agricultural soil	ng/kg	85	10–390	700	100–1700	52	8–350	10	6–21	35	18–75	76	10–530
Natural soil	ng/kg	180	24–1100	1500	200–4900	120	18–900	24	13–61	83	41–220	170	24–1500

Urban soil	ng/kg	330	39–1500	2700	300–6700	200	30–1300	40	23–81	130	71–290	300	39–2100
Sludge treated soil	ng/kg	1,300,000	130,000–3,100,000	170,000	17,000–480,000	NA	NA	170	20–530	60	30–180	1500	94–5100
Air	ng/m^3	0.10	0.01–0.5	0.70	0.08–2	0.04	0.005–0.2	0.007	0.004–0.011	0.042	0.022–0.091	0.1	0.01–0.6

Mode value and 95% interval are shown. *NA*, Not available.

Modified from Gottschalk, F., Lassen, C., Kjoelholt, J., Christensen, F., & Nowack, B. (2015). Modeling flows and concentrations of nine engineered nanomaterials in the Danish environment. *International Journal of Environmental Research and Public Health, 12*(5), 5581–5602. https://doi.org/10.3390/ijerph120505581.

filtration). Massari et al. (2014) studied the behavior of Ag, TiO_2, and CeO_2 ENPs during a lab-scale incineration of solid paint waste and found that ENPs were mainly in fly ash and on the surface of bottom ash. Less than 1 mg/kg leaching of Ti ENPs was also detected in the bottom ash. Waste incineration process may cause chemical transformation of ENPs to alter their environmental impacts. Previous studies have shown that incineration process can influence the transformations of Ag NPs in biosolid to strongly affect soil microbial diversity if the ash is land applied (Forstner et al., 2020; Impellitteri et al., 2013). Superparamagnetic NPs, anthropogenically generated from the high-temperature incineration of iron contained waste, can pose risks once they are released into soils (Funari et al., 2018).

The groundwater close to waste disposal facilities may be contaminated by the release of ENPs (Baumann, 2010). The concerns of ENPs released from landfills into groundwater are rising. Lee et al. demonstrated the bentonite clay liner can block the release of Ag ENPs from landfill (Lee et al., 2021). Saheli et al. (2017) evaluated the diffusion potential of MWCNTs through a simulated high-density polyethylene (HDPE) geomembrane and found that their permeation was relatively low with an estimated diffusion coefficient less than 5.1×10^{-15} m^2/s. However, there are many open landfills and dumps worldwide without liners or other protections. In these landfills, ENPs can easily enter the soil and groundwater systems. In addition, activities of ENP manufacturing facilities, stormwater runoff, and irrigation may also become the pathways for ENPs entering soils and groundwater.

6.3.3.4 *Runoff and others*

As mentioned previously, NNPs are originated from geochemical and biotic processes and abundant in soils and groundwater. Typical NNP concentrations in groundwater are in the lower mg/L level. Freshwater sediment constitutes one of the largest environmental sinks for NNPs. Only few published studies have measured ENPs in sediments. Because of the complexity of sediment-water matrices, it requires thoroughly computational knowledge to determine the residence time and subsequent fate, transport, and transformation of ENPs. Factors such as aggregation/disaggregation, sorption to larger particles and NNPs, solubility, and interaction with NOM should be coupled with mass flow models in order to accurately describe the behaviors and fate of ENPs in soils and groundwater.

Beside the point source release of ENPs from technical compartments (landfill and biosolid), nonpoint source release of ENPs into soil is also important. However, nonpoint sources such as stormwater runoff and agricultural runoff have received much less attention due to limited investigations and measurements. Some studies have identified the occurrence of nano-TiO_2 in road-environment samples, such as road dust, stormwater sludge, and curbside soil, which can be traced back to ENP-containing products including coating, paint, and pigments (Adamiec, 2017). Wang et al. (2020) collected runoff samples from two bridges in urban area. The samples were digested with acid and extracted with $Na_4P_2O_7$ to create dispersed particles of different size fractions. TiO_2 ENPs were finally identified and quantified as 0.14–0.2 mg/L, which is higher than those measured in surface water.

ENPs are used in many applications in agriculture, such as nano-fertilizers, nano-pesticides, and nano-biosensors. They are often incorporated with fertilizers to improve nutrient efficiency, develop slow release fertilizers, and improve the accessibility of poorly available nutrients (Liu & Lal, 2015). Nano-pesticides and herbicides can offer slow degradation and controlled release of active ingredients for effective agricultural uses over long time. ENPs have also been applied to upgrade sensors to provide better detection to inform rapid decision to enhance crop yields by monitoring water, land, fertilizers, and pesticides (Usman et al., 2020). However, it is inevitable that these ENPs would find their ways into agricultural soils and runoff and thus become nonpoint source pollution of ENPs to the environment. Moreover, most of the studies on agricultural applications of ENPs mainly focus on their performances and benefits while limited information about their release is available. Although applications of ENPs can improve agricultural productions, firm understanding of the occurrences, transport, and impacts of agricultural applied ENPs remains a major challenge in agricultural and environmental sciences (Usman et al., 2020).

Intentional environmental applications of ENPs also offer a promising option for groundwater and soil remediation. nZVI (nano-Fe^0) is an excellent electron donor and reducing agent to adsorb, reduce, and immobilize various environmental pollutants including heavy metals, nutrients, and organics such as chlorinated, nitroaromatic, and phenol compounds (Fu et al., 2014; Wang et al., 2019). Based on the existing literature, the injected nZVI has to be aqueous stabilized to avoid aggregation and blocking of the pore space. Many surface modification techniques have been demonstrated to be effective to stabilize nZVI and facilitate its distribution in the contaminated sites during in situ applications. However, the long-term fate of the nZVI and the surface modifiers in soils and groundwater remains unclear. More field studies are suggested to investigate the occurrences, fate, and transport of ENPs in soil and groundwater systems.

6.4 Analytical techniques

6.4.1 Separation and sample preparation

6.4.1.1 Digestion

Digestion of environmental samples, especially solid samples, is the first step for measuring the concentrations of ENPs. Common oxidizing acids like nitric acid alone or in combination with hydrogen peroxide or hydrochloric acid to digest the samples under heat treatment or other processes, such as microwave-assisted process (Laborda et al., 2016). Alkaline reagents such as tetramethylammonium hydroxide are also used to degrade the organic matters in samples. Furthermore, enzymatic digestions are employed to solubilize biological materials through degrading proteins and plant cell walls (Shrivastava et al., 2019). Both alkaline and enzymatic digestion strategies preserve the cores of inorganic ENPs, allowing the direct detection, quantification, and size characterization of the ENPs in complex environmental samples.

6.4.1.2 Centrifugation

Centrifugation is the simplest approach to extract ENPs from aqueous samples. It requires high centrifugal speeds and long times to separate ENPs from supernatants, which may not guarantee complete separations (Laborda et al., 2016). Moreover, some unwanted super molecules, such as organic matters and proteins, may accompany with ENPs. Thus preparative centrifugation alone is not considered as effective technique for the fractionation of ENPs (Tsao et al., 2011). The directly available pool of ENPs in soils can be collected from soil pore water with subsequent characterization of ENPs. Centrifugation is the most commonly used method to extract pore water from soils (Rodrigues et al., 2016).

6.4.1.3 Filtration, ultrafiltration, and dialysis

Filtration with 0.45 μm membrane is no longer used to distinguish dissolved and particulate fractions. Instead, serials of filtration in combined with ultrafiltration has been used for size fractionation of ENPs in water. Dialysis and ultrafiltration are also applied with different materials and nominal molecular-weight cut-offs (Bakshi et al., 2015). Dialysis is based on pure diffusion of concentration gradients, which takes long time to achieve equilibrium than ultrafiltration. Ranging from 10^3-10^6 Da (Dalton, unified atomic mass unit), ENPs can be separated from complex sample matrices based on molecular-weight cut-offs (Laborda et al., 2016). It should be noted that ENPs may interact with the membrane surface during filtration or dialysis, depending on their composition and surface functionality (Ladner et al., 2012).

6.4.1.4 Liquid phase extraction

ENPs can be extracted from solid and aqueous samples by chemical solutions or organic solvents. $Ca(NO_3)_2$, KNO_3, hexane, EDTA, nonionic surfactant, thiosulfate, $Na_4P_2O_7$, and many other organic and inorganic chemicals are used for ENP extraction alone or in combination with filtration (Laborda et al., 2016; Tsao et al., 2011; Wang et al., 2020). Selective extraction of metal-based ENPs in the presence of corresponding cations can be achieved by certain complexing agents, which is critical for extracting ENPs from field samples (Laborda et al., 2016).

6.4.1.5 *Solid phase extraction*

Ion exchange resins and magnetic NPs can be used for solid phase extraction of ENPs. Anionic exchange resins have been used for the selective extraction of noble metal ENPs through noncovalent reversible reaction (Li et al., 2012). Functionalized magnetic NPs have also been applied for the extraction of Au ENPs under certain conditions, depending on the specific needs of the quantification methods (Su et al., 2014).

6.4.2 Detection and characterization

6.4.2.1 *Electron microscopy*

Electron microscopy is one of the most powerful visualization techniques to characterize ENPs for their sizes, shapes, and morphologies. SEM, TEM, atomic force microscopy (AFM), and scanning transmission electron microscopy (STEM) with high resolutions are common electron microscopy equipment. Energy dispersive X-ray spectroscopy (EDS) with distribution mapping option is usually combined with SEM and TEM to quantify surface elements heavier than boron. TEM can be also equipped with electron energy-loss spectroscopy (EELS), which can provide structural and chemical oxidation down to the atomic level. High-angle annular dark field (HAADF) is also a method of mapping samples in STEM to achieve atomic resolution images (Laborda et al., 2016).

6.4.2.2 *Particle size and surface characteristics*

Dynamic light scattering (DLS) is a useful technique to quantify ENPs via measuring intensity of scattered light from NP suspensions. Mobility and hydrodynamic diameters (HDD) are measured based on autocorrelated functions to determine the size distribution of ENPs. Similar to DLS, NP tracking analysis (NTA), ultraviolet–visible (UV–vis) spectroscopy and fluorescence correlation spectroscopy (for fluorescent ENPs only) are also used to determine the size distributions of ENPs (Rodrigues et al., 2016).

Continuous separation and size fractionation techniques have shown great promise for ENPs in complex environmental samples. Asymmetric flow field-flow fractionation (AF4) can achieve high size resolution separation and is commercially available. Two electrophoretic techniques, gel electrophoresis (GE) and capillary electrophoresis (CE), are also used for separation and characterization of ENPs based on the migration of charged species under the influence of electric field. Hydrodynamic chromatography and other liquid chromatography techniques, such as size-exclusion chromatography (SEC), are applied to separate ENPs with different size ranges by flowing into the chromatography column. These size fractionation techniques are usually coupled with in-line detection systems.

Zeta potential is measured to determine the surface charge of suspended ENPs. Point of zero charge is measured according to the changes of zeta potential with solution pH, which can be used to determine the highest aggregation of ENPs (Wang et al., 2019; Wang, Gao, et al., 2016). The Brunauer–Emmett–Teller (BET) method is widely applied to determine the specific surface area of ENPs. The adsorption and desorption of gas molecules (N_2 or CO_2) onto ENPs are monitored to determine specific surface area and pore size distributions. Thermogravimetric analysis is a method that can measure the surface functional groups and thermostability of ENPs.

6.4.2.3 *Chemical composition*

Atomic absorption spectrometry (AAS), graphite furnace atomic absorption spectrometry (GFAAS), X-ray fluorescence (XRF), inductively coupled plasma optical emission spectroscopy (ICP-OES), and inductively coupled plasma mass spectrometry (ICP-MS) are commonly used for detection and quantification of the chemical composition of ENPs after they are acid digested. The selection and application of these technologies are based on the portability, detection limits, sample accessibility, and other factors. Single particle inductively coupled plasma mass spectrometry (SP-ICP-MS) is developed specifically for NPs to determine their number concentrations and the elemental mass content of one single NP (Meermann & Nischwitz, 2018). ICP-MS and SP-ICP-MS have several advantages over other analytical techniques in detection limit (10 ng/g), spatial resolution (15–50 μm), and analytical depth (200 μm) (Schultz et al., 2015).

6.4.2.4 *Chemical speciation*

X-ray, as an energy source, is widely applied in analytical instrument for identification and speciation of ENPs. X-ray diffraction (XRD) is a powerful quantitative method to reveal crystallinity and particle size of ENPs. X-ray photoelectron (XPS) is a surface-sensitive spectroscopic technique to characterize ENP surface and coating. Synchrotron radiation is advanced by the high energy X-ray from the acceleration of electrons at nearly the speed of light. Synchrotron radiation-based techniques, such as XRF and X-ray absorption spectroscopy (XAS), are widely used to understand the chemical transformation and speciation of ENPs in complex environmental samples. Two types of X-ray adsorption spectroscopy, extended X-ray absorption fine structure (EXAFS) and X-ray near-edge absorption spectroscopy (XANES), are combined with XRF to provide high spatial resolution analysis of ENPs. Both of them can define the chemical environment of a single element in relation to the number and type of neighbor atoms, interatomic distances, and structural features.

Other qualitative analysis techniques have been applied to investigate the interactions of ENPs in soil samples. Raman spectroscopy can provide information related to chemical bond vibration frequencies and molecular symmetry of ENP structures (Bakshi et al., 2015). Attenuated total reflectance-Fourier transform infrared spectroscopy (ATR-FTIR) is used to characterize chemical bonding between ENPs and organic matters, as well as to identify the physicochemical binding mechanisms. Nuclear magnetic resonance (NMR) enables the assessment of the dynamics and 3D structures of ENPs and their composites (Schultz et al., 2015).

A tabulated summary of separation, sample preparation, detection, and characterization methods for ENPs in complex matrices is provided in Table 6.3.

6.5 Risks and impacts

6.5.1 Transformations

Once released into the natural environment, ENPs may undergo various transformation processes altering their potential risks and impacts due to the changes in physicochemical properties. As shown in Fig. 6.8, physical, chemical, and biological transformations are the main processes inducing the risks of ENPs in the environment (Abbas et al., 2020).

Physical transformation mainly includes agglomeration and sedimentation/deposition influencing the fate and transport in soil and water systems. In both water and soil systems, the mobility and bioavailability of ENPs are affected by pH, ionic strength, and the interactions with NOM. These environmental conditions also influence the adsorption of ENPs to other substances. For example, ENPs can associate with heavy metals and pesticides to damage root membrane of plants and increase the bioaccumulation of containments in the food chain (Servin & White, 2016).

ENPs may undergo a series of chemical transformations that can induce their toxicity. For example, redox reactions of ENPs occur with the existence of electron donors (reducing agent) or acceptor (oxidizing agent) in the environment. Chemical transformation can increase the dissolution and water solubility of ENPs to release toxic ions (Colvin, 2003). Photochemically induced reactions are also important to the environmental exposure of photoactive ENPs, such as TiO_2, ZnO, and CeO_2. These ENPs can trigger the process of photo-activation to form unpaired electrons and generate reactive oxygen species (Rajput et al., 2020). The toxicity of TiO_2 ENPs to *Escherichia coli* increases dramatically upon irradiation with sunlight for 30 minutes (Maurer-Jones et al., 2013).

Biological transformation primarily influences ENPs by biodegradation and biomodification. Extracellular enzymes, extracellular polymeric substances (EPS), and soil microbes can interact with ENPs. Bacteria, fungi, and other microorganisms, or their enzymatic components, can degrade ENPs under favorable conditions. For example, white-rot basidiomycete fungi can

TABLE 6.3 Separation, detection, and characterization methods for ENPs in complex matrices.

Analysis	Method	Abbreviation	Purpose or information obtained	Detection limit or technical limitation	References
Separation	Digestion	NA	Sample pretreatment to separate ENPs from background matrix. Usually target at total concentration measurement	Possible modification of ENPs properties	Laborda et al. (2016)
	Centrifugation	NA	Separate ENPs from aqueous suspension or collect pore water from soil	Incomplete separation, high energy, and long time	Laborda et al. (2016)
	Filtration	NA	Separation and sample preparation	0.45 μm no longer distinguish dissolved and particulate fractions	Schultz et al. (2015)
	Ultrafiltration	NA	Enhanced separation with ENPs size fractions	Possible artifacts due to particle–membrane interactions	Schultz et al. (2015)
	Dialysis	NA	Aqueous separation based on pure diffusion of concentration gradients	ENPs could interact with the membrane	Shrivastava et al. (2019)
	Liquid phase extraction	NA	Sample preparation	Extraction chemicals and recovery varies by sample matrix	Li et al. (2012)
	Solid phase extraction	NA	Separate NMs from background matrix components	Possible modification of NM properties	Li et al. (2012)

Electron microscopy	Scanning electron microscopy	SEM	Visualization, morphology, size, and aggregation analysis. SEM usually equip with EDS for elemental mapping	1 nm–1 μm image resolution or μg/g in environmental samples	Shrivastava et al. (2019)
	Transmission electron microscopy	TEM	Morphology and aggregation analysis. Equipped with EELS for surface interaction analysis	>0.1 nm image resolution or μg/g in environmental samples	Koh et al. (2009)
	Scanning transmission electron microscopy	STEM	Equipped with HAADF in atomic resolution for crystalline analysis	<0.1 nm image resolution or μg/g in environmental samples	Koh et al. (2009)
	Atomic force microscopy	AFM	Aggregation, shape, size, surface morphology, speciation structure	~0.1 nm resolution or ppb in environmental samples. Overestimation of particle dimensions.	Shrivastava et al. (2019)
Particle size and other characteristics	Dynamic light scattering	DLS	Size distribution in aqueous suspension by measuring HDD	Detection in range of 3 nm–μm. Poor size resolution. Inability to distinguish size of NPs within heteroaggregates in unknown samples.	Bhattacharjee, (2016)
	Asymmetric flow field-flow fractionation	AF4	Size separation with detection and characterization equipped with in-line detector	Detection in ng/L. Need method development to minimize membrane interaction for each sample.	López-Lorente et al. (2011), Zhang et al. (2018)

(*Continued*)

TABLE 6.3 (Continued)

Analysis	Method	Abbreviation	Purpose or information obtained	Detection limit or technical limitation	References
	Gel electrophoresis	GE	separation and characterization of ENPs	Average detection size in nm	López-Lorente et al. (2011)
	Capillary electrophoresis	CE	Separation and characterization of ENPs	Average detection size in nm	López-Lorente et al. (2011)
	Size-exclusion chromatography	SEC	Separation and characterization of ENPs	Wide dynamic ranges covering particles from 5 to >300 nm	Zhang et al. (2018)
	Zeta potential	Zeta	Measure the surface charge properties	Require good suspension of ENPs	Bhattacharjee, (2016)
	Brunauer–Emmett–Teller (BET)	BET	Measure the specific surface area	Surface area in m^2/g	Tiede et al. (2008)
Chemical composition	Atomic absorption spectrometry	AAS	Measure the total concentration of metal elements	μg/L. Require light source change for different element.	Frankowski et al. (2011), Shrivastava et al. (2019)
	Inductively coupled plasma optical emission spectroscopy	ICP-OES	Measure the total concentration of metal elements	μg/L	Funari et al. (2018), Shrivastava et al. (2019)
	Inductively coupled plasma mass spectrometry	ICP-Ms	Identify and measure the total concentration of metal elements in environmental samples	ng/L	Frankowski et al. (2011), Meermann & Nischwitz, (2018)

	Single particle ICP-MS	SP-ICP-MS	Measure number concentration of ENPs suspensions, and the elemental mass content of single particle	ng/L. Unable to distinguish homoaggregates. Need prior knowledge of primary ENP.	Montaño et al. (2016)
	X-ray diffraction	XRD	Analyzing crystalline and speciation structure	Larger crystalline samples (>1 mg) required	Shrivastava et al. (2019)
Chemical speciation	X-ray fluorescence or synchrotron micro-X-rays fluorescence	XRF or μ-XRF	Understand the chemical transformation and speciation	XRF can detect in μg/g and nm–μm depth. μ-XRF can detect in range of 10^{-16} g with detection limit below ng/g.	Shrivastava et al. (2019)
	X-ray absorption spectroscopy or synchrotron X-ray absorption spectroscopy	XAS or μ-XAS	Understand the chemical transformation and speciation, which usually equipped with	XAS can detects in μg/g and nm–μm depth. μ-XAS can detect in range of 10^{-16} g with detection limit below ng/g.	Shrivastava et al. (2019)
	Extended X-ray absorption fine structure spectroscopy	EXAFS	Understand the chemical transformation and speciation of ENPs, which is equipped into μ-XAS system	Minimum of 10^9 total effective single counts for a whole spectrum	Heald, (2015)
	X-ray absorption near-edge spectroscopy	XANES	Understand the chemical transformation and speciation of ENPs, which is equipped into μ-XAS system	Minimum of 10^7 total effective single counts for a whole spectrum	Heald, (2015)
	X-ray photoelectron spectroscopy	XPS	Characterize ENP surfaces, such as speciation and functional groups in organic coatings	Selective for surface in 1–12 nm depth with detection limit of 0.1–1 atomic percent	Shard, (2020)

(Continued)

TABLE 6.3 (Continued)

Analysis	Method	Abbreviation	Purpose or information obtained	Detection limit or technical limitation	References
	Nuclear magnetic resonance	NMR	Characterize surface ligands on ENPs and aqueous surface area of ENPs	<1.5 μg/dosage	Pan et al. (2013), Xing et al. (2016)
	Raman spectroscopy	Raman	Imaging of ENPs location and associations in environmental matrices	Identify ENPs in nm range	Vandenabeele et al. (2012)
	Attenuated total reflectance–Fourier transform infrared spectroscopy	ATR-FTIR	Chemical composition and functional group analysis. Analysis is possible for both solid and liquid samples.	~0.1 weight percentage	Xing et al. (2016)

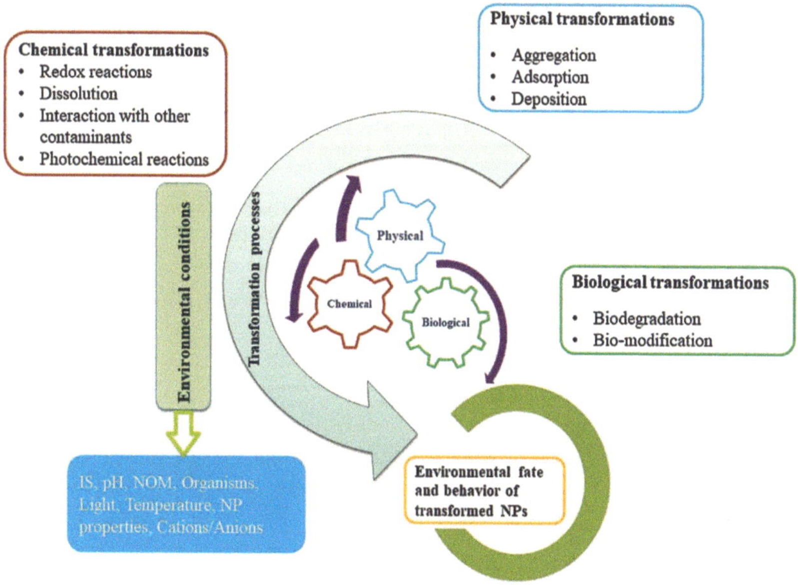

FIGURE 6.8 Schematic diagram showing transformations of ENPs in the environment. ENPs may undergo physically, chemically, and biologically mediated transformations in the environments. *Modified from Abbas, Q., Yousaf, B., Amina, Ali, M., Munir, M., El-Naggar, A., Rinklebe, J., & Naushad, M. (2020). Transformation pathways and fate of engineered nanoparticles (ENPs) in distinct interactive environmental compartments: A review.* Environment International, 138. *https://doi.org/10.1016/j.envint.2020.105646.*

enzymatically degrade hydroxylated carbon ENPs by oxidation process and mineralized them into CO_2 in 32 weeks (Schreiner et al., 2009). EPS and microbes from fungi and algae are capable of modifying metal and metal oxide ENPs by changing their oxidation states and ligand functional groups (Abbas et al., 2020; Amde et al., 2017).

6.5.2 Risks and impacts

Bacteria serve as components for integral environmental functions. This ecological function has made bacteria a primary focus of toxicity studies to date, including use of a wide range of model organisms (e.g., *E. coli*, *Bacillus subtilis*, and *Pseudomonas aeruginosa*) in toxicity studies to assess the impacts of ENPs (Maurer-Jones et al., 2013). ENPs also show evidence on reducing soil microbial diversity. Studies have evaluated the microbial activities of WWTP sludge-amended soils and shown that Ag NP exposure significantly affected 38 bacteria in both sludge and incorporated soils (Forstner et al., 2020; Wang, Menzies, et al., 2016).

ENPs in soils can closely interact with plants, such as uptake and accumulation in plant biomass and plant cell walls. These interactions will greatly affect the fate and transport of ENPs and food safety. Evidence has shown that ENPs can inhibit plant growth at different developmental stages such as seed germination and root elongation (Ma et al., 2010, 2014). Determination of the phytotoxicity of metal/metal oxide ENPs is complex due to the potential dissolution of metallic ions from these ENPs and their effect to plant growth.

Soil earthworms comprise a critical component of the ecosystem and are often used as a model organism in nanotoxicity studies. Earthworms can accumulate gold ENPs in soils at low dosages. Amde et al. (2017) found that Ag ENPs had 100% reproductive failure on earthworm with 1000 mg/g of dry soil.

ENPs can enter human body via oral and skin exposures. The potential exposure of ENPs in groundwater can deteriorate the quality of drinking water. ENPs in edible plants and organisms may be transferred and bioaccumulated in the food chain. With increasing production of ENPs and their consumer products, more ENPs are released into soils and groundwater to increase their potential exposures to the public.

6.6 Case study of Ag ENPs

Silver engineered nanoparticles (Ag ENPs), one of the most important metal ENPs, are widely used and applied in industrial and commercial products. During the manufacturing, use, and disposal of Ag ENP–containing products, Ag ENPs are inevitably released into the environment. This section presents a case study on the environmental sources, occurrences, impacts, and transformations of Ag ENPs.

Ag ENPs are widely used in the fields of medical, antibiotics, electronics, cosmetics, and textiles. China has the largest Ag ENP production and consumption. Based on a national nano-silver market report in China, Ag ENPs were mainly applied in seven use categories: textile (15.9% of the national total production), cosmetics (15.9%), plastics (10.2%), paints (23.3%), medical supplies (17.7%), ceramic agent (12.4%), and electronic equipment (4.6%) (Ding et al., 2019). Ag ENPs are first released from anthropogenic processes such as production, manufacturing, and consumption into different environmental compartments (i.e., air, soil, sediment, surface water, and groundwater) and waste disposal facilities (e.g., sewage treatment plants, WIPs, landfills, and recycling sites). Ag ENPs in waste disposal facilities may be released into the environmental compartments after processing.

From a life-cycle perspective of Ag ENP–containing products, predicted environmental concentrations of Ag ENPs can be obtained from the probabilistic material flow analysis. The modeled regional concentration modes of Ag ENPs are 0.9 ng/L in surface water, 20 ng/kg in soil, 6.9 μg/kg in sediment, and negligible in air (Ding et al., 2019). Moreover, studies presented that over 90% of the bulk Ag ENPs, primarily originating from production and consumption, end up in WWTP, landfill, and WIP. The waste from those disposal facilities would be primary sinks for the release of Ag ENPs.

Primarily, wastewater containing Ag ENPs from manufacturing and consumer products enters the WWTPs. According to a field analysis of representative WWTPs in Germany, 96.4% of Ag ENPs from wastewater influent was removed by WWTPs, even with high influent concentrations of Ag ENPs, resulting in effluent concentrations of 0.7 − 11.1 ng/L over the seasons (Li et al., 2016, 2019). Wimmer et al. (2019) confirmed that 90%–95% of the Ag ENPs were retained in the WWTP sludge. A one-dimensional model was developed, which can predict Ag ENP concentrations in most part of real environmental measurements. The Ag ENP contents of WWTP sludge in China was reported to be 2.72 mg/kg in average and most of them end up in soil by landfill, incineration, land application, and building materials (Chen et al., 2020). Most of the Ag in WWTP sludge is relatively stable and in the form of silver sulfide (Ag_2S); however, thiosulfate ($S_2O_3^{2-}$) can cause dramatic leaching of Ag due to the replacement of Ag_2S to soluble $Ag_2S_2O_3$ (Chen et al., 2020). Based on the investigation over different biosolid-amended soil samples, Ag has the concentration of 25–40 mg/kg in the surface layers (0–30 cm) of soils, which is much higher than 10–13 mg/kg in other soil depths (30–100 cm) (Yang et al., 2014).

Most of the solid products containing Ag ENPs end up in landfills at the end of their usages. The amount of Ag ENPs in consumer products and their leachability are essential to their occurrences and fate in landfills. Studies have shown that the average Ag ENP content of textile consumer products such as body suit, car sheet, socks, and face mask are in the range of 1–3 μg/g. These products in landfills would result in 4–65 μg/L of Ag release based on the toxicity characteristic leaching procedure (TCLP, USEPA) test (Limpiteeprakan & Babel, 2016). On the other hand, medical cloth resulted in much higher TCLP Ag concentration at 2900 μg/L, which is on the same order as the toxicity characterization limit (USEPA) of 5000 μg/L (Benn et al., 2010). The release of Ag ENPs release from landfill depends on the types of products. Moreover, other factors may also strongly influence the release of Ag ENPs from landfills to soils and groundwater. NTA provides a cost-effective method to determine particle aggregation, mass distribution, and stability of ENPs in landfill leachates (Benn et al., 2010). Although landfill clay liner and HDPE geomembrane can serve as safeguards to mitigate the leaching of Ag ENPs into soils and groundwater, in situ monitoring of Ag ENPs in landfills and soils is still necessary and challenging.

Various instrument and tools have been used to characterize Ag ENPs in soil and groundwater systems. XAS, ICP-OES, and ICP-Ms are usually used as the quantitative techniques to measure the total mass concentrations of Ag in the aqueous samples after digestion (Wimmer et al., 2019). Extraction and separation of Ag ENPs from solid samples are important prerequisite for SEM and TEM to determine their size, spatial distribution, and morphology

(Li et al., 2019, 2020). XANES and EXAFS have also been employed to determine the chemical speciation and transformation of Ag ENPs in soils and biosolids (Ma et al., 2010, 2014; Wang, Menzies, et al., 2016).

The environmental impacts of Ag ENPs are closely related with their dissolution and transformation in the environment. The oxidative dissolution of Ag ENPs in aerated soils and water can trigger Ag^0 to be converted into silver ions (Ag^+), where the oxidation of silver by H_2O_2 or H_2O is explained by Eqs. (6.1)–(6.3):

$$Ag(ENP)_n + H_2O_2 \rightarrow Ag(ENP)_{n-2} + 2Ag^+ + 2OH^- \tag{6.1}$$

$$Ag(ENP)_n + H_2O^- + H_2O \rightarrow Ag(ENP)_{n-2} + 2Ag^+ + 3OH^- \tag{6.2}$$

$$Ag^+ + n(\text{ligand}) \rightarrow (Ag(\text{ligand})_n)^+ \tag{6.3}$$

Ag^+ released from particle dissolution may be adsorbed on the surface of particles or complexes in soils and water, which influences its bioavailability and toxicity. Ag ENPs, under the condition of high sulfide content, have a high tendency to form silver sulfide (Ag_2S) NPs by direct particle fluid reaction in biosolids and soils (Fig. 6.9) (Liu et al., 2011; Wang, Menzies, et al., 2016). The potential bioavailability of Ag in soils is low but may increase markedly with elevated Cl, likely due to the formation of soluble AgCl complexes in soil pore water (Liu & Lal, 2015). Van der Ploeg et al. (2014) found that Ag ENP exposure can affect earthworm populations and growth. As deposited Ag ENPs may be accumulated in top soils, they may present potential risks to plants, animals, and human beings.

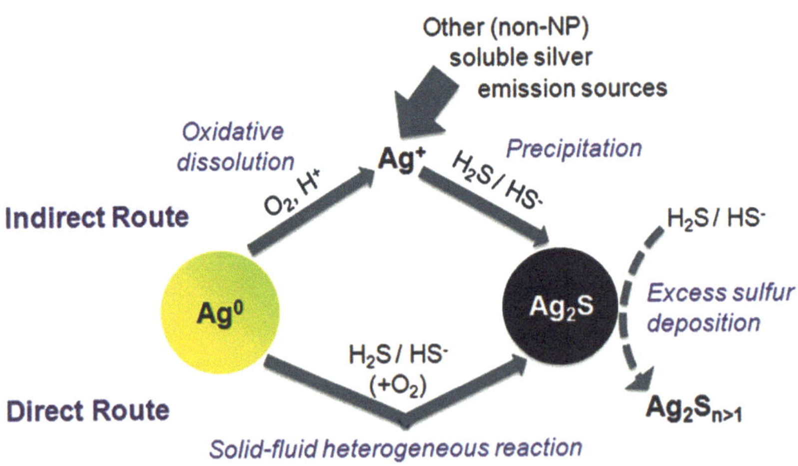

FIGURE 6.9 Kinetics and mechanisms of Ag ENP transformation with oxysulfidation. *Modified from Liu, J., Pennell, K.G., & Hurt, R.H. (2011). Kinetics and mechanisms of nanosilver oxysulfidation.* Environmental Science & Technology, 45*(17), 7345–7353. https://doi.org/10.1021/es201539s.105646.*

6.7 Conclusions and perspectives

This chapter provides a general overview of the properties, productions, and commercial uses of ENPs. Based on the global abundance and popularity of ENPs, their potential releases, occurrences, sources, and impacts in soils and groundwater system are systematically reviewed. As an emerging environmental contaminant, ENPs have much lower concentrations than naturally occurring NPs in the soil and groundwater systems. Most of ENPs in the environment are life cycle orientated, where the major sources include production facilities, waste disposal facilities, and other nonpoint sources. Thus ENPs are released into soils and farther into groundwater through complex pathways. Many analytical techniques have been developed and applied in the sampling, separation, and characterization of ENPs in soils and groundwater. Environmental conditions such as pH, ionic strength, and natural components (e.g., NOM, natural colloids, and living organisms) play vital roles in controlling the physical and chemical transformations of ENPs, affecting their potential risks and impacts to the ecosystems.

Although progresses have been made toward better understanding of the environmental fate, transport, transformation, and impacts of ENPs, the research on environmentally emerging ENPs in soils and groundwater is still challenging because of their relatively low environmental concentration and difficult seperation from naturally occurring NPs. When it comes to the studies of ENPs in soils, methodologies are even more limited due to the lack of customized analytical techniques and the complexity of soil systems. The development of new analytical tools and the adaption of existing analytical techniques are in critical needs in the research of ENPs in soils and groundwater. Moreover, establishing advanced experimental and computational methods to investigate the fate, transport, transformation, and interactions of ENPs in different environmental components under natural conditions is also crucial. These not only lead to much better understanding and prediction of the fate of ENPs in soils and groundwater, but also inform the development of more reliable evaluations on their long-term impacts and ecotoxicity.

References

Abbas, Q., Yousaf, B., Amina., Ali, M. U., Munir, M. A. M., El-Naggar, A., Rinklebe, J., & Naushad, M. (2020). Transformation pathways and fate of engineered nanoparticles (ENPs) in distinct interactive environmental compartments: A review. *Environment International, 138*. Available from https://doi.org/10.1016/j.envint.2020.105646.

Adamiec, E. (2017). Road environments: Impact of metals on human health in heavily congested cities of Poland. *International Journal of Environmental Research and Public Health, 14*(7). Available from https://doi.org/10.3390/ijerph14070697.

Amde, M., Liu, J. f, Tan, Z. Q., & Bekana, D. (2017). Transformation and bioavailability of metal oxide nanoparticles in aquatic and terrestrial environments. A review. *Environmental Pollution, 230*, 250–267. Available from https://doi.org/10.1016/j.envpol.2017.06.064.

Bakshi, S., He, Z. L., & Harris, W. G. (2015). Natural nanoparticles: Implications for environment and human health. *Critical Reviews in Environmental Science and Technology*, *45*(8), 861–904. Available from https://doi.org/10.1080/10643389.2014.921975.

Baumann, T. (2010). *Nanoparticles in groundwater - Occurrence and applications. Nanoparticles in the water cycle: Properties, analysis and environmental relevance* (pp. 23–34). Berlin Heidelberg: Springer. Available from https://doi.org/10.1007/978-3-642-10318-6_3.

Benn, T., Cavanagh, B., Hristovski, K., Posner, J. D., & Westerhoff, P. (2010). The release of nanosilver from consumer products used in the home. *Journal of Environmental Quality*, *39* (6), 1875–1882. Available from https://doi.org/10.2134/jeq2009.0363.

Bernd, G., Fred, K., Barry, P., Ralf, K., Michael, S., Henning, W., Arnim, V. G., & Fadri, G. (2018). Risks, release and concentrations of engineered nanomaterial in the environment. *Scientific Reports*. Available from https://doi.org/10.1038/s41598-018-19275-4.

Bhattacharjee, S. (2016). DLS and zeta potential—what they are and what they are not? *Journal of Controlled Release*, *235*, 337–351. Available from https://doi.org/10.1016/j.jconrel.2016.06.017.

Chen, Y., Mao, Y., Song, M., Yin, Y., Liu, G., & Cai, Y. (2020). Occurrence and leaching of silver in municipal sewage sludge in China. *Ecotoxicology and Environmental Safety*, *189*. Available from https://doi.org/10.1016/j.ecoenv.2019.109929.

Colvin, V. L. (2003). The potential environmental impact of engineered nanomaterials. *Nature Biotechnology*, *21*(10), 1166–1170. Available from https://doi.org/10.1038/nbt875.

Ding, R., Li, L., Yang, P., Luo, L., Li, L., & Wang, Q. (2019). Assessing the environmental occurrence and risk of nano-silver in Hunan, China using probabilistic material flow modeling. *Science of the Total Environment*, *658*, 1249–1255. Available from https://doi.org/10.1016/j.scitotenv.2018.12.254.

Dulger, M., Sakallioglu, T., Temizel, I., Demirel, B., Copty, N. K., Onay, T. T., Uyguner-Demirel, C. S., & Karanfil, T. (2016). Leaching potential of nano-scale titanium dioxide in fresh municipal solid waste. *Chemosphere*, *144*, 1567–1572. Available from https://doi.org/10.1016/j.chemosphere.2015.10.037.

Fawole, O. G., Cai, X. M., & Mackenzie, A. R. (2016). Gas flaring and resultant air pollution: A review focusing on black carbon. *Environmental Pollution*, *216*, 182–197. Available from https://doi.org/10.1016/j.envpol.2016.05.075.

Forstner, C., Orton, T. G., Wang, P., Kopittke, P. M., & Dennis, P. G. (2020). Wastewater treatment processing of silver nanoparticles strongly influences their effects on soil microbial diversity. *Environmental Science and Technology*, *54*(21), 13538–13547. Available from https://doi.org/10.1021/acs.est.0c01312.

Frankowski, M., Zioła-Frankowska, A., Kurzyca, I., Novotný, K., Vaculovič, T., Kanický, V., Siepak, M., & Siepak, J. (2011). Determination of aluminium in groundwater samples by GF-AAS, ICP-AES, ICP-MS and modelling of inorganic aluminium complexes. *Environmental Monitoring and Assessment*, *182*(1-4), 71–84. Available from https://doi.org/10.1007/s10661-010-1859-8.

Fu, F., Dionysiou, D. D., & Liu, H. (2014). The use of zero-valent iron for groundwater remediation and wastewater treatment: A review. *Journal of Hazardous Materials*, *267*, 194–205. Available from https://doi.org/10.1016/j.jhazmat.2013.12.062.

Funari, V., Mantovani, L., Vigliotti, L., Tribaudino, M., Dinelli, E., & Braga, R. (2018). Superparamagnetic iron oxides nanoparticles from municipal solid waste incinerators. *Science of the Total Environment*, *621*, 687–696. Available from https://doi.org/10.1016/j.scitotenv.2017.11.289.

Giese, B., Klaessig, F., Park, B., Kaegi, R., Steinfeldt, M., Wigger, H., Von Gleich, A., & Gottschalk, F. (2018). Risks, release and concentrations of engineered nanomaterial in the environment. *Scientific Reports*, *8*(1), 1–18. Available from https://doi.org/10.1038/s41598-018-19275-4.

Gottschalk, F., Lassen, C., Kjoelholt, J., Christensen, F., & Nowack, B. (2015). Modeling flows and concentrations of nine engineered nanomaterials in the Danish environment. *International Journal of Environmental Research and Public Health, 12*(5), 5581–5602. Available from https://doi.org/10.3390/ijerph120505581.

Gottschalk, F., Sun, T., & Nowack, B. (2013). Environmental concentrations of engineered nanomaterials: Review of modeling and analytical studies. *Environmental Pollution, 181*, 287–300. Available from https://doi.org/10.1016/j.envpol.2013.06.003.

Heald, S. M. (2015). Strategies and limitations for fluorescence detection of XAFS at high flux beamlines. *Journal of Synchrotron Radiation, 22*(2), 436–445. Available from https://doi.org/10.1107/S1600577515001320.

Hochella, M. F., Mogk, D. W., Ranville, J., Allen, I. C., Luther, G. W., Marr, L. C., McGrail, B. P., Murayama, M., Qafoku, N. P., Rosso, K. M., Sahai, N., Schroeder, P. A., Vikesland, P., Westerhoff, P., & Yang, Y. (2019). Natural, incidental, and engineered nanomaterials and their impacts on the Earth system. *Science (New York, N.Y.), 363*(6434). Available from https://doi.org/10.1126/science.aau8299.

Ijaz, I., Gilani, E., Nazir, A., & Bukhari, A. (2020). Detail review on chemical, physical and green synthesis, classification, characterizations and applications of nanoparticles. *Green Chemistry Letters and Reviews, 13*(3), 59–81. Available from https://doi.org/10.1080/17518253.2020.1802517.

Impellitteri, C. A., Harmon, S., Silva, R. G., Miller, B. W., Scheckel, K. G., Luxton, T. P., Schupp, D., & Panguluri, S. (2013). Transformation of silver nanoparticles in fresh, aged, and incinerated biosolids. *Water Research, 47*(12), 3878–3886. Available from https://doi.org/10.1016/j.watres.2012.12.041.

Janković, N. Z., & Plata, D. L. (2019). Engineered nanomaterials in the context of global element cycles. *Environmental Science: Nano, 6*(9), 2697–2711. Available from https://doi.org/10.1039/c9en00322c.

Jarvie, H. P., Al-Obaidi, H., King, S. M., Bowes, M. J., Lawrence, M. J., Drake, A. F., Green, M. A., & Dobson, P. J. (2009). Fate of silica nanoparticles in simulated primary wastewater treatment. *Environmental Science and Technology, 43*(22), 8622–8628. Available from https://doi.org/10.1021/es901399q.

Keller, A. A., & Lazareva, A. (2013). Predicted releases of engineered nanomaterials: From global to regional to local. *Environmental Science & Technology Letters*, 65–70. Available from https://doi.org/10.1021/ez400106t.

Kim, B., Murayama, M., Colmane, B. P., & Hochella, M. F. (2012). Characterization and environmental implications of nano- and larger TiO2 particles in sewage sludge, and soils amended with sewage sludge. *Journal of Environmental Monitoring, 1129*. Available from https://doi.org/10.1039/c2em10809g.

Koh, A. L., Bao, K., Khan, I., Smith, W. E., Kothleitner, G., Nordlander, P., Maier, S. A., & McComb, D. W. (2009). Electron Energy-Loss Spectroscopy (EELS) of surface plasmons in single silver nanoparticles and dimers: Influence of beam damage and mapping of dark modes. *ACS Nano, 3*(10), 3015–3022. Available from https://doi.org/10.1021/nn900922z.

Laborda, F., Bolea, E., Cepriá, G., Gómez, M. T., Jiménez, M. S., Pérez-Arantegui, J., & Castillo, J. R. (2016). Detection, characterization and quantification of inorganic engineered nanomaterials: A review of techniques and methodological approaches for the analysis of complex samples. *Analytica Chimica Acta, 904*, 10–32. Available from https://doi.org/10.1016/j.aca.2015.11.008.

Ladner, D. A., Steele, M., Weir, A., Hristovski, K., & Westerhoff, P. (2012). Functionalized nanoparticle interactions with polymeric membranes. *Journal of Hazardous Materials, 211–212*, 288–295. Available from https://doi.org/10.1016/j.jhazmat.2011.11.051.

Lee, Y. S., Kim, Y. M., Lee, J., & Kim, J. Y. (2021). Evaluation of silver nanoparticles (AgNPs) penetration through a Clay Liner in Landfills. *Journal of Hazardous Materials*, *404*, 124098. Available from https://doi.org/10.1016/j.jhazmat.2020.124098.

Li, L., Leopold, K., & Schuster, M. (2012). Effective and selective extraction of noble metal nanoparticles from environmental water through a noncovalent reversible reaction on an ionic exchange resin. *Chemical Communications*, *48*(73), 9165–9167. Available from https://doi.org/10.1039/c2cc34838a.

Li, L., Stoiber, M., Wimmer, A., Xu, Z., Lindenblatt, C., Helmreich, B., & Schuster, M. (2016). To what extent can full-scale wastewater treatment plant effluent influence the occurrence of silver-based nanoparticles in surface waters? *Environmental Science and Technology*, *50* (12), 6327–6333. Available from https://doi.org/10.1021/acs.est.6b00694.

Li, L., Wang, Q., Yang, Y., Luo, L., Ding, R., Yang, Z. G., & Li, H. P. (2019). Extraction method development for quantitative detection of silver nanoparticles in environmental soils and sediments by single particle inductively coupled plasma mass spectrometry. *Analytical Chemistry*, *91*(15), 9442–9450. Available from https://doi.org/10.1021/acs.analchem.8b05575.

Li, M., Liu, H. L., Dang, F., Hintelmann, H., Yin, B., & Zhou, D. (2020). Alteration of crop yield and quality of three vegetables upon exposure to silver nanoparticles in sludge-amended soil. *ACS Sustainable Chemistry and Engineering*, *8*(6), 2472–2480. Available from https://doi.org/10.1021/acssuschemeng.9b06721.

Li, X., Rui, M., Song, J., Shen, Z., & Zeng, H. (2015). Carbon and graphene quantum dots for optoelectronic and energy devices: A review. *Advanced Functional Materials*, *25*(31), 4929–4947. Available from https://doi.org/10.1002/adfm.201501250.

Limpiteeprakan, P., & Babel, S. (2016). Leaching potential of silver from nanosilver-treated textile products. *Environmental Monitoring and Assessment*, *188*(3), 1–12. Available from https://doi.org/10.1007/s10661-016-5158-x.

Liné, C., Larue, C., & Flahaut, E. (2017). Carbon nanotubes: Impacts and behaviour in the terrestrial ecosystem—A review. *Carbon*, *123*, 767–785. Available from https://doi.org/10.1016/j.carbon.2017.07.089.

Liu, J., Pennell, K. G., & Hurt, R. H. (2011). Kinetics and mechanisms of nanosilver oxysulfidation. *Environmental Science & Technology*, *45*(17), 7345–7353. Available from https://doi.org/10.1021/es201539s.

Liu, R., & Lal, R. (2015). Potentials of engineered nanoparticles as fertilizers for increasing agronomic productions. *Science of the Total Environment*, *514*, 131–139. Available from https://doi.org/10.1016/j.scitotenv.2015.01.104.

López-Lorente, A. I., Simonet, B., & Valcárcel, M. (2011). Electrophoretic methods for the analysis of nanoparticles. *TrAC Trends in Analytical Chemistry*, *30*(1), 58–71. Available from https://doi.org/10.1016/j.trac.2010.10.006.

Ma, R., Levard, C., Judy, J. D., Unrine, J. M., Durenkamp, M., Martin, B., Jefferson, B., & Lowry, G. V. (2014). Fate of zinc oxide and silver nanoparticles in a pilot wastewater treatment plant and in processed biosolids. *Environmental Science and Technology*, *48*(1), 104–112. Available from https://doi.org/10.1021/es403646x.

Ma, X., Geiser-Lee, J., Deng, Y., & Kolmakov, A. (2010). Interactions between engineered nanoparticles (ENPs) and plants: Phytotoxicity, uptake and accumulation. *Science of the Total Environment*, *408*(16), 3053–3061. Available from https://doi.org/10.1016/j.scitotenv.2010.03.031.

Massari, A., Beggio, M., Hreglich, S., Marin, R., & Zuin, S. (2014). Behavior of TiO2 nanoparticles during incineration of solid paint waste: A lab-scale test. *Waste Management*, *34*(10), 1897–1907. Available from https://doi.org/10.1016/j.wasman.2014.05.015.

Maurer-Jones, M. A., Gunsolus, I. L., Murphy, C. J., & Haynes, C. L. (2013). Toxicity of engineered nanoparticles in the environment. *Analytical Chemistry*, *85*(6), 3036–3049. Available from https://doi.org/10.1021/ac303636s.

Meermann, B., & Nischwitz, V. (2018). ICP-MS for the analysis at the nanoscale-a tutorial review. *Journal of Analytical Atomic Spectrometry*, *33*(9), 1432–1468. Available from https://doi.org/10.1039/c8ja00037a.

Mehrabi, K., Nowack, B., Arroyo Rojas Dasilva, Y., & Mitrano, D. M. (2017). Improvements in nanoparticle tracking analysis to measure particle aggregation and mass distribution: A case study on engineered nanomaterial stability in Incineration Landfill Leachates. *Environmental Science and Technology*, *51*(10), 5611–5621. Available from https://doi.org/10.1021/acs.est.7b00597.

Methner, M., Beaucham, C., Crawford, C., Hodson, L., & Geraci, C. (2012). Field application of the nanoparticle emission assessment technique (NEAT): Task-based air monitoring during the processing of engineered nanomaterials (ENM) at four facilities. *Journal of Occupational and Environmental Hygiene*, *9*(9), 543–555. Available from https://doi.org/10.1080/15459624.2012.699388.

Mitrano, D. M., Limpiteeprakan, P., Babel, S., & Nowack, B. (2016). Durability of nano-enhanced textiles through the life cycle: Releases from landfilling after washing. *Environmental Science: Nano*, *3* (2), 375–387. Available from https://doi.org/10.1039/c6en00023a.

Mitrano, D. M., Mehrabi, K., Dasilva, Y. A. R., & Nowack, B. (2017). Mobility of metallic (nano)particles in leachates from landfills containing waste incineration residues. *Environmental Science: Nano*, *4*(2), 480–492. Available from https://doi.org/10.1039/c6en00565a.

Montaño, M. D., Olesik, J. W., Barber, A. G., Challis, K., & Ranville, J. F. (2016). Single Particle ICP-MS: Advances toward routine analysis of nanomaterials. *Analytical and Bioanalytical Chemistry*, *408*(19), 5053–5074. Available from https://doi.org/10.1007/s00216-016-9676-8.

Nowack, B., & Bucheli, T. D. (2007). Occurrence, behavior and effects of nanoparticles in the environment. *Environmental Pollution*, *150*(1), 5–22. Available from https://doi.org/10.1016/j.envpol.2007.06.006.

Pan, B., Zhang, D., Li, H., Wu, M., Wang, Z., & Xing, B. (2013). Increased adsorption of sulfamethoxazole on suspended carbon nanotubes by dissolved humic acid. *Environmental Science & Technology*, *47*(14), 7722–7728. Available from https://doi.org/10.1021/es4008933.

Park, B., Donaldson, K., Duffin, R., Tran, L., Kelly, F., Mudway, I., Morin, J. P., Guest, R., Jenkinson, P., Samaras, Z., Giannouli, M., Kouridis, H., & Martin, P. (2008). Hazard and risk assessment of a nanoparticulate cerium oxide-based diesel fuel additive—A case study. *Inhalation Toxicology*, *20*(6), 547–566. Available from https://doi.org/10.1080/08958370801915309.

Part, F., Berge, N., Baran, P., Stringfellow, A., Sun, W., Bartelt-Hunt, S., Mitrano, D., Li, L., Hennebert, P., Quicker, P., Bolyard, S. C., & Huber-Humer, M. (2018). A review of the fate of engineered nanomaterials in municipal solid waste streams. *Waste Management*, *75*, 427–449. Available from https://doi.org/10.1016/j.wasman.2018.02.012.

Part, F., Zaba, C., Bixner, O., Zafiu, C., Lenz, S., Martetschläger, L., Hann, S., Huber-Humer, M., & Ehmoser, E. K. (2020). Mobility and fate of ligand stabilized semiconductor nanoparticles in landfill leachates. *Journal of Hazardous Materials*, *394*. Available from https://doi.org/10.1016/j.jhazmat.2020.122477.

Picca, R. A., Sportelli, M. C., Hötger, D., Manoli, K., Kranz, C., Mizaikoff, B., Torsi, L., & Cioffi, N. (2015). Electrosynthesis and characterization of ZnO nanoparticles as inorganic component in organic thin-film transistor active layers. *Electrochimica Acta*, *178*, 45–54. Available from https://doi.org/10.1016/j.electacta.2015.07.122.

Praetorius, A., Scheringer, M., & Hungerbühler, K. (2012). Development of environmental fate models for engineered nanoparticles—A case study of TiO2 nanoparticles in the rhine river. *Environmental Science and Technology*, *46*(12), 6705–6713. Available from https://doi.org/10.1021/es204530n.

Rajput, V., Minkina, T., Sushkova, S., Behal, A., Maksimov, A., Blicharska, E., Ghazaryan, K., Movsesyan, H., & Barsova, N. (2020). ZnO and CuO nanoparticles: A threat to soil organisms, plants, and human health. *Environmental Geochemistry and Health*, *42*(1), 147–158. Available from https://doi.org/10.1007/s10653-019-00317-3.

Rodrigues, S. M., Trindade, T., Duarte, A. C., Pereira, E., Koopmans, G. F., & Römkens, P. F. A. M. (2016). A framework to measure the availability of engineered nanoparticles in soils: Trends in soil tests and analytical tools. *Trends in Analytical Chemistry*, *75*, 129–140. Available from https://doi.org/10.1016/j.trac.2015.07.003.

Saheli, P. T., Rowe, R. K., Petersen, E. J., & O'Carroll, D. M. (2017). Diffusion of multiwall carbon nanotubes through a high-density polyethylene geomembrane. *Geosynthetics International*, *24*(2), 184–197. Available from https://doi.org/10.1680/jgein.16.00025.

Sakallioglu, T., Bakirdoven, M., Temizel, I., Demirel, B., Copty, N. K., Onay, T. T., Uyguner Demirel, C. S., & Karanfil, T. (2016). Leaching of nano-ZnO in municipal solid waste. *Journal of Hazardous Materials*, *317*, 319–326. Available from https://doi.org/10.1016/j.jhazmat.2016.05.094.

Saleh, T. A. (2020). Nanomaterials: Classification, properties, and environmental toxicities. *Environmental Technology and Innovation*, *20*. Available from https://doi.org/10.1016/j.eti.2020.101067.

Schreiner, K. M., Filley, T. R., Blanchette, R. A., Bowen, B. B., Bolskar, R. D., Hockaday, W. C., Masiello, C. A., & Raebiger, J. W. (2009). White-rot basidiomycete-mediated decomposition of C60 fullerol. *Environmental Science and Technology*, *43*(9), 3162–3168. Available from https://doi.org/10.1021/es801873q.

Schultz, C., Powell, K., Crossley, A., Jurkschat, K., Kille, P., Morgan, A. J., Read, D., Tyne, W., Lahive, E., Svendsen, C., & Spurgeon, D. J. (2015). Analytical approaches to support current understanding of exposure, uptake and distributions of engineered nanoparticles by aquatic and terrestrial organisms. *Ecotoxicology (London, England)*, *24*(2), 239–261. Available from https://doi.org/10.1007/s10646-014-1387-3.

Sengani, M., Grumezescu, A. M., & Rajeswari, V. D. (2017). Recent trends and methodologies in gold nanoparticle synthesis – A prospective review on drug delivery aspect. *OpenNano*, *2*, 37–46. Available from https://doi.org/10.1016/j.onano.2017.07.001.

Servin, A. D., & White, J. C. (2016). Nanotechnology in agriculture: Next steps for understanding engineered nanoparticle exposure and risk. *NanoImpact*, *1*, 9–12. Available from https://doi.org/10.1016/j.impact.2015.12.002.

Shard, Alexander G. (2020). *X-ray photoelectron spectroscopy. Characterization of Nanoparticles* (pp. 349–371). Elsevier.

Shrivastava, M., Srivastav, A., Gandhi, S., Rao, S., Roychoudhury, A., Kumar, A., Singhal, R. K., Jha, S. K., & Singh, S. D. (2019). Monitoring of engineered nanoparticles in soil-plant system: A review. *Environmental Nanotechnology, Monitoring and Management*, *11*. Available from https://doi.org/10.1016/j.enmm.2019.100218.

Singh, S., Thiyagarajan, P., Mohan Kant, K., Anita, D., Thirupathiah, S., Rama, N., Tiwari, B., Kottaisamy, M., & Ramachandra Rao, M. S. (2007). Structure, microstructure and physical properties of ZnO based materials in various forms: Bulk, thin film and nano. *Journal of Physics D: Applied Physics*, *40*(20), 6312–6327. Available from https://doi.org/10.1088/0022-3727/40/20/S15.

Stamou, I., & Antizar-Ladislao, B. (2016). A life cycle assessment of the use of compost from contaminated biodegradable municipal solid waste with silver and titanium dioxide nanoparticles. *Journal of Cleaner Production*, *135*, 884–891. Available from https://doi.org/10.1016/j.jclepro.2016.06.150.

Su, S., Chen, B., He, M., Xiao, Z., & Hu, B. (2014). A novel strategy for sequential analysis of gold nanoparticles and gold ions in water samples by combining magnetic solid phase extraction with inductively coupled plasma mass spectrometry. *Journal of Analytical Atomic Spectrometry*, *29*(3), 444–453. Available from https://doi.org/10.1039/c3ja50342a.

Sun, T. Y., Bornhöft, N. A., Hungerbühler, K., & Nowack, B. (2016). Dynamic probabilistic modeling of environmental emissions of engineered nanomaterials. *Environmental Science and Technology*, *50*(9), 4701–4711. Available from https://doi.org/10.1021/acs.est.5b05828.

Tian, Y., Gao, B., Silvera-Batista, C., & Ziegler, K. J. (2010). Transport of engineered nanoparticles in saturated porous media. *Journal of Nanoparticle Research*, *12*(7), 2371–2380. Available from https://doi.org/10.1007/s11051-010-9912-7.

Tiede, K., Boxall, A. B. A., Tear, S. P., Lewis, J., David, H., & Hassellöv, M. (2008). Detection and characterization of engineered nanoparticles in food and the environment. *Food Additives and Contaminants*, *25*(7), 795–821. Available from https://doi.org/10.1080/02652030802007553.

Tsao, T. M., Chen, Y. M., & Wang, M. K. (2011). Origin, separation and identification of environmental nanoparticles: A review. *Journal of Environmental Monitoring*, *13*(5), 1156–1163. Available from https://doi.org/10.1039/c1em10013k.

Usman, M., Farooq, M., Wakeel, A., Nawaz, A., Cheema, S. A., Rehman, H. U., Ashraf, I., & Sanaullah, M. (2020). Nanotechnology in agriculture: Current status, challenges and future opportunities. *Science of the Total Environment*, *721*. Available from https://doi.org/10.1016/j.scitotenv.2020.137778.

USEPA, Nanotechnology White Paper, 100/B-07/001, February 2007.

Van der Ploeg, M. J. C., Handy, R. D., Waalewijn-Kool, P. L., van den Berg, J. H. J., Herrera Rivera, Z. E., Bovenschen, J., Molleman, B., Baveco, J. M., Tromp, P., Peters, R. J. B., Koopmans, G. F., Rietjens, I. M. C. M., & Van den Brink, N. W. (2014). Effects of silver nanoparticles (NM-300K) on *Lumbricus rubellus* earthworms and particle characterization in relevant test matrices including soil. *Environmental Toxicology and Chemistry*, *33*(4), 743–752. Available from https://doi.org/10.1002/etc.2487.

Vandenabeele, P., Jehlička, J., Vítek, P., & Edwards, H. G. M. (2012). On the definition of Raman spectroscopic detection limits for the analysis of biomarkers in solid matrices. *Planetary and Space Science*, *62*(1), 48–54. Available from https://doi.org/10.1016/j.pss.2011.12.006.

Walser, T., Limbach, L. K., Brogioli, R., Erismann, E., Flamigni, L., Hattendorf, B., Juchli, M., Krumeich, F., Ludwig, C., Prikopsky, K., Rossier, M., Saner, D., Sigg, A., Hellweg, S., Günther, D., & Stark, W. J. (2012). Persistence of engineered nanoparticles in a municipal solid-waste incineration plant. *Nature Nanotechnology*, *7*(8), 520–524. Available from https://doi.org/10.1038/nnano.2012.64.

Wang, J., Nabi, M. M., Mohanty, S. K., Afrooz, A. N., Cantando, E., Aich, N., & Baalousha, M. (2020). Detection and quantification of engineered particles in urban runoff. *Chemosphere*, *248*. Available from https://doi.org/10.1016/j.chemosphere.2020.126070.

Wang, M., Gao, B., & Tang, D. (2016). Review of key factors controlling engineered nanoparticle transport in porous media. *Journal of Hazardous Materials*, *318*, 233–246. Available from https://doi.org/10.1016/j.jhazmat.2016.06.065.

Wang, P., Menzies, N. W., Dennis, P. G., Guo, J., Forstner, C., Sekine, R., Lombi, E., Kappen, P., Bertsch, P. M., & Kopittke, P. M. (2016). Silver nanoparticles entering soils via the wastewater-sludge-soil pathway pose low risk to plants but elevated Cl concentrations increase Ag bioavailability. *Environmental Science and Technology*, *50*(15), 8274–8281. Available from https://doi.org/10.1021/acs.est.6b01180.

Wang, S., Zhao, M., Zhou, M., Li, Y. C., Wang, J., Gao, B., Sato, S., Feng, K., Yin, W., Igalavithana, A. D., Oleszczuk, P., Wang, X., & Ok, Y. S. (2019). Biochar-supported nZVI (nZVI/BC) for contaminant removal from soil and water: A critical review. *Journal of Hazardous Materials*, *373*, 820–834. Available from https://doi.org/10.1016/j.jhazmat.2019.03.080.

Weir, A., Westerhoff, P., Fabricius, L., Hristovski, K., & Von Goetz, N. (2012). Titanium dioxide nanoparticles in food and personal care products. *Environmental Science and Technology*, *46*(4), 2242–2250. Available from https://doi.org/10.1021/es204168d.

Wimmer, A., Markus, A. A., & Schuster, M. (2019). Silver nanoparticle levels in river water: Real environmental measurements and modeling approaches—A comparative study. *Environmental Science and Technology Letters*, *6*(6), 353–358. Available from https://doi.org/10.1021/acs.estlett.9b00211.

Xing, B., Vecitis, C. D., & Senesi, N. (2016). *Challenges Facing the Environmental Nanotechnology Research Enterprise. Engineered nanoparticles and the environment: biophysicochemical processes and toxicity*. John Wiley & Sons.

Yang, X., Wan, Y., Zheng, Y., He, F., Yu, Z., Huang, J., Wang, H., Ok, Y. S., Jiang, Y., & Gao, B. (2019). Surface functional groups of carbon-based adsorbents and their roles in the removal of heavy metals from aqueous solutions: A critical review. *Chemical Engineering Journal*, *366*, 608–621. Available from https://doi.org/10.1016/j.cej.2019.02.119.

Yang, Y., Wang, Y., Westerhoff, P., Hristovski, K., Jin, V. L., Johnson, M. V. V., & Arnold, J. G. (2014). Metal and nanoparticle occurrence in biosolid-amended soils. *Science of the Total Environment*, *485–486*(1), 441–449. Available from https://doi.org/10.1016/j.scitotenv.2014.03.122.

Zhang, H., Freitas, D., Kim, H. S., Fabijanic, K., Li, Z., Chen, H., Mark, M. T., Molina, H., Martin, A. B., Bojmar, L., et al. (2018). Identification of distinct nanoparticles and subsets of extracellular vesicles by asymmetric flow field-flow fractionation. *Nature Cell Biology*, *20*(3). Available from https://doi.org/10.1038/s41556-018-0040-4.

Chapter 7

Fate and transport of engineered nanoparticles in soils and groundwater

Huimin Sun[1], Shi Zhou[1], Yanji Jiang[1], Xianglong Xi[1], Yuehui Tan[1], Guangcai Zhang[1], Nan Jiang[1], Ting Zhou[1], Xianqiang Yin[1], Mei Wang[2] and Bin Gao[3]

[1]*College of Natural Resources and Environment, Northwest A&F University, Yangling, P.R. China,* [2]*School of Water Conservancy and Environment, Zhengzhou University, Zhengzhou, P.R. China,* [3]*Department of Agricultural and Biological Engineering, University of Florida, Gainesville, FL, United States*

Chapter Outline

Emerging Contaminants in Soil and Groundwater Systems. DOI: https://doi.org/10.1016/B978-0-12-824088-5.00003-3

7.1 Overview

7.1.1 Classification and applications of manufactured nanomaterials

As shown in the previous chapter, nanoparticles (NPs) include natural nanoparticles (NNPs) and man-made nanoparticles. Both intentional and unintentional human activities can produce nanoscale materials that can be further classified into incidental nanoparticles (INPs) and engineered nanoparticles (ENPs). Fig. 7.1 provides some examples of NNPs, INPs, and ENPs. As shown in Fig. 7.1, ENPs are used in a wide range, covering the energy sector, telecommunications sector, computers, agrochemicals, and personal care

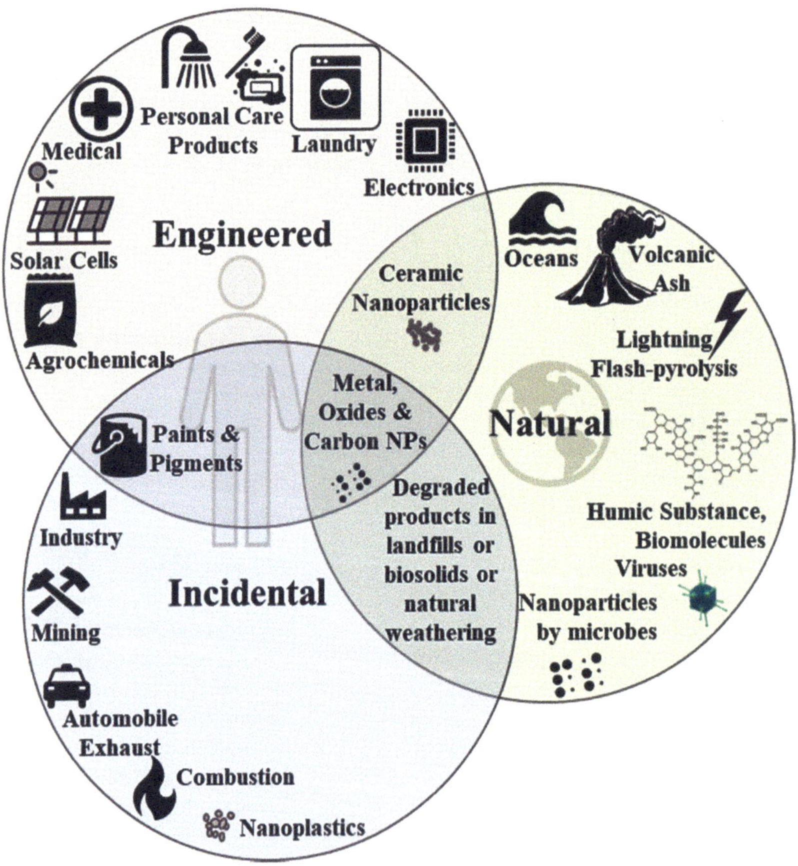

FIGURE 7.1 Classification of nanomaterials. *From Malakar, A., Kanel, S. R., Ray, C., Snow, D. D., & Nadagouda, M. N. (2021). Nanomaterials in the environment, human exposure pathway, and health effects: A review.* Science of the Total Environment, 759, *143470–143488. https://doi.org/10.1016/j.scitotenv.2020.1434704.*

products. The use of ENPs is increasing every day, causing more and more ENPs to enter the soil and groundwater systems (Malakar et al., 2019).

The most commonly encountered ENPs in the environment can be categorized into different groups based on their compositions. Detailed on the definition and classification of various ENPs can be found in the previous chapter. Here, we only give a brief overview of the applications of some of the ENPs that have been heavily investigated for their fate and transport in soils and groundwater.

Carbon ENPs such as graphene and its derivatives, carbon nanotubes (CNTs), and fullerenes (e.g., C_{60}) have many potential applications (Malakar et al., 2021). For example, CNTs have an annual output of approximately 270 tons and are used in microelectronics, medicine, and hydrogen fuel cell technology (Malakar et al., 2021). In addition, they have also used in water treatment (Kanel et al., 2016; Pokhrel et al., 2017). Graphene oxide (GO) is an oxidized derivative of graphene and is considered to be a recently developed nanomaterial that has promising applications in many fields (Chen et al., 2020). Another group of carbon ENPs is composed of polymer organic molecules, such as chitosan, cellulose, polylactic acid, polyacrylonitrile, polyhydroxyalkanoate, and poly (lactic-glycolic acid copolymer) (Hauser et al., 2019). These ENPs can have various morphologies, but are usually nanospheres or nanocapsules (Khan et al., 2019). Because they are biodegradable and biocompatible, polymeric ENPs can be used in biomedical applications as a safe and effective delivery tool for antigens, proteins, drugs, and vaccines (Han et al., 2018; Hashem & Gonzalez-Bulnes, 2020).

Metal-based ENPs including nanosized metals and metal oxides are also among the most popular nanomaterials. Metal ENPs are usually produced by a bottom-up method and are strictly controlled for customized forms. Their nanocrystalline structures can be defect-free. These ENPs are widely used in optoelectronics (Gad & Hegazy, 2019). Because they are highly reactive (zero valent), metal ENPs have also been applied in environmental remediation (Zou et al., 2016). As one of the most promising environmental remediation agent, nanosized zero-valent iron (NZVI) has been directly applied to the soils and groundwater to remove various contaminants (Zou et al., 2016). Other zero-valent metal nanomaterials such as gold (Au) and gold-based ENPs are often used in sensor applications (Xiao et al., 2020). Nanosized metal/metalloid oxides have strong physical and chemical stability, which makes them good modifiers of other materials for enhanced properties and applications (Malakar et al., 2021).

7.1.2 ENPs in soils and groundwater

The likelihood that ENPs end up in the soil and groundwater environment depends largely on the amount of industrial production and how they are used. As shown in Chapter 6, Occurrences and Impacts of Engineered

Nanoparticles in Soils and Groundwater, ENPs released into the environment may found different pathways to enter the soils and groundwater (Fig. 7.2). Various estimates of the number of different ENPs currently being produced have been published (Bogdal et al., 2010; Castellaneta et al., 2012; He et al., 2014; Hendren et al., 2011). Giese et al. estimated that 17,080 tons of nano-Ag, CeO_2, and SiO_2 were released in the environmental compartment in 2017 through the use phase (2.5–3 times more than in the technical compartment) and 7 tons in the treatment phase (Giese et al., 2018). However, such estimates of the annual production of different ENPs are highly uncertain and vary considerably (often as much as two to three orders of magnitude) not only between studies, but also between different assumptions for the same study. For example, Hendren et al. estimate annual production of titanium dioxide NPs in the United States at between 7800 and 38,000 tons (Hendren et al., 2011), while global production estimates are lower (Castellaneta et al., 2012). Despite the wide range of these estimates, it is clear that some ENPs are produced in tons per year and therefore have the potential to be released into the environment including soils in large quantities.

As nanomaterial production continues to increase, the number and diversity of ENPs released into the environment will lead to a rapid increase of

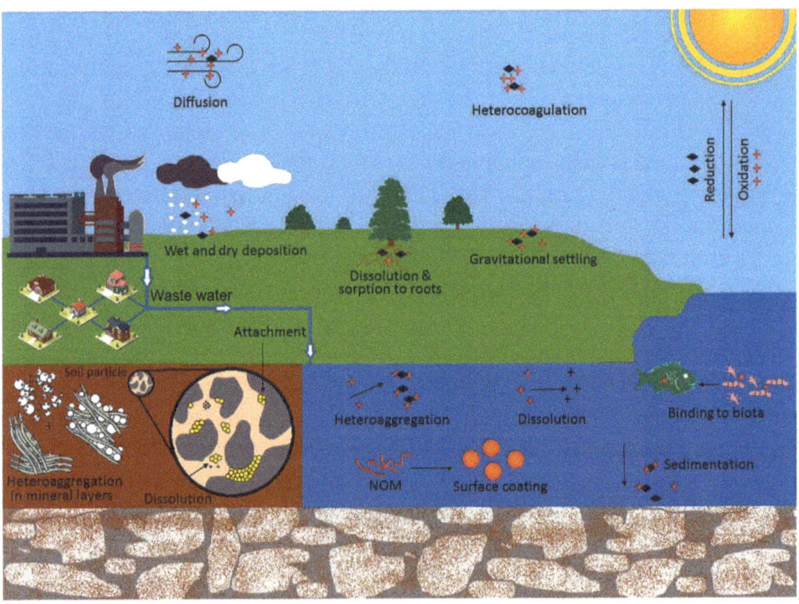

FIGURE 7.2 Pathways and transformations of ENPs in the atmosphere, aquatic, and terrestrial environment. *From Abbas, Q., Yousaf, B., Amina, Ali, M. U., Munir, M. A. M., El-Naggar, A., Rinklebe, J., & Naushad, M. (2020). Transformation pathways and fate of engineered nanoparticles (ENPs) in distinct interactive environmental compartments: A review.* Environment International, 138, *105646–105663. https://doi.org/10.1016/j.envint.2020.1056463.*

ENPs in soils and groundwater (Zhang et al., 2020). The soil and groundwater systems can serve as an important sink of ENPs in the environment. It is estimated that the soil receives of 8%–28% a year of the ENPs produced worldwide (Keller et al., 2013). Nano-pesticides contain metal-based NPs can be one of the main sources of ENPs in soils and groundwater. In addition, sewage sludge applied to land can be another important source of ENPs in soils and groundwater. The fate and transport of these ENPs in subsurface are affected by multiple factors, which are the focus of this chapter.

7.1.3 Risks of ENPs in soils and groundwater

The entry of ENPs into soils and groundwater may have significant impact on the function of soil biota. Plant is an important part of soil ecosystem, and its growth is closely related to soil environment. If the particle size of ENPs is smaller than the cell wall pore size of plant organs, they can be transferred to the plant system through roots, stomata, and leaves, and can migrate and accumulate in different parts of the body, even affect the growth and development of plants. Navarro et al. showed NPs while reacting with the elements of the cell wall may initiate the forming of new and larger pores that can increase the permeability of the cell wall (Navarro et al., 2008). In addition, ENPs can also have negative effects on soil microorganisms through inhibition of soil enzyme activities or direct toxicity. Rousk et al. demonstrated that CuO and ZnO ENPs were toxic to bacterial growth in mineral soils (Rousk et al., 2012). Moreover, ENPs also affected soil enzyme activities. Jośko et al. found that, in a soil with nano-CuO, the dehydrogenase activity increased twofold in relation to the control soil (7.4 mg TPF/kg), while the activity of dehydrogenase was significantly decreased after adding nano Cu (Jośko et al., 2019).

ENPs in groundwater can be dissolved, agglomerated, and migrated, and it has strong adsorption capacity, which can absorb other harmful substances. Because groundwater is one of the main sources of water supplies (e.g., drinking water and irrigation) world widely, the presence of ENPs can impose potential risks on public health. Moreover, ENPs in soils can groundwater can also enter the food chain, accumulate and produce toxic effects, and threaten human health. Previous studies have demonstrated that ENPs not only have toxic effects on organ system in human body, but also may affect normal reproduction and fetal development, thus affecting human healthy life (Oomen et al., 2018; Prajitha et al., 2019). Because of the high adsorption ability, ENPs can be carriers of other pollutants and enter the organism in the form of particulate pollutant complexes, which increases the toxicity of ENPs to organisms. Jia et al. found that ZnO ENPs and Pb^{2+}coexposure enhanced the Pb accumulation in all major organs of mice to cause damages (Jia et al., 2017). It is thus important to understand the fate and transport of ENPs in soils and groundwater.

7.2 Transformation and stability of subsurface ENPs

7.2.1 Transformation of ENPs in the subsurface environment

ENPs in the subsurface environment may undergo various physical, chemical, and biological transformation processes (Abbas et al., 2020). The dynamic transformations of ENPs in the environment are shown in Fig. 7.3. After entering the subsurface, different ENP transformation processes may take place, such as homo- or hetero-agglomeration, dissolution/sedimentation, sulfidation, adsorption, oxidation, chemical reduction, and biologically mediated reactions, which would greatly affect their fate and transport in soils and groundwater (Meseret et al., 2017). These transformation processes play an important role in controlling the fate and transport of ENPs in soils and groundwater. The most important physicochemical characteristics of ENPs that are affected by their transformation in the subsurface environment include particle size, surface area, zeta potential or surface charge, colloidal stability, and core-shell composition (Bhatt & Tripathi, 2011).

7.2.2 Stability of ENPs in the subsurface environment

Several different natural colloids such as clay minerals, Fe and Mn hydrous oxides, organic matter and extracellular polymeric substances are presented in the soil and groundwater environment (Luo et al., 2017), which may strongly affect the agglomeration stability of ENPs in the subsurface environment. The phenomenon of agglomeration in a natural system is driven by Brownian movement, differential settling, and fluid motion (Peng et al., 2010; Shi et al., 2020). Natural colloids are ubiquitous in the soil and groundwater systems, the interactions between these colloids and ENPs can either enhance or suppress the agglomeration/stability of ENPs depending upon the environmental conditions (Ma et al., 2018).

In natural groundwater and soils, some metallic elements (e.g., Ag, Cu, and NZVI) and metal oxide (e.g., Fe oxides) ENPs are susceptible to the oxidation−reduction process (Mudunkotuwa et al., 2012). On one hand, redox reactions form an insoluble coating on ENP surface to strengthen their stability in the environment. On the other hand, these reactions can also decrease the stability of ENPs by the development of an active surface layer to facilitate the interactions between the particles and other subsurface components (Amde et al., 2017). Furthermore, surface atom oxidation promotes the dissolution of ENPs.

The organic substances of soils also influence the adsorption of NPs onto soil surface and hence enhance the stability of NPs. ENPs can also be dissolved in soil pore water and their dissolution largely depends on the type and physicochemical properties of soils (Ahmed et al., 2021).

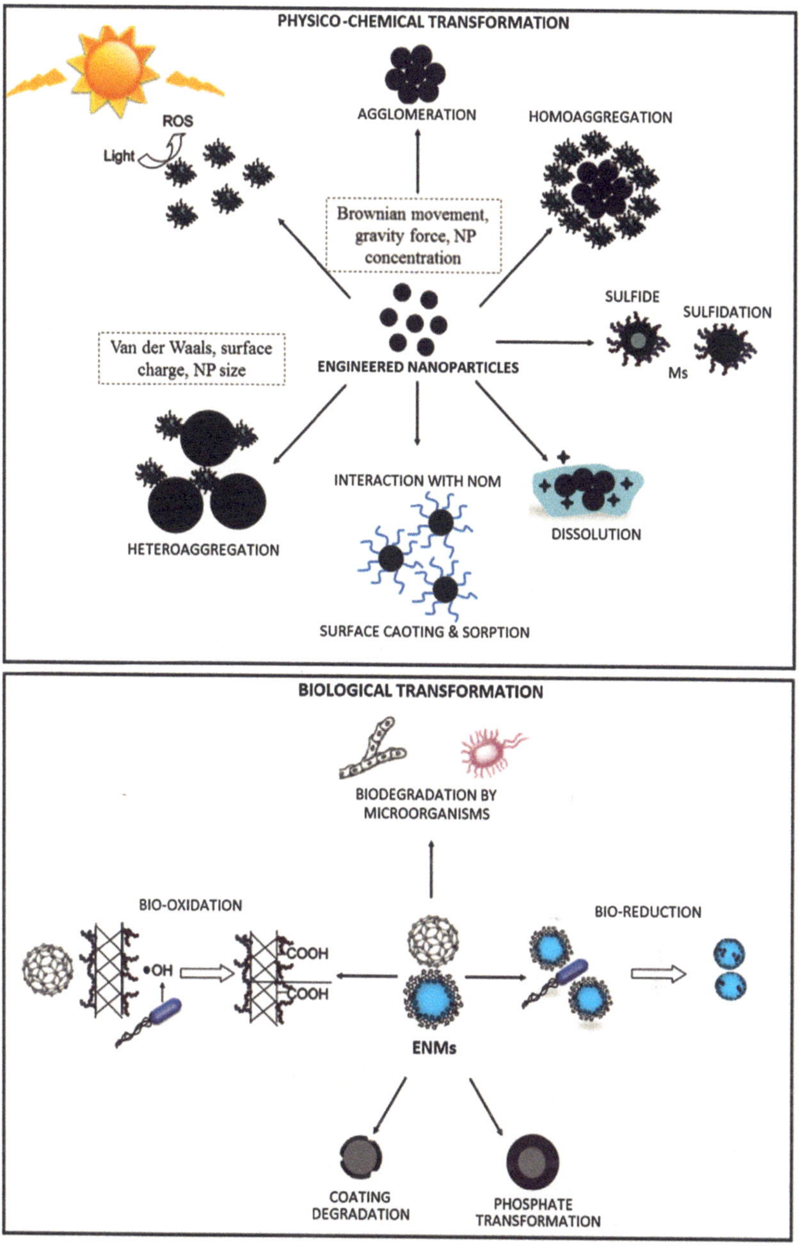

FIGURE 7.3 Physicochemical and biological transformations of ENPs. *From Abbas, Q., Yousaf, B., Amina, Ali, M. U., Munir, M. A. M., El-Naggar, A., Rinklebe, J., & Naushad, M. (2020). Transformation pathways and fate of engineered nanoparticles (ENPs) in distinct interactive environmental compartments: A review.* Environment International, 138, *105646–105663. https://doi.org/10.1016/j.envint.2020.1056463.*

7.3 Retention, transport, and release

The transport behaviors of ENPs in the subsurface can be regarded as functions of their own properties (such as particle size, surface properties, and shape), physical and chemical properties of porous media (such as porosity, saturation, and type), physical and chemical properties of pore water [such as pH, ionic strength (IS), ionic type, and flow rate] and other environmental factors (such as temperature, organic matter, and clay minerals) (Sun et al., 2015; Wang et al., 2016). The existing researches on the fate and transport of ENPs in the soil and groundwater system mainly focus on the above-mentioned influencing factors, have changed from the single influencing factor to the synergistic effect of multiple influencing factors (Fig. 7.4), and a lot of research results have been obtained (Wang et al., 2016).

7.3.1 Effects of physicochemical properties of ENPs

7.3.1.1 Particle size

The physicochemical characteristics of ENPs are very important to their fate and transport in soils and groundwater. Compared with primary particles, aggregates have a smaller ratio of total surface area to volume, and the active sites are more accessible (Espinasse et al., 2018). Thus the reactivity and bioavailability of the aggregated ENPs in the subsurface are reduced (Amde

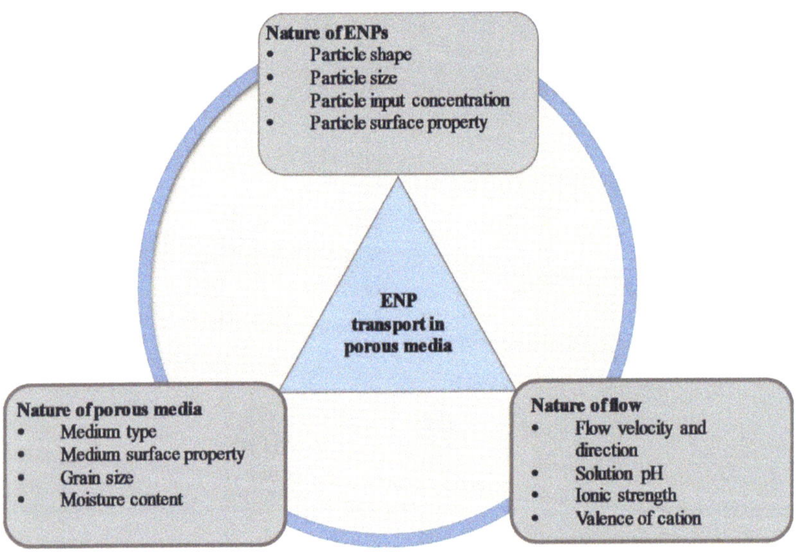

FIGURE 7.4 Influential factors of the fate and transport of ENPs in soils and groundwater. *From Wang, M., Gao, B., & Tang, D. (2016). Review of key factors controlling engineered nanoparticle transport in porous media.* Journal of Hazardous Materials, 318, *233–246. https://doi.org/10.1016/j.jhazmat.2016.06.0651.*

et al., 2017; Lowry et al., 2012). For example, nanoscale Fe_2O_3 (40 nm), titanium dioxide (216 nm), and silver (60 nm) particles aggregate into micron-sized aggregates after being released into different environmental waters, and their fluidity in the environment is reduced (Chekli et al., 2015). The time scale of ENPs size growth is mainly determined by the intrinsic physicochemical properties of particles and atmospheric environmental conditions (John et al., 2017).

Some scholars have shown the change of particle size of NPs when studying the transport characteristics of ENPs, and the particle size increased during the transport in porous media (Wang, Gao, et al., 2012). Specifically, the mean and standard deviation of particle size of ENPs increased with the increase of transport distance. This means that NP transport is accompanied by polymerization processes. Sun et al found that the particle size out of the porous media column was much greater than the particle size out before entering the column ($R_{effulent} > R_{retentate} > R_{influent}$), and with the increase of transport distance, the particle size in the column becomes larger and larger, and the dispersion degree between particle size also shows an increasing trend (Fig. 7.5) (Sun et al., 2015). The simultaneous polymerization and transport will complicate the environmental behavior of ENPs in soils and

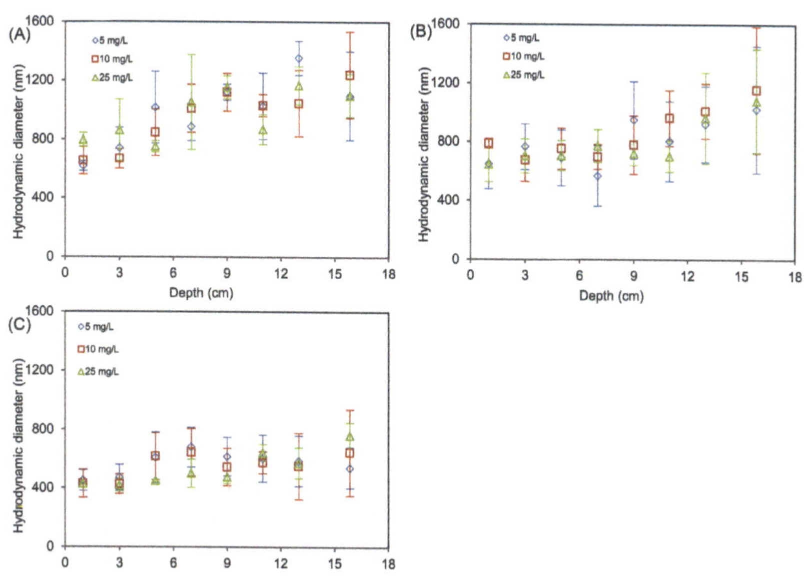

FIGURE 7.5 Size distribution of retained GO in coarse (A), medium (B), and fine (C) sand columns. Data are expressed as means ± standard deviation ($n = 4-9$). *From Sun, Y., Gao, B., Bradford, S. A., Wu, L., Chen, H., Shi, X., & Wu, J. (2015). Transport, retention, and size perturbation of graphene oxide in saturated porous media: Effects of input concentration and grain size.* Water Research, 68, *24–33. https://doi.org/10.1016/j.watres.2014.09.02530.*

groundwater. Some other scholars have demonstrated the phenomenon of "particle size selective" filtering when studying the fate and transport of ENPs in porous media and found that the average particle size decreased with the increase of transport distance (Braun et al., 2015; Wang et al., 2015; Xia et al., 2015). They have analyzed the particle size of $R_{effentate}$ out of the media, $R_{retentate}$ stuck in the media, and $R_{influent}$ in the original flow, and found that the particle size from large to small is: $R_{effulent} > R_{retentate} > R_{influent}$. Wang et al. found that porous media can filter the large-size particles preferentially (Wang et al., 2015). In their study, the small-size particles were the first to migrate out of the porous media, forming the first penetration peak, while the large-size particles lagged out and forming the second penetration peak (Fig. 7.6). At high IS, the particle size distribution of ENPs is more heterogeneous, and the "particle size selective" filtering of NPs by porous media is more significant (Wang, Park, et al., 2017).

It can be seen that during their transport in porous media, ENP size will change in space and time and their heterogeneity may further expand or decrease under different subsurface environment conditions. This makes the fate and transport of ENPs in soils and groundwater hard to be accurately predicted.

7.3.1.2 Particle shape

Previous studies have examined the effects of particle shape on ENP transport in porous media (Zhang, Zhu, et al., 2016). The exact results of convective boundary conditions and velocity slip for ENPs of four different geometries (sphere, brick, cylinder, and sheet) have been studied (Iftikhar et al., 2021). The results indicate that particle shape has strong influences on the transport behaviors of ENPs in porous media. Theoretical tests of the formulas show that spherical NPs have a greater effect on the temperature and velocity curves than ENPs of other shapes (Wang, Park, et al., 2017). Therefore there are great differences in the transport among ENPs of different shapes in soils and groundwater. However, the exact mechanisms of the shape effect on ENP transport in the subsurface is still unclear. Most of the previous studies of the fate and transport of ENPs have ignored the shape effect and treated the NPs as spherical ones regardless of their actual shapes (e.g., cylindered CNTs and GO nanosheets) (Prabhavathy & Arivuoli, 2021).

7.3.1.3 Particle concentration

Under constant experimental conditions, the initial NP concentration has a considerable effect on their transport and retention in porous media (He & Zhao, 2007). Related studies have shown that the mobility of the same ENPs in porous media varies at different NP concentrations (Wu et al., 2019). There are contradictory results on effect of particle concentration on ENP transport in soils, which show a promotion of particle transport with

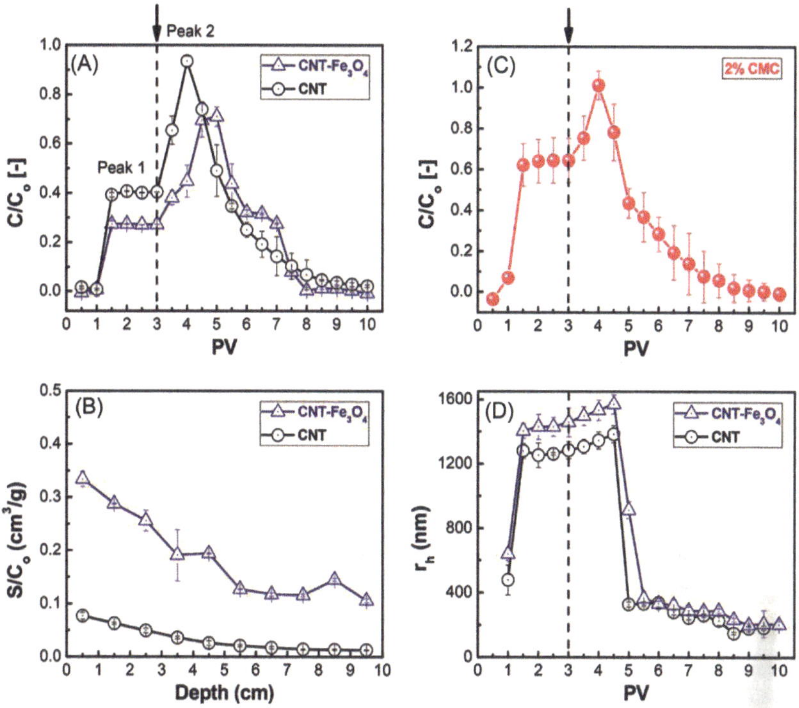

FIGURE 7.6 Breakthrough curves (BTCs; A) and retention profiles (RPs; B) of CNT-Fe_3O_4 NHs and individual CNT, respectively, in water-saturated sand columns at 1 mM NaCl and pH 7.3 (in 2% CMC). Breakthrough curve describes the normalized effluent concentration of NMs, C/Co (where Co is the initial influent concentration of CNT-Fe_3O_4 NHs or CNT) as a function of pore volume (PV); and retention profile shows the normalized solid-phase retention concentration of NMs, S/Co (where S is retention amount of CNT-Fe_3O_4 NHs or CNT per gram dry sand) as a function of distance from the column inlet. For column transport experiments, 3 PVs (marked by *dash line* in A) of CNT-Fe_3O_4 NH or CNT influent were injected in the column followed by elution with 7 PVs of NM-free background electrolyte solution (without CMC). Other column transport parameters were the same including: CNT-Fe_3O_4 NH or CNT influent concentration = 10 mg/L, influent pH = 7.3; background electrolyte = 1 mM NaCl; average sand grain size = 360 μm; and Darcy velocity = 0.44 cm/min. For comparison, the breakthrough curve of 2% CMC alone under the experimental condition (1 mM NaCl and pH 7.3) identical to that of NHs/CNT is shown in (C). Figure (D) shows the average hydrodynamic radius (r_h) of CNT-Fe_3O_4 NHs and CNT in the effluents (collected from A). The *error bars* represent the standard deviations from duplicate experiments. Note different *y*-axis scales in figure. *From Wang, D., Park, C., Masud, A., Aich, N., & Su, N. (2017). Carboxymethylcellulose mediates the transport of carbon nanotube—Magnetite nanohybrid aggregates in water-saturated porous media.* Environmental Science and Technology, *51(21), 12405–12415. https://doi.org/10.1021/acs. est.7b0403712409.*

increasing concentration and an intensification of particle retention in porous media at higher concentrations (Phenrat et al., 2009; Xia et al., 2019). For example, the effect of input particle concentration on the transport and retention of AgNPs in soil has been systematically investigated (He et al., 2019). The breakthrough curve (BTC) particle number and retention profiles (RPs) of silver nanoparticles (AgNPs) with different input particle concentrations ($C_0 = 2.5$, 5.0, and 10 mg/L) were observed and simulated under TEM with size = 15.0 nm and surface coating = 0.5% polyvinylpyrrolidone (PVP) (Fig. 7.7). The normalized emission concentration of radionuclides (C/C_0) increases with the increase of input concentration (C_0). Higher C_0 correspond to steeper slope of the BTC, indicating that high concentration of AgNPs can promote the down gradient mobility (He et al., 2019). This concentration-dependent transport behavior of AgNPs can be reasonably explained by the blocking effect (Bradford & Bettahar, 2006). That is, at higher C_0, the attachment sites are occupied rapidly over time. Liang observed a shape shift of RPs from super-exponential to nonmonotone to homogeneous when increasing from 1 to 5 and then to 10 mg/L. The development of nonmonotonic RPs may be attributed to time-dependent blockage (Liang et al., 2013). Similar results have been reported in other publications that the input particle concentration has a significant effect on the mobility of ENPs in porous media (Kasel et al., 2013; Sun et al., 2015).

Most existing studies have been conducted at low particle concentrations (30 < mg/L), whereas ENPs, especially the ones for remediation, may have much higher particle concentrations in soils and groundwater (Phenrat et al., 2009). Moreover, many studies have examined particle transport in distilled water or low IS, which is not related to natural groundwater chemistry. Particle transport and retention have been performed abroad by injecting

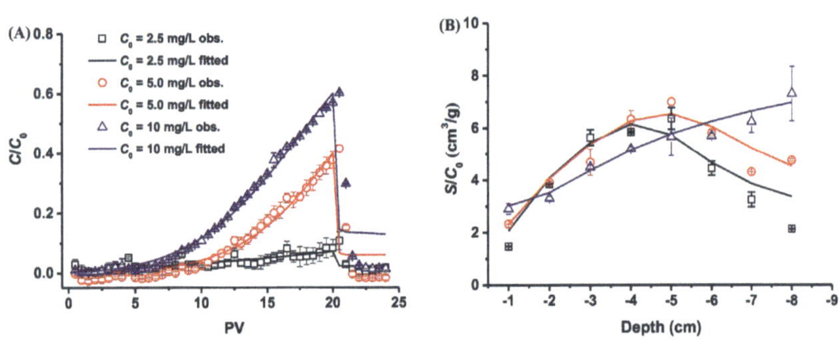

FIGURE 7.7 Observed and simulated BTCs (A) and RPs (B) for AgNPs at different input concentrations (C_0). The BTCs are plotted as normalized effluent concentration (C/C_0) of AgNPs as a function of PVs. The RPs are plotted as normalized solid-phase concentration (S/C_0) versus depth from the column inlet. *Open symbols* are experimental data and *solid lines* are fitted results using the two-species model since nonmonotonic RPs occur. *Error bars* represent the standard deviations in duplicate experiments. *From He, J., Wang, D., & Zhou, D. (2019). Transport and retention of silver nanoparticles in soil: Effects of input concentration, particle size and surface coating.* Science of the Total Environment, 648, *102–108. https://doi.org/10.1016/j.scitotenv.2018.08.136107.*

stable NZVI suspensions at various initial concentrations (0.1, 0.5, and 2 g/L) into saturated sand columns. The results show that the deposition of NZVIs was significant at higher initial particle concentrations. The maximum relative concentration (C_{max}/C_0) decreased from 0.601 to 0.267 when the initial concentration of PAA-NZVI was increased from 0.1 to 2 g/L. However, in the case of PVP-NZVI, C_{max}/C_0 decreased from 0.307 to 0.07. Usually, increasing the initial concentration of each polymer-stabilized NZVI leads to more retention in porous media (Esfahani et al., 2014). Phenrat et al also noted that the transport of NZVI stabilized by polystyrene sulfonate was significantly lower at higher concentrations compared to lower concentrations (Phenrat et al., 2008).

7.3.1.4 Particle surface properties

Previous studies have shown that the transport and retention of ENPs in porous media highly depend on the surface properties of particles (e.g. morphology, surface charge and functional group distribution) (Dong et al., 2019; Wang et al., 2018). For example, carbonaceous nanomaterials are usually negatively charged and can easily migrate through soils that are usually negatively charged (Park et al., 2016). Xia et al. experimentally investigated the transport of GO, reduced GO (RGO), and polystyrene nanoplastics (PSNP) in saturated quartz sand (Fig. 7.8) (Xia et al., 2021). Under the experimental conditions, all three NPs were negatively charged. In selected background solutions, all the three carbon-based nanomaterials exhibited considerable mobility compared to metal/metal oxide based nanomaterials (Afrooz et al., 2016). The order of their mobility is GO > RGO > PSNP because of the decrease in negative surface charge and increase in hydrophobicity. Compared with GO and RGO, PSNP has less negative charge, which also leads to a weaker electrostatic repulsion between PSNP and quartz sand, thus promoting the retention of PSNP on the sand surface. Moreover, NPs with different charge properties are likely to interact with each other once they are released into the soils and groundwater. For example, iron oxide NPs can inhibit the transport of negatively charged bacteria, plastics, and hydroxyapatite NPs because positively charged iron oxide NPs are preferentially deposited on porous media and induce aggregation of other types of particles through electrostatic attraction (Cai et al., 2013).

The type of functional groups on the surface of ENPs determines their functionality, and NPs with different functional groups exhibit significantly different transport and retention properties in saturated media (Xia et al., 2019). A comparison of GO and RGO revealed that, after reduction, the carbon to oxygen ratio of GO increased from 2.09 in the original to 2.97 in RGO, while the content of epoxy/hydroxyl, carboxyl, and carbonyl groups on the surface of GO decreased significantly. As a result, the mobility of RGO in saturated media is lower than that of GO. The oxygen functional

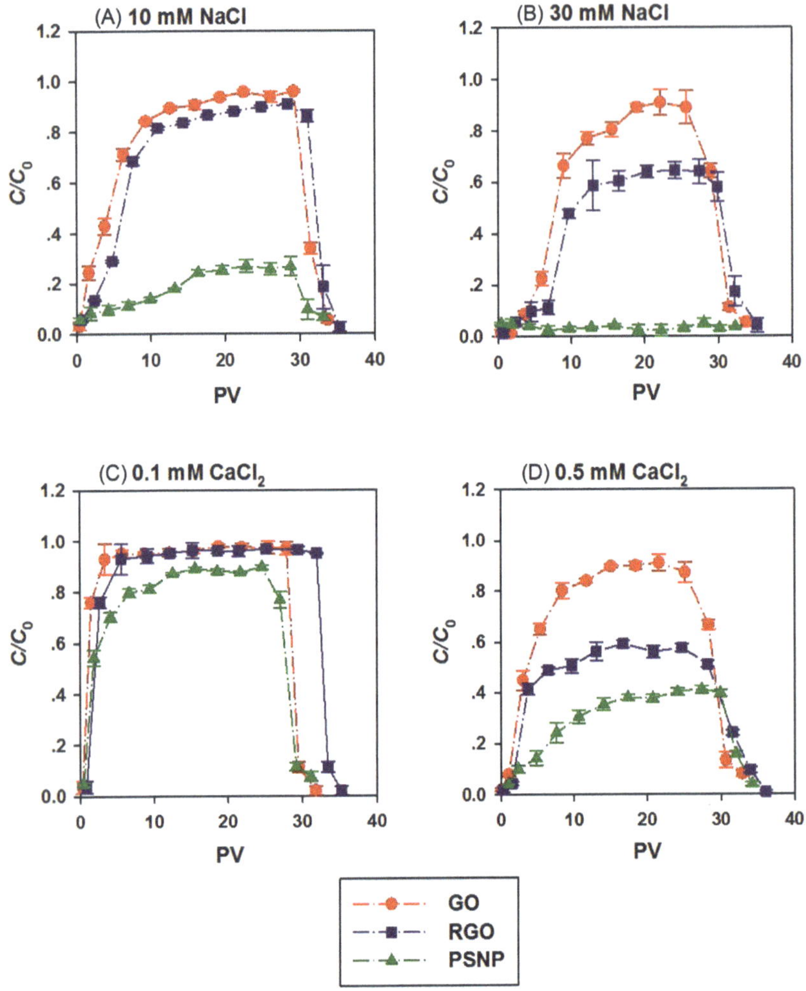

FIGURE 7.8 Breakthrough curves of GO, RGO, and PSNP in the single transport system in (A) 10 mM NaCl; (B) 30 mM NaCl; (C) 0.1 mM $CaCl_2$, and (D) 0.5 mM $CaCl_2$ electrolyte solutions. *From Xia, T., Lin, Y., Li, S., Yan, N., Xie, Y., He, M., Guo, X., & Zhu, L. (2021). Co-transport of negatively charged nanoparticles in saturated porous media: Impacts of hydrophobicity and surface O-functional groups.* Journal of Hazardous Materials, 409, *124477. https://doi.org/10.1016/j.jhazmat.2020.124477S14.*

group of RGO is less than that of GO, so the surface negative charge of RGO is also less, and thus its mobility is lower. Previous studies have also examined the effect of different types of surface functional groups on the fate and transport of ENPs in soils and groundwater (Esfahani et al., 2014; Yecheskel et al., 2019). Plastic nanoparticles are considered as emerging contaminants, and their transport behavior in the soil-groundwater environment has been studied in detail, showing that polystyrene NPs with surface functional groups have different mobility profiles (Shaniv et al., 2020).

Most ENPs incorporated into commercial products are surface modified, which can strongly influence their surface chemistry and thus their fate and transport in soils and groundwater (Baddar et al., 2019; Levard et al., 2012). For example, AgNPs are usually coated with capping agents such as ligands (e.g., citrate) and polymers (e.g., polyvinylpyrrolidone and polyvinylpyrrolidone). An indoor column experiment was conducted in this regard to systematically investigate the effect of surface coatings on the transport and retention of AgNPs in soil (He et al., 2019). The results demonstrate that the presence of surface coatings (i.e., PVP and citrate) further promoted the transport and decreased the retention of AgNPs in soil (Fig. 7.9), probably because they effectively blocked the solid phase sites that could otherwise be used for AgNP retention. Wen et al. investigated the effect of two different coating types on NP transport in porous media (Wen et al., 2020). They found that polymer surface coatings have a significant effect on NP transport in porous media with spatial repulsion being the main force leading to more distant transport. These results suggest that surface modification of ENPs with polymers, polyelectrolytes or surfactants can greatly enhance their mobility in the subsurface (Liu et al., 2016; Zhao et al., 2016). The modifiers attach to ENPs surface and generate or enhance repulsive forces, including permeation repulsion, electrostatic double-layer repulsion and electrostatic repulsion, to counter van der Waals forces and any magnetic attraction, thus affecting their stability and their fate and transport in soils and groundwater (Sangani et al., 2018).

7.3.2 Effects of physicochemical properties of porous media

7.3.2.1 Grain size and medium heterogeneity

Many studies have found that the smaller the particle size of porous media, the more ENPs deposited on the surface of the media (Sun et al., 2015;

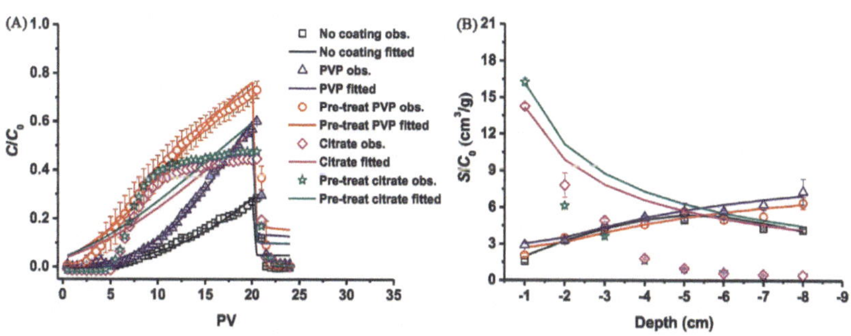

FIGURE 7.9 Observed and simulated BTCs (A) and RPs (B) for AgNPs with different surface coatings. *From He, J., Wang, D., & Zhou, D. (2019). Transport and retention of silver nanoparticles in soil: Effects of input concentration, particle size and surface coating.* Science of the Total Environment, 648, *102–108. https://doi.org/10.1016/j.scitotenv.2018.08.136107.*

Wang, Gao, et al., 2017). The pore surface area increases with the decrease of the particle size of the medium, and the attachment sites provided by the medium increase accordingly (Kamrani et al., 2018). In addition, the smaller the particle size of the medium, the easier it is to form small pores, which increases the possibility of the medium retaining ENPs through physical blocking (Xin et al., 2016). Kamrani et al. used quartz sand with different particle sizes as the medium to explore the transport and retention of carbon dots (CDs) (Kamrani et al., 2018). The results show that the CDs outflow decreases with the decrease of quartz sand particle size (Fig. 7.10), mainly because the decrease of quartz sand particle size increases the mass transfer rate of CDs to the medium surface. CDs have more contact with the reserved position on quartz particles, and the residence time increases. In addition, the decrease of grain size of quartz sand increases the specific surface area of the medium, increases the available adsorption sites, and decreases the effective porosity, which is due to the increase of dead-end pores. The study of Lv et al. shows that the retention rate of nano-TiO_2 particles in sand columns increases with the decrease of the medium particle size under three kinds of IS (Lv et al., 2016). Hou et al. reported that the mobility of poly vinylpyrrolidone-coated AgNPs can be increased by increasing the grain size of quartz sand (Hou et al., 2017).

Most of the previous studies on the influence of medium particle size on the transport of ENPs in porous media have used packed columns with relatively homogeneous particle size range, and ignored the influence of spatial structure heterogeneity caused by the difference of medium particle size.

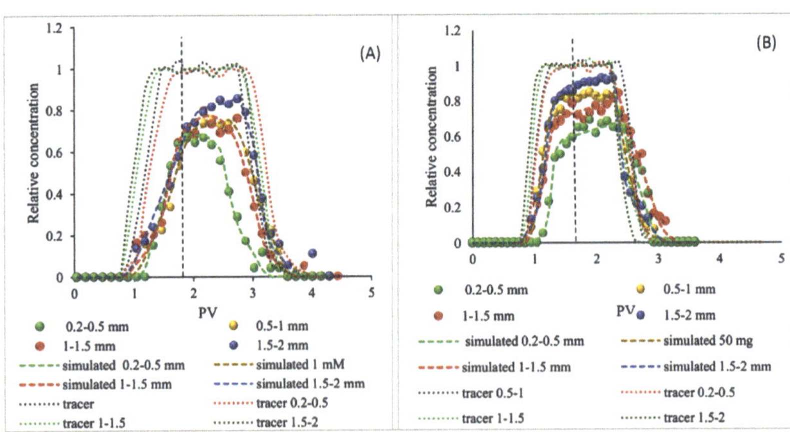

FIGURE 7.10 Breakthrough curves (BTCs) and fitted simulation (R2 > 0.9) for different grain diameters in (A) downward flow and (B) upward flow. *From Kamrani, S., Rezaei, M., Kord, M., & Baalousha, M. (2018). Transport and retention of carbon dots (CDs) in saturated and unsaturated porous media: Role of ionic strength, pH, and collector grain size.* Water Research, 133, *338–347. https://doi.org/10.1016/j.watres.2017.08.045344.*

Only few studies have tested the transport of ENPs in heterogeneous porous media and found that spatial heterogeneity plays an important role in controlling the mobility of ENPs in porous media (Chen et al., 2018; Lv et al., 2016). Some large pore channels in porous media are often caused by the particle size difference of soil aggregates and the objective existence of soil stratification, which form preferential flow paths to dominate the transport of ENPs in soils and groundwater (Chen et al., 2018). Freeze thaw alternation and dry wet alternation can also change soil structures to create macropores and preferential flow paths, and thus affect the transport of ENPs (Liu et al., 2021; Xu et al., 2021). Pore scale heterogeneity and its spatial structure can also play important roles in the fate and transport of ENPs in soils and groundwater (Yin et al., 2019). Therefore more attention should be paid to the study of pore scale heterogeneity.

7.3.2.2 *Medium surface roughness*

The surface roughness of porous media has an obvious effect on the transport of ENPs in soils and groundwater. ENPs deposition in porous media with rough surfaces has been reported previously (Shen et al., 2011). The energy barrier between particles and porous media is weakened or eliminated by the high surface roughness, resulting in the deposition of particles at the primary minima (Henry et al., 2011). Surface roughness can also increase the dominant effect of viscous torque on the fluid resistance experienced by particles in the flow field (Bradford et al., 2013), thus enhancing particle adhesion or inhibiting particle separation from the surface. Liang et al. Experimentally studied the retention and release of AgNP using two sands with very different surface roughness properties in a range of solution pH and/or IS (Liang et al., 2020). The transport of AgNPs in the relatively smooth sand was significantly stronger than that in the coarse sand. When the solution decreased from 40 mM NaCl to deionized water, and then the pH of the solution increased from 6.5 to 10, it was observed that the retained AgNPs were completely released from the relatively smooth sand. On the contrary, less than 40% of the residual AgNPs were released from the coarse sand in a similar process (Figs. 7.11 and 7.12). Another study on the transport of carbon black nanoparticles (CBNP) in two soils with different surface roughness also showed that CBNP mainly remained in the surface roughness area (Lohwacharin et al., 2015). The adhesion of roughness enhanced particles may be due to the increase of contact area through the concave part of a single asperity or a single asperity, as well as the increase of surface charge inhomogeneity related to roughness (Rasmuson et al., 2017).

7.3.2.3 *Medium surface impurity*

Chemical impurities (e.g., oxides of Al, Fe, and Mn) can cause local physical or chemical inhomogeneity on the surface of the media, which directly leads

FIGURE 7.11 Atomic force microscopy images of (A) sand and (B) glass beads surfaces. *From Liang, Y., Zhou, J., Dong, Y., Klumpp, E., Šimůnek, J., & Bradford, S. A. (2020). Evidence for the critical role of nanoscale surface roughness on the retention and release of silver nanoparticles in porous media.* Environmental Pollution, 258, *113803–113811. https://doi.org/10.1016/j.envpol.2019.11380312.*

to the difference in interaction forces between ENPs and the media (Zhang et al., 2019). This is mainly because the metallic impurity on the surface of porous media can generate strong electrostatic interactions with ENPs. Liu et al. found that the retention of functionalized graphene in the column filled with natural sand was more prominent than that in the column filled with acid pickled sand (Fig. 7.13) (Liu, Gao, Wu, Yang, et al., 2013). Wang et al. demonstrated that hematite coating of porous media greatly increased the retention of GO in porous media (Wang, Shen, et al., 2017). This is because of the strong electrostatic attraction between positively charged mineral impurities and NPs on the surface of the media, which inhibits transport. Similarly, biofilm coating can also increase medium surface impurity to affect the interactions between ENPs and porous media. The mechanism of ENPs deposition on hydrophilic or hydrophobic biofilm coatings is fundamentally different, due to the different hydration characteristics of the biofilms (Mitzel et al., 2016). Natural colloids as a surface impurity can also affect the fate and transport of ENPs in soils and groundwater. Liang et al. proved that the existence of soil colloid can inhibit ENP deposition by changing the surface morphology of soil particles and the shallow primary minimum interaction, thus promoting the release and cotransport of ENPs in the subsurface (Liang et al., 2021). The presence of these impurities on the surface of porous media also brings certain surface roughness. The concave

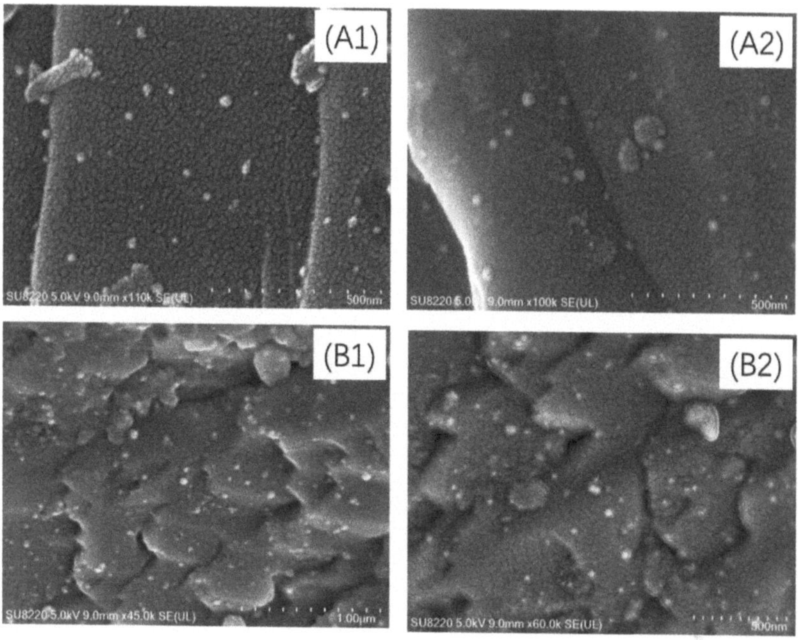

FIGURE 7.12 SEM images of retained AgNPs on relatively smooth sand (A) and rough sand (B) surfaces. *Bright dots* are AgNPs (Liang et al., 2020). *From Liang, Y., Zhou, J., Dong, Y., Klumpp, E., Šimůnek, J., & Bradford, S. A. (2020). Evidence for the critical role of nanoscale surface roughness on the retention and release of silver nanoparticles in porous media.* Environmental Pollution, 258, *113803–113811. https://doi.org/10.1016/j.envpol.2019.11380312.*

and protrusion of nanoscale have opposite effects on the transport of ENPs in the subsurface (Shen et al., 2018; Zhang, Bradford, et al., 2016). In unsaturated porous media, the increase of surface roughness can improve the surface friction and the interception effect of pore capillary, which can inhibit the transport of ENPs in porous media (Gao et al., 2008).

7.3.2.4 Environmental temperature

Temperature has different effects on the retention and migration of ENPs in porous media (Feng et al., 2019; Wang, Gao, et al., 2017; Yuan et al., 2020). The influence of temperature on the adsorption behavior of ENPs in the column system may be due to the different interaction dynamics and behavior between ENPs and the media. In general, temperature has a significant effect on the interception and migration of ENPs in porous media. When the temperature is lowered, the recovery rate increases significantly (Figs. 7.14 and 7.15) (Wang, Gao, et al., 2017; Yuan et al., 2020). When combined with particle size, temperature also has different effects on the migration of ENPs in porous media (Bras et al., 2010; Wang, Gao, et al., 2017). The recovery rate of ENPs in coarse media increases as the temperature decreases. In porous

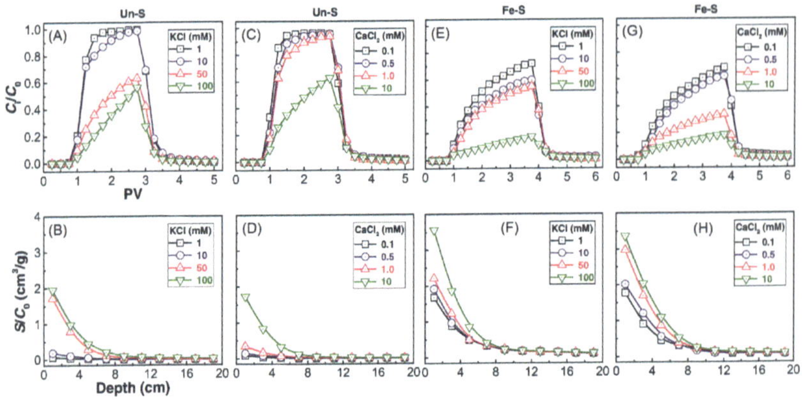

FIGURE 7.13 Retention-excavation experiments (the first-set experiments): breakthrough curves (BTCs; A, C, E, and G) and retention profiles (RPs; B, D, F, and H) of GONMs (10 mg/L at pH 7.0) under different KCl (1, 10, 50, and 100 mM; A, B, E, and F) and $CaCl_2$ concentrations (0.1, 0.5, 1.0, and 10 mM; C, D, G, and H) from Un-S (A–D) and Fe-S (E–H), respectively. In this figure (A, C, E, and G), the normalized concentration (C_i/C_0) of GONMs is plotted as a function of eluted pore volume (PV). In this figure (B, D, F, and H), the normalized solid-phase concentration (S/C_0) of retained GONMs is plotted as a function of distance from the column inlet. *From Wang, D., Shen, C., Jin, Y., Su, C., Chu, L., & Zhou, D. (2017). Role of solution chemistry in the retention and release of graphene oxide nanomaterials in uncoated and iron oxide-coated sand.* Science of the Total Environment, 579, *776–785. https://doi.org/10.1016/j.scitotenv.2016.11.02977.*

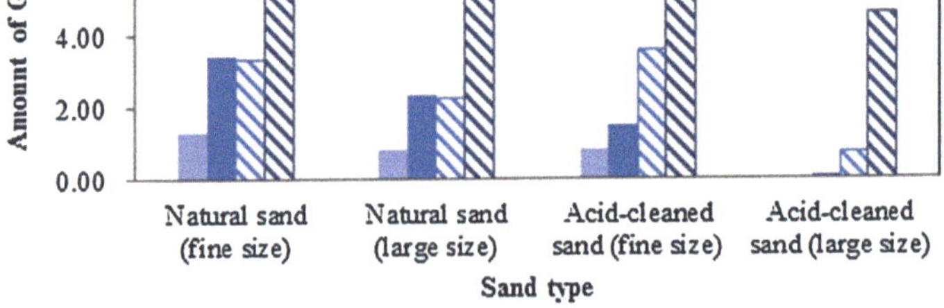

FIGURE 7.14 Effects of temperature on GO sorption onto sand. *From Wang, M., Gao, B., Tang, D., Sun, H., Yin, X., & Yu, C. (2017). Effects of temperature on graphene oxide deposition and transport in saturated porous media.* Journal of Hazardous Materials, 331, *28–35. https://doi.org/10.1016/j.jhazmat.2017.02.01433.*

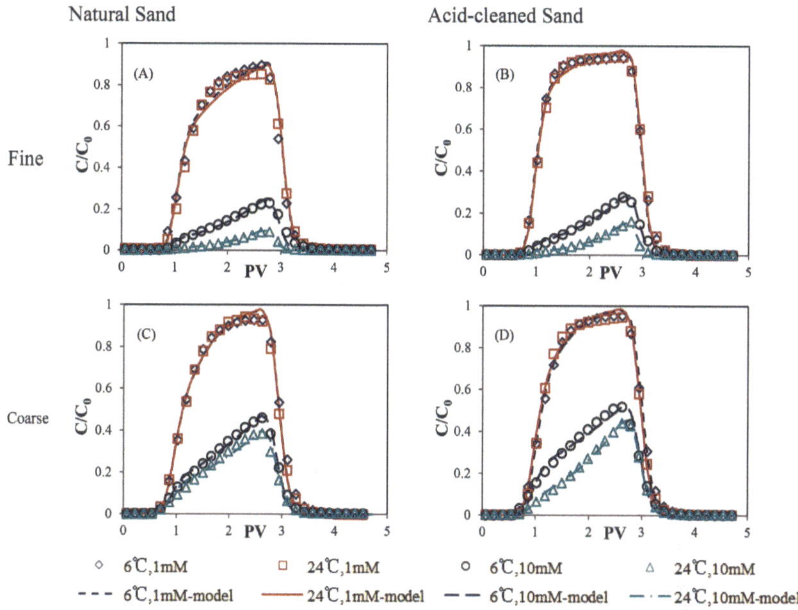

FIGURE 7.15 Transport of GO at different ISs (1 and 10 mM) and temperatures (6°C and 24°C) in saturated columns packed with (A) fine-size natural sand, (B) fine-size acid-cleaned sand, (C) coarse-size natural sand, and (D) coarse-size acid-cleaned sand. *From Wang, M., Gao, B., Tang, D., Sun, H., Yin, X., & Yu, C. (2017). Effects of temperature on graphene oxide deposition and transport in saturated porous media.* Journal of Hazardous Materials, 331, *28–35. https://doi.org/10.1016/j.jhazmat.2017.02.01433.*

media with smaller particle size, lowering temperature is more significant to improve the mobility of ENPs (Wang, Gao, et al., 2017). In addition, the retention rate of ENPs in acid-cleaned sand columns is slightly more dependent on temperature than that in natural sand columns (Wang, Gao, et al., 2017; Yuan et al., 2020). Under the same conditions, the ENPs recovery rate of clean sand is generally higher than that of natural sand (Feng et al., 2019; Wang, Gao, et al., 2017).

7.3.3 Effects of flow hydraulic properties and solution chemistry

The physical and chemical properties of fluids strongly affect the retention, transport and release of ENPs in soils and groundwater. The physical properties mainly refer to the hydrodynamic conditions of the flow in porous media, including velocity and flow direction. Chemical properties mainly refer to moisture content, solution pH, IS, and cation valence, which can change the interaction between ENPs and the surrounding environment (Wang et al., 2016).

7.3.3.1 Velocity and flow direction

A number of studies have shown that the mobility of ENPs in porous media is positively correlated with the flow velocity, that is, the mobility increases with the increase of the flow velocity (Braun et al., 2015; Liang et al., 2013). When all other conditions are the same, the flow rate controls ENPs to reach favorable deposition locations in porous media (Braun et al., 2015). Compared with AgNPs stabilized by surfactants, in the presence of higher IS, divalent cations and fine particles, and at lower water speeds, polymer-stabilized AgNPs are effective in transporting silver ENPs in porous media (Adrian et al., 2018). The higher the flow rate, the easier it is for the retained NPs to be separated from the porous medium under the action of a relatively large hydrodynamic force (Adrian et al., 2018). AgNPs show relatively high fluidity in loam sand, especially under low IS conditions and high flow rates (Figs. 7.16 and 7.17). The flow direction may also affect the transport of

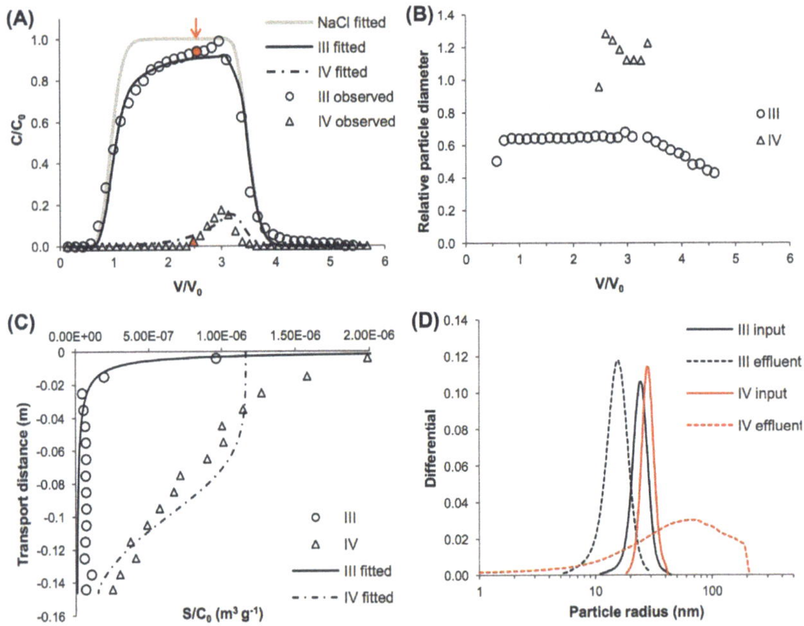

FIGURE 7.16 Observed and modeled BTCs and observed BTC of NaCl tracer (A), RPs (B), effluent mean particle size (C), and particle size distribution of input samples and representative effluent samples (D) of AgNPs in loamy sand with varied flow velocities. The experiments were conducted with 10 mM $NaNO_3$. III: 0.2 cm/min, IV: 0.0033 cm/min. *Highlighted data points* in the BTCs signify the end of the AgNP pulse and the *arrow* indicates the end of the tracer pulse. Particle diameters and size distributions were only determined when the sample concentration exceeded the detection limit of the flow field-flow fractionation detection. *From Braun, A., Klumpp, E., Azzam, R., & Neukum, C. (2015). Transport and deposition of stabilized engineered silver nanoparticles in water saturated loamy sand and silty loam.* Science of the Total Environment, 535, *102–112. https://doi.org/10.1016/j.scitotenv.2014.12.023109.*

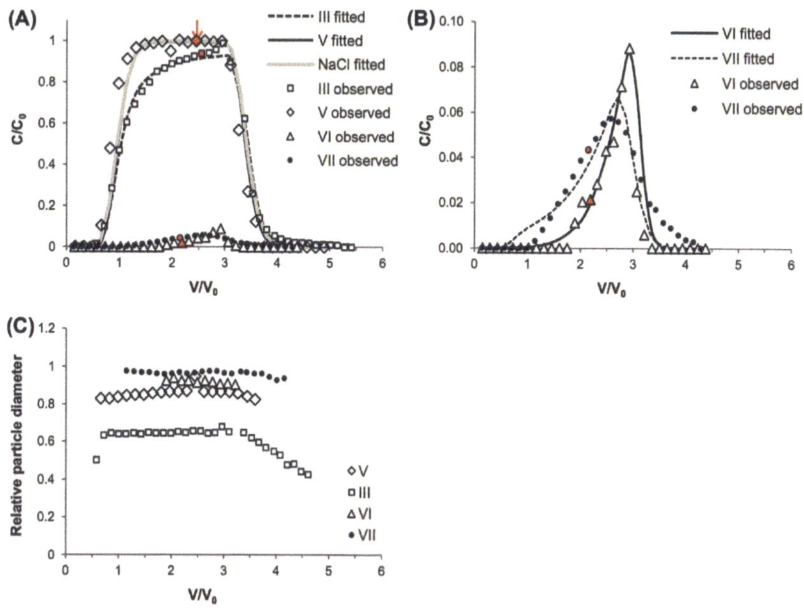

FIGURE 7.17 Observed and modeled BTCs of AgNPs and observed BTC of NaCl tracer in loamy sand under various IS conditions. (A) V: H_2O, III: 10 mM $NaNO_3$, VI: 50 mM $NaNO_3$, VII: 1 mM $CaNO_3$, enlarged view of the BTCs of VI and VII (B), and effluent relative mean particle sizes (C). *Highlighted data points* in the BTCs signify the end of the AgNP pulse and the *arrow* indicates the end of the tracer pulse. *From Braun, A., Klumpp, E., Azzam, R., & Neukum, C. (2015). Transport and deposition of stabilized engineered silver nanoparticles in water saturated loamy sand and silty loam.* Science of the Total Environment, 535, *102–112. https://doi.org/10.1016/j.scitotenv.2014.12.023109.*

ENPs in porous media. The change of flow direction can remobilize some previously retained ENPs, especially those retained by pore straining (Tian et al., 2012).

7.3.3.2 Moisture content

The soil moisture content determines the degree of soil saturation. "Saturated soil" means that all soil pores are filled with water. Under saturated conditions, the underground environment is a two-phase space, where the solid–water interface interaction and pore strain mainly control the particle transport process. On the contrary, "unsaturated soil" means that the air phase exists in the pores of the soil. Additional mechanisms for the retention and transport of ENPs in unsaturated porous media may be introduced, such as air–water interface adhesion and water film strain (Chen et al., 2010). Under unsaturated conditions, colloids and ENPs may also remain in the solid–water–air triple position (Lu et al., 2021; Zevi et al., 2012). As the results, the retention rate of ENPs in saturated porous media is often much

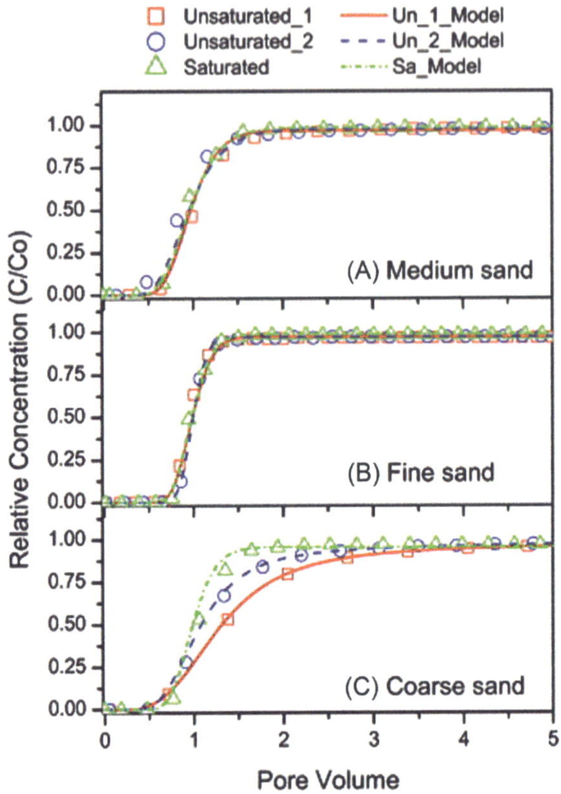

FIGURE 7.18 Transport of SDBS-dispersed SWNTs in sand columns under different volumetric moisture content conditions: (A) medium-sand columns with moisture contents of 0.40 (labeled as saturated), 0.21 (labeled as unsaturated 1), and 0.14 (labeled as unsaturated 2), (B) fine-sand columns with moisture contents of 0.38 (labeled as saturated), 0.34 (labeled as unsaturated 1), and 0.29 (labeled as unsaturated 2), and (C) coarse-sand columns with moisture contents of 0.40 (labeled as saturated), 0.10 (labeled as unsaturated 1), and 0.07 (labeled as unsaturated 2). *From Tian, Y., Gao, B., & Ziegler, K. J. (2011). High mobility of SDBS-dispersed single-walled carbon nanotubes in saturated and unsaturated porous media.* Journal of Hazardous Materials, 186*(2–3), 1766–1772. https://doi.org/10.1016/j.jhazmat.2010.12.0721769.*

lower than that in unsaturated porous media (Liu, Gao, Wu, Morales, et al., 2013). However, there are some exceptions. Tian et al. have shown that dispersed single-walled CNTs have a high degree of mobility in both saturated and unsaturated porous media and show retention only under conditions of extremely low water content (Fig. 7.18) (Tian et al., 2011). The authors believe that the high mobility of NPs in unsaturated porous media can be attributed to the fact that the surfactant-dispersed single-walled CNTs are not attached to the air–water interface. Fang et al. also found that the reduction of water saturation has no effect on the retention of ENPs and excluded the adhesion mechanism of the air–water interface attachment mechanisms due to the repulsive interaction between negatively charged air–water interaction (Fang, Xu, et al., 2013).

7.3.3.3 Solution pH

After entering the subsurface environment, ENPs may accumulate and precipitate into the sediments, but they may also be stable and transported in the water flow, depending on the properties of ENPs and the conditions of water chemistry, such as pH, IS, dissolved organic matter content, and property (Lin et al., 2010). In addition, pH can affect the fate and transport of ENPs entering soils and groundwater (Tan et al., 2018). Therefore pH is one of the critical environmental factors for ENPs retention and transport in porous media (Dibyanshu & Raychoudhury, 2019). While several studies have shown that the retention and transport of ENPs in porous media are influenced by solution pH (Dibyanshu & Raychoudhury, 2019; Rastghalam et al., 2020); more research efforts have been spent on the effect of pH on the stability of ENPs (Cupi et al., 2016; Van Koetsem et al., 2015). In theory, when pH is at point of zero charge or isoelectric point, the stability of the colloidal system is the smallest (i.e., coagulation/flocculation is the largest). This is because the net effect of the surface charge and the rate of transition are zero (Zhang, Hu, et al., 2016). Under normal conditions, ENPs are highly unstable. When their zeta-potential values are between 0 and 5 (negative or positive), agglomerations are formed rapidly. When the zeta-potential values of ENPs are higher than ± 30 mV, they are highly stable (Tan et al., 2018). The surfaces of ENPs are modified with different polymers or other inorganic NPs to stabilize the suspended ENPs (Kango et al., 2013). The coating material on ENPs may be affected with pH change since the capping agents surface functional groups ($-COOH$, $-OH$, $-NH_2$, $-N-(CH_3)_2$, $-Si-(CH_3)_3$ or $-SO_3$) can protonate or deprotonate at specific pH which in turn influence the charge of zeta potential (Tan et al., 2018). Kim et al. found that the charge of zero valent iron ENPs under the capping agent was affected by the pH change of the aqueous solutions (Kim et al., 2012).

7.3.3.4 Ionic strength and cation valence

It is well known that solution IS has strong effect on the fate and transport of ENPs in soils and groundwater. With the increase of IS, the ionic electrical double layer (EDL) in the adjacent medium is compressed, resulting in the decrease of the potential and thus increase of ENPs deposition in porous media (Adrian et al., 2018). Because the space repulsion between particles is overcome by gravity, this process can also promote the rapid aggregation of ENPs in the subsurface environment (Abbas et al., 2020). So IS is one of the most important parameters for ENPs retention, transport, and mobilization in soils and groundwater. Adrian et al. showed that in the presence of divalent

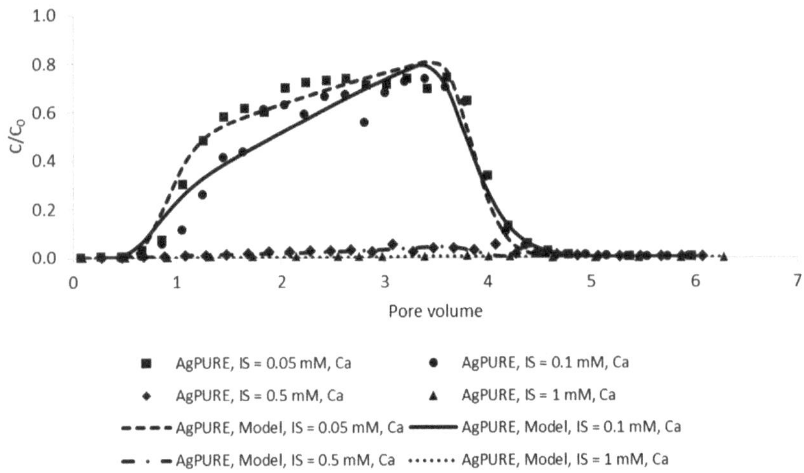

FIGURE 7.19 Effect of ionic strength on AgPURE transport in the absence of fines. Pore water velocity = 8.64 m/d, d50 = 0.7 mm. *From Adrian, Y. F., Schneidewind, U., Bradford, S. A., Simunek, J., Fernandez-Steeger, T. M., & Azzam, R. (2018). Transport and retention of surfactant- and polymer-stabilized engineered silver nanoparticles in silicate-dominated aquifer material.* Environmental Pollution, 236, *195–207. https://doi.org/10.1016/j.envpol.2018.01.011201.*

cations and fine-grained particles, increasing solution IS increased the retention of polymer-stabilized Ag ENPs (AgPURE) and delay was observed in their transport at low water velocity (Fig. 7.19) (Adrian et al., 2018). The cation valence also has an important influence on the stability and transport of ENPs in porous media. Multivalent cations such as Ca^{2+} and Mg^{2+} are much more effective than more than monovalent cations such as Na^{+} and K^{+} to reduce the transport of ENPs in soils and groundwater (Wang et al., 2016).

7.3.4 Soil colloids

Soil colloids including abiotic and bio-colloids (e.g., bacteria) can easily interact with ENPs to change their distribution, migration, and bioavailability in the subsurface environment (Nowack & Bucheli, 2007). Studies have shown that the soil organic colloids such as humus can change the surface properties of ENPs to not only promote their migration in soils but also enhance their release and mobility (Yin et al., 2019). Previous research has also examined the cotransport of inorganic colloids and ENPs in soils. Wang et al. found that, when hydroxyapatite and goethite NPs comigrated (Wang et al., 2015), larger particles were deposited preferentially, while smaller particles were easier to migrate (Table 7.1). Cai et al. found that, when sand was covered with polystyrene latex or bacterial colloids (Cai et al., 2016), the transport of TiO_2 ENPs was significantly accelerated. It is worth noting

TABLE 7.1 Average hydrodynamic diameters (DH) of HANPs – GNPs samples in the influents, effluents, and retentates in the column at different depths under different pHs and NOM concentrations and types.

D_H of HANPs – GNPs samples (nm)							
pH	SRHA	SRFA	Influent	Effluent	Retentate (cm)		
	(mg/L)	(mg/L)			0–1	1–2	2–3
6.5			2835 ± 677 a	984 ± 183 a	3196 ± 529 a	1836 ± 173 b	731 ± 70 a
7.5			1656 ± 117 b	810 ± 171 a	2215 ± 378 b	2404 ± 206 a	834 ± 67 a
10.5			401 ± 16 c	325 ± 22 b	658 ± 195 c	445 ± 240 c	585 ± 12 b
7.6	0.1		387 ± 15 c	359 ± 10 b	483 ± 67 c	356 ± 71 b	421 ± 78 b
7.6	1		282 ± 5.9 d	264 ± 7.1 b	458 ± 93 c	348 ± 63 b	352 ± 98 b
7.6	10		282 ± 7.6 d	257 ± 7.1 b	362 ± 74 c	361 ± 47 b	351 ± 56 b
7.7		0.1	472 ± 20 c	395 ± 8.8 b	574 ± 205 c	565 ± 129 b	448 ± 217 b
7.7		1	293 ± 4.9 d	290 ± 4.1 b	411 ± 81 c	400 ± 52 b	383 ± 19 b
7.7		10	287 ± 4.0 d	273 ± 4.4 b	385 ± 41 c	369 ± 80 b	331 ± 70 b

Mean values ± standard deviations in each vertical column followed by the same lowercase letters are not significantly different using Tukey's HSD test at $p < 0.05$.
From Wang, D., Yan, J., & Jaisi, D. P. (2015). Effect of size-selective retention on the cotransport of hydroxyapatite and goethite nanoparticles in saturated porous media. *Environmental Science and Technology, 49*(14), 8461–8470. https://doi.org/10.1021/acs.est.5b01210.

that some research has focused on the significance of ENGs in the cotransport and deposition of bacteria in porous media. For instance, the cotransport of nano and micron plastic particles with bacteria has been studied, and the results show that the transport of plastic particles and bacteria is very high under 10 mM NaCl and 1 mM $CaCl_2$ conditions (He et al., 2018). In addition, nano and micron plastic particles can promote the transport of bacteria in porous media under different environmental conditions. Findings from another study show that the transport of hAdV35 in saturated porous media was severely hampered, mostly by the presence of TiO_2 ENPs (Syngouna et al., 2017). Similarly, the retention and transport of *Escherichia coli* in quartz sand may be greatly affected by the presence of CNTs (Yang et al., 2013).

7.3.5 Heavy metals and organic pollutants

The long-distance migration of ENPs in the soils and groundwater provides the possibility of their cooperative migration with other chemical pollutants (Nowack & Bucheli, 2007). After entering the subsurface environment, ENPs can serve as a carrier to carry pollutants further into the soils and groundwater (Prédélus et al., 2017). Laboratory column experiments have been conducted to investigate the cooperative migration behavior of various ENPs (e.g., GO) and pollutants (Liu, Gao, Wu, Morales, et al., 2013). For example, GO can remobilize heavy metal pollutants such as divalent lead (Pb^{2+}), divalent cadmium (Cd^{2+}), divalent copper (Cu^{2+}), and hexavalent uranium (U^{6+}) trapped in porous media, and thus promote their migration in soils (Jiang et al., 2018). On the other hand, heavy metals in the soils can also strongly affect the fate and transport of ENPs (Liu et al., 2020). It has also been found that polychlorinated biphenyls (PCBs) are easily fixed to buckminsterfullerene (C_{60}) for cotransport, affecting the retention and transport of both PCBs and the NPs in soils (Wang, Huang, et al., 2012; Zhang et al., 2011). Similar phenomena have also between observed for the cotransport of CNTs and phenanthrene, carbon dioxide nanotubes and phenanthrene, and carbon dioxide nanotubes and oxytetracycline (Fang, Shan, et al., 2013; Fang et al., 2017). These results demonstrated that the interactions between ENPs and chemical pollutants such as heavy metals and organic pollutants can strongly their fate and transport in soils and groundwater (Hofmann & von der Kammer, 2009).

7.4 Modeling

7.4.1 Derjaguin–Landau–Verwey–Overbeek theory

Derjaguin–Landau–Verwey–Overbeek (DLVO) theory has been widely applied in describing the stability and interactions of ENPs in soils and groundwater in the literature (Misra et al., 2019). The original DLVO theory is a theory on the stability of colloids (Lyklema et al., 1999). It was proposed by Derjaguin and Landau (1941) and by Verwey and Overbeek (1948) respectively. Hence, the theory is named after the initials of the four scientists. The theory quantitatively explains the aggregation of aqueous dispersion systems and describes the forces of charged surfaces interacting through a liquid medium. It combines the van der Waals diffusive force with the effect of electrostatic repulsion caused by a so-called double elcounterion layer (i.e., EDL). The theory is based on the assumption that the EDL and van der Waals forces are independent and can therefore be superimposed or added at each interaction distance between the two particles (Fig. 7.20).

The DVLO theory studies the stability and coalescence of particles including colloids and ENPs from the perspective of the interaction between the potential energy of repulsion and attraction. The theory holds that there are two kinds of interaction forces between charged particles, namely electrostatic repulsion when EDLs overlap (Fig. 7.21) and long-range van der Waals attraction, determining the stability (Bajpai, 2018). According to the theory, the total potential energy V_T^D between particles in the dispersed system is the sum of the repulsive potential energy

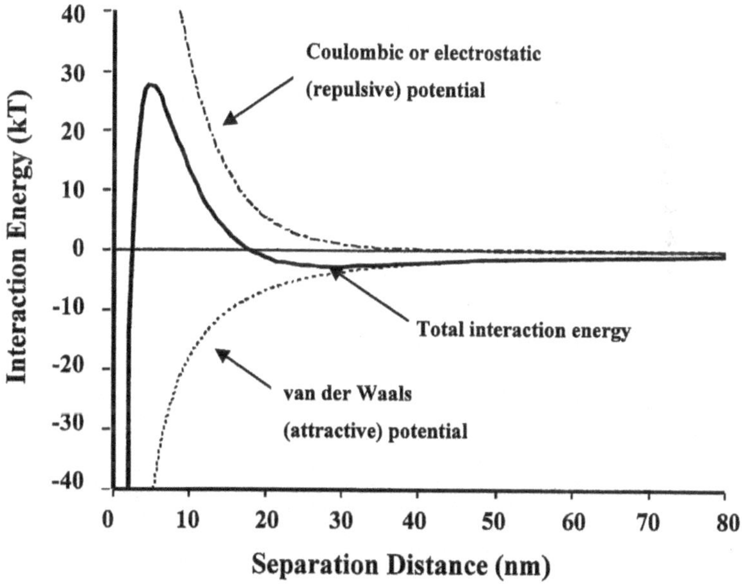

FIGURE 7.20 Relationship between distance and total potential energy between particles. *From Adair, J. H., Suvaci, E., & Sindel, J. (2001).* Surface and colloid chemistry. *Elsevier. https://doi.org/10.1016/B0-08-043152-6/01622-36.*

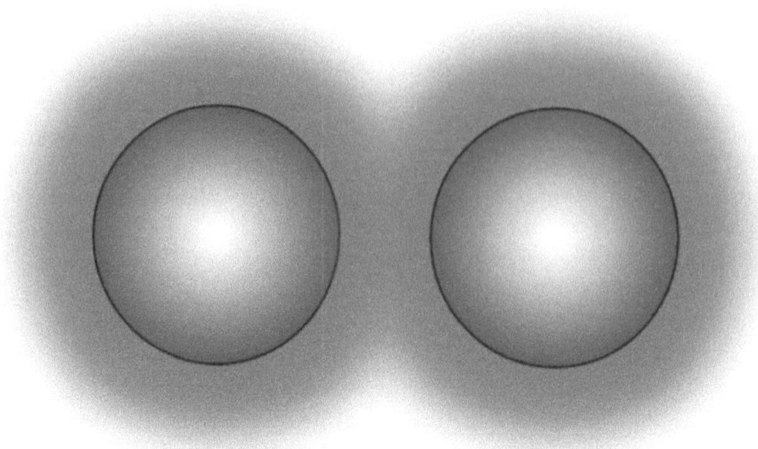

FIGURE 7.21 Interaction between spheres 1 and 2 due to overlapping of their electrical double layers. *From Ohshima, H. (2014). In H. Ohshima & K. Makino (Eds.),* Colloid and interface science in pharmaceutical research and development. *Elsevier. https://doi.org/10.1016/B978-0-444-62614-1.00001-68.*

(electrostatic repulsive potential energy) V_{ER} and attractive potential energy (long-range van der Waals potential energy) V_{WA}:

$$V_T^D = V_{ER} + V_{WA} \quad (7.1)$$

The main influencing factors of the DLVO potential energy include particle size, shape, distance between particles, Hamaker constant, particle surface potential, and the IS of the electrolyte. The attraction between colloidal/nanosized particles is essentially van der Waals forces, but the particles are aggregates of many molecules. The attraction between these particles is thus the sum of the attraction of all molecules in the particles. Electrostatic repulsion exists between colloidal/nanosized particles, resulting in repulsive potential energy. The repulsive potential energy between these particles depends on not only their size and shape but also the thermodynamic potential of the surface. It is exponentially related to the ionic atmosphere radius of the micelles and the shortest distance between the particles.

The classical DLVO theory has been used in the past for a long time to explain the agglomeration and dispersion behavior between colloidal and nanosized particles (Yang et al., 2016). When water is used as the liquid medium, there are many special forces between hydrophilic and hydrophobic particles, affecting their stability. These forces are not considered in the classical DLVO theory, making the theory cannot explain some of the agglomeration and dispersion behaviors. The extended DLVO (XDLVO) theory has been developed to satisfy the need (Hoek & Agarwal, 2006; Van Oss, 2008).

$$V_T^D = V_{ER} + V_{WA} + V_{SR} + V_{HR} + V_{HA} + V_{MA} \quad (7.2)$$

where V_{SR} is the space repulsive potential energy, V_{HR} is the hydration interaction repulsive energy, V_{HA} is the hydrophobic interaction attractive energy, and V_{MA} is the magnetic attractive potential energy. Details of the applications of DLVO/XDLVO theory in predicting the stability of ENPs in soils and groundwater can be found in many of recent research papers (Jiang et al., 2021; Wu et al., 2020).

7.4.2 Filtration of ENPs in the subsurface environment

As porous media, soils in the subsurface can always serve as natural filters to remove colloidal and nanosized particles from water flow (Grolimund & Borkovec, 2006). Colloid filtration theory (CFT) is one of the most commonly used theories to describe the filtration of ENPs in the subsurface environment (Harvey & Garabedlan, 1991). CFT has practical implications for chromatographic separation, particle (deep bed) filtration of colloids and colloid-related pollutants in water and wastewater treatment, migration and destination in subsurface environments, and natural filtration of microorganisms such as bacteria, viruses, and protozoa (Zhang et al., 2020).

In the subsurface, the transport of ENPs through saturated homogeneous soils is mainly controlled by particle advection, hydrodynamic dispersion, and deposition (filtration) (Wang et al., 2016). For relatively low particle concentration and medium IS, the inhibition and ripening effect is not obvious, and ENP release in the subsurface is usually negligible (Dragoman et al., 2017). Under these conditions, at a soil depth x and time t, the concentrations of ENPs in groundwater flow $C(x, t)$ and detained particles $S(x, t)$ can be described by the one-dimensional advection dispersion equation of the first-order dynamic deposition/filtration term (Grolimund et al., 1998; Kretzschmar et al., 1997).

$$\frac{\partial C}{\partial t} + \frac{\rho b}{\varepsilon}\frac{\partial S}{\partial t} = D\frac{\partial^2 C}{\partial x^2} - v\frac{\partial C}{\partial x} \quad (7.3)$$

$$\frac{\rho_b}{\varepsilon}\frac{\partial S}{\partial t} = kC \quad (7.4)$$

where v is particle velocity, D is the hydrodynamic dispersion coefficient, ε is the porosity of the bed, $\rho_b$is the bulk density of porous media, k is the coefficient of particle deposition rate. The latter is related to the commonly used single collector removal efficiency η.

$$k = \frac{3(1-\varepsilon)v}{2d_c}\eta \quad (7.5)$$

where d_c is the diameter of the collector particle and η is defined as the product of single collector efficiency (η_0) and attachment efficiency (α). Based on the single collect efficiency theory, η_0 is contributed by efficiencies from Brownian diffusion (η_D), gravity sedimentation (η_G), and interception (η_I), and can be calculated using following equation (Tufenkji & Elimelech, 2004):

$$\eta_0 = 2.4A_s^{1/3}N_R^{-0.081}N_{P_e}^{-0.715}N_{\mathrm{vdw}}^{0.052} + 0.55A_sN_R^{1.675}N_A^{0.125} + 0.22N_R^{-0.24}N_G^{1.11}N_{\mathrm{vdw}}^{0.053} \tag{7.6}$$

where N_R is an aspect ratio, N_{Pe} is the Peclet number, N_A is the attraction number, N_{vdw} is the van der Waals number, N_G is the gravity number, and A_S is a porosity-dependent parameter calculated by Eq. (7.7).

$$A_s = \frac{2 - 2(1-n)^{5/3}}{2-3(1-f)^{1/3} + 3(1-f)^{5/3} - 2(1-f)^2} \tag{7.7}$$

The attachment efficiency (α) is generally obtained by two methods. One is through the calculation based on experimental results of particle BTCs. Another is to set up advection-dispersion model equations and fit the attachment coefficient value.

For most practical applications (such as chromatographic separation and particle filtration), the system is considered to be in a stable state and the effects of hydrodynamic dispersion are negligible. It is also assumed that the particle deposition rate is constant in space and time. In other words, a single value of k is specified. Under these conditions, for continuous particle injection with concentration C_0 (at $x = 0$) and time period t_0, the solutions of Eqs. (7.8) and (7.9) for the initial particle-free soils are:

$$C(x) = C_0 \exp\left[-\frac{k}{v}x\right] \tag{7.8}$$

$$S(x) = \frac{t_0 \varepsilon k}{\rho_b} C(x) = \frac{t_0 \varepsilon k C_0}{\rho_b} \exp\left[-\frac{k}{v}x\right] \tag{7.9}$$

In recent years, many studies have reported the discrepancy between the theoretical prediction of CFT and the actual experimental results. For example, according to the CFT, the smaller the particle size is, the more intense the Brownian motion is, and the greater the chance of collision between particles and porous media is. A considerable number of studies have found an experimental phenomenon contrary to the prediction, that is, the larger the particle size, the weaker the migration (He et al., 2019; Wang, Kim, et al., 2012; May & Li, 2013). Furthermore, there is research about RP shows super exponential rule that is inconformity about the theory (Tufenkji & Elimelech, 2005).

7.4.3 One-dimensional convective dispersion model for saturated porous media

For ENP transport in a steady uniform flow through isotropic homogeneous porous media without a concentration gradient or advection perpendicular to the primary flow direction, the mass balance equation for particles in aqueous phase can be written as:

$$\frac{\partial C}{\partial t} = D\frac{\partial^2 C}{\partial x^2} - v\frac{\partial C}{\partial x} - \frac{\rho b}{n}\frac{\partial S}{\partial t} \tag{7.10}$$

where C is particle concentration in aqueous phase at a time t and a distance χ from the column inlet, D is hydrodynamic dispersion coefficient, v is pore-water velocity, ρb is the dry bulk density of the porous media, n is the porosity of the porous media, and S is particles concentration in solid phase. The mass balance equation for particles in solid phase is expressed as Eq. (7.11):

$$\frac{\rho_b}{n}\frac{\partial s}{\partial t} = k_d C - \frac{\rho_b}{n}k_r S \tag{7.11}$$

where k_d is deposition rate coefficient, and k_r is release (i.e., detachment) rate coefficient (Bradford et al., 2003, 2004; Tufenkji & Elimelech, 2004).

7.4.4 One-dimensional convective dispersion model for unsaturated porous media

As indicated in previous sections, ENPs in unsaturated porous media may be retained at multiple sites including solid–water, air–water, and air–water solid interfaces (Chen et al., 2008; Keller et al., 2020). The transport of ENPs in unsaturated porous media is thus often described with dual-site or multisite deposition-release models that assume reversible deposition (i.e., deposition and release) of ENPs at two or more sites. The governing mass balance equations in the case of uniform one-dimensional flow of a dual-site particle transport model can be written as (Schijven et al., 2002):

$$\frac{\partial C}{\partial t} = D\frac{\partial^2 C}{\partial x^2} - v\frac{\partial c}{\partial x} - \frac{\rho_b}{n}\frac{\partial s_1}{\partial t} - \frac{p_b}{n}\frac{\partial s_2}{\partial t} \tag{7.12}$$

$$\frac{\rho_b}{n}\frac{\partial S_1}{\partial t} = k_{d,1}C - \frac{\rho_b}{n}k_{r,1}S_1 \tag{7.13}$$

$$\frac{p_b}{n}\frac{\partial s_2}{\partial t} = k_{d,2}C - \frac{\rho_b}{n}k_{r,2}S_2 \tag{7.14}$$

where S_1 and S_2 are solid phase particle concentration at the two retention sites. $k_{d,1}$ and $k_{r,1}$ are deposition and release rate constants at the first site, respectively. $k_{r,2}$ and $k_{d,2}$ are deposition and release rate constants at the second site, respectively.

7.5 Conclusions and perspectives

This chapter not only summarizes the transformation and stability of ENPs in soils and groundwater, but also outlines the key factors that control the retention, transport, and release of ENPs in porous mediation under various conditions. In particular, the chapter provides a comprehensive review on how the physicochemical properties of media, NPs, and flow can affect the fate and transport of ENPs in the soil and groundwater systems. The commonly used mathematical models of ENP stability, filtration, and transport in porous media are also briefly introduced. Although previous studies have greatly expanded the knowledge of ENP fate and transport in soils and groundwater, there are still many areas that require further investigations.

Significant past research has been devoted to the fate and environmental risks associated with the growing use of ENPs. It is clear from many studies that most of the ENPs can retained the soils, however, some fractions may find their ways into groundwater. The current literature provides a great deal of information on the effects of soil hydrochemistry on the accumulation and deposition of ENPs, which are related to the migration of ENPs into soils and groundwater. However, most of the previous studies are based on short-term laboratory experiment. Owing to the limitation of experimental site and cost, there is a lack of long-term and field-scale studies on the fate and transport of ENPs in the soil and groundwater environment. In addition, more attention should be paid to the fate and behavior of ENPs in natural environment and their impact on soil sustainability.

So far, there is a serious lack of information on the long-term impact of ENPs on the soil and groundwater environment. Some studies have reported the toxic effects of ENPs on many organisms (Bardos et al., 2018; Galdames et al., 2020; Patil et al., 2016). However, some studies have also pointed out that ENPs have no significant effect on microbial community in soils (Fajardo et al., 2013; Hou et al., 2019; Lacalle et al., 2020). Considering the above two situations, short-term and local studies often cannot fully reveal the impact of ENPs on biological populations, communities, and even the whole ecosystem. Therefore it is necessary to conduct more systematic studies to assess the long-term application risk of ENPs in the ecosystem including soil and groundwater systems.

In view of the insufficient understanding of the long-term risks of ENPs in soils and groundwater, it is necessary to conduct additional research to formulate universal application standards, detection methods, and subsequent management rules and regulations. In addition, effective and reliable prediction models of the fate and transport of ENPs in natural soils and groundwater are in great need. We need to identify and select the most influential features that affect the transport of ENPs in soils and groundwater. Owing to the complex interaction between these influential features, it is necessary to generate new knowledge in large databases through data mining technology.

So far, machine learning has been applied to a series of different engineering and environmental fields, but it has not been widely used to simulate the interaction between ENPs and soils (Sangani et al., 2019).

References

Abbas, Q., Yousaf, B., Amina., Ali, M. U., Munir, M. A. M., El-Naggar, A., Rinklebe, J., & Naushad, M. (2020). Transformation pathways and fate of engineered nanoparticles (ENPs) in distinct interactive environmental compartments: A review. *Environment International*, *138*, 105646–105663. Available from https://doi.org/10.1016/j.envint.2020.105646.

Adrian, Y. F., Schneidewind, U., Bradford, S. A., Simunek, J., Fernandez-Steeger, T. M., & Azzam, R. (2018). Transport and retention of surfactant- and polymer-stabilized engineered silver nanoparticles in silicate-dominated aquifer material. *Environmental Pollution*, *236*, 195–207. Available from https://doi.org/10.1016/j.envpol.2018.01.011.

Afrooz, A. R. M. N., Das, D., Murphy, C. J., Vikesland, P., & Saleh, N. B. (2016). Co-transport of gold nanospheres with single-walled carbon nanotubes in saturated porous media. *Water Research*, *99*, 7–15. Available from https://doi.org/10.1016/j.watres.2016.04.006.

Ahmed, B., Rizvi, A., Ali, K., Lee, J., Zaidi, A., Khan, M. S., & Musarrat, J. (2021). Nanoparticles in the soil–plant system: A review. *Environmental Chemistry Letters*, *19*, 1545–1609. Available from https://doi.org/10.1007/s10311-020-01138-y.

Amde, M., Liu, J., Tan, Z., & Bekana, D. (2017). Transformation and bioavailability of metal oxide nanoparticles in aquatic and terrestrial environments. A review. *Environmental Pollution*, *230*, 250–267. Available from https://doi.org/10.1016/j.envpol.2017.06.064.

Baddar, Z. E., Matocha, C. J., & Unrine, J. M. (2019). Surface coating effects on the sorption and dissolution of ZnO nanoparticles in soil. *Environmental Science: Nano*, *6*(8), 2495–2507. Available from https://doi.org/10.1039/c9en00348g.

Bajpai, P. (2018). Chapter 19—Colloid and surface chemistry. In P. Bajpai (Ed.), *Biermann's handbook of pulp and paper* (3rd ed.). Elsevier. Available from https://doi.org/10.1016/B978-0-12-814238-7.00019-2.

Bardos, P., Merly, C., Kvapil, P., & Koschitzky, H.-P. (2018). Status of nanoremediation and its potential for future deployment: Risk-benefit and benchmarking appraisals. *Remediation*, *28*(3), 43–56. Available from https://doi.org/10.1002/rem.21559.

Bhatt, I., & Tripathi, B. N. (2011). Interaction of engineered nanoparticles with various components of the environment and possible strategies for their risk assessment. *Chemosphere*, *82*(3), 308–317. Available from https://doi.org/10.1016/j.chemosphere.2010.10.011.

Bogdal, C., Scheringer, M., Schmid, P., Bläuenstein, M., Kohler, M., & Hungerbühler, K. (2010). Levels, fluxes and time trends of persistent organic pollutants in Lake Thun, Switzerland: Combining trace analysis and multimedia modeling. *Science of the Total Environment*, *408*(17), 3654–3663. Available from https://doi.org/10.1016/j.scitotenv.2010.04.038.

Bradford, S. A., & Bettahar, M. (2006). Concentration dependent transport of colloids in saturated porous media. *Journal of Contaminant Hydrology*, *82*(1–2), 99–117. Available from https://doi.org/10.1016/j.jconhyd.2005.09.006.

Bradford, S. A., Bettahar, M., Simunek, J., & Van Genuchten, M. T. (2004). Straining and attachment of colloids in physically heterogeneous porous media. *Vadose Zone Journal*, *3*(2), 384–394. Available from https://doi.org/10.2113/3.2.384.

Bradford, S. A., Simunek, J., Bettahar, M., Van Genuchten, M. T., & Yates, S. R. (2003). Modeling colloid attachment, straining, and exclusion in saturated porous media. *Environmental Science and Technology*, *37*(10), 2242–2250. Available from https://doi.org/10.1021/es025899u.

Bradford, S. A., Torkzaban, S., & Shapiro, A. (2013). A theoretical analysis of colloid attachment and straining in chemically heterogeneous porous media. *Langmuir: The ACS Journal of Surfaces and Colloids*, *29*(23), 6944–6952. Available from https://doi.org/10.1021/la4011357.

Bras, A., Henriques, F. M. A., & Cidade, M. T. (2010). Effect of environmental temperature and fly ash addition in hydraulic lime grout behaviour. *Construction and Building Materials*, *24* (8), 1511–1517. Available from https://doi.org/10.1016/j.conbuildmat.2010.02.001.

Braun, A., Klumpp, E., Azzam, R., & Neukum, C. (2015). Transport and deposition of stabilized engineered silver nanoparticles in water saturated loamy sand and silty loam. *Science of the Total Environment*, *535*, 102–112. Available from https://doi.org/10.1016/j.scitotenv.2014.12.023.

Cai, L., Peng, S., Wu, D., & Tong, M. (2016). Effect of different-sized colloids on the transport and deposition of titanium dioxide nanoparticles in quartz sand. *Environmental Pollution*, *208*, 637–644. Available from https://doi.org/10.1016/j.envpol.2015.10.040.

Cai, L., Tong, M., Ma, H., & Kim, H. (2013). Cotransport of titanium dioxide and fullerene nanoparticles in saturated porous media. *Environmental Science and Technology*, *47*(11), 5703–5710. Available from https://doi.org/10.1021/es400256d.

Castellaneta, S., Piccinno, E., Frezza, E., Oliva, M., Fontana, C., Indrio, F., Capriari, T., Cavallo, L., & Francavilla, R. (2012). CO35 spontaneous normalization of anti-tissue transglutaminase antibody levels is common in children with type 1 diabetes mellitus: A ten years experience. *Digestive and Liver Disease*, *44*, S256. Available from https://doi.org/10.1016/S1590-8658(12)60663-X.

Chekli, L., Zhao, Y. X., Tijing, L. D., Phuntsho, S., Donner, E., Lombi, E., Gao, B. Y., & Shon, H. K. (2015). Aggregation behaviour of engineered nanoparticles in natural waters: Characterising aggregate structure using on-line laser light scattering. *Journal of Hazardous Materials*, *284*, 190–200. Available from https://doi.org/10.1016/j.jhazmat.2014.11.003.

Chen, L., Sabatini, D. A., & Kibbey, T. C. G (2008). Role of the air–water interface in the retention of TiO_2 nanoparticles in porous media during primary drainage. *Environmental Science & Technology*, *42*(6), 1916–1921.

Chen, C., Shang, J., Zheng, X., Zhao, K., Yan, C., Sharma, P., & Liu, K. (2018). Effect of physicochemical factors on transport and retention of graphene oxide in saturated media. *Environmental Pollution*, *236*, 168–176. Available from https://doi.org/10.1016/j.envpol.2018.01.026.

Chen, L., Li, J., Chen, Z., Gu, Z., Yan, L., Zhao, F., & Zhang, A. (2020). Toxicological evaluation of graphene-family nanomaterials. *Journal of Nanoscience and Nanotechnology*, *20*(4), 1993–2006. Available from https://doi.org/10.1166/jnn.2020.17364.

Cupi, D., Hartmann, N. B., & Baun, A. (2016). Influence of pH and media composition on suspension stability of silver, zinc oxide, and titanium dioxide nanoparticles and immobilization of Daphnia magna under guideline testing conditions. *Ecotoxicology and Environmental Safety*, *127*, 144–152. Available from https://doi.org/10.1016/j.ecoenv.2015.12.028.

Derjaguin, B., & Landau, L. (1941). Theory of the stability of strongly charged lyophobic sols and of the adhesion of strongly charged particles in solutions of electrolytes. *Acta Physicochim: USSR*, *14*, 633–662.

Dibyanshu., & Raychoudhury, T. (2019). Co-transport behavior of nano-ZnO particles in the presence of metal-nanoparticles through saturated porous media. *Journal of Environmental Chemical Engineering*, *7*(3), 103103–103110. Available from https://doi.org/10.1016/j.jece.2019.103103.

Dong, S., Gao, B., Sun, Y., Guo, H., Wu, J., Cao, S., & Wu, J. (2019). Visualization of graphene oxide transport in two-dimensional homogeneous and heterogeneous porous media. *Journal of Hazardous Materials*, *369*, 334–341. Available from https://doi.org/10.1016/j.jhazmat.2019.02.042.

Dragoman, R. M., Grogg, M., Bodnarchuk, M. I., Tiefenboeck, P., Hilvert, D., Dirin, D. N., & Kovalenko, M. V. (2017). Surface-engineered cationic nanocrystals stable in biological buffers and high ionic strength solutions. *Chemistry of Materials*, *29*(21), 9416–9428. Available from https://doi.org/10.1021/acs.chemmater.7b03504.

Esfahani, A. R., Firouzi, A. F., Sayyad, G., & Kiasat, A. R. (2014). Transport and retention of polymer-stabilized zero-valent iron nanoparticles in saturated porous media: Effects of initial particle concentration and ionic strength. *Journal of Industrial and Engineering Chemistry*, *20*(5), 2671–2679. Available from https://doi.org/10.1016/j.jiec.2013.10.054.

Espinasse, B. P., Geitner, N. K., Schierz, A., Therezien, M., Richardson, C. J., Lowry, G. V., Ferguson, L., & Wiesner, M. R. (2018). Comparative persistence of engineered nanoparticles in a complex aquatic ecosystem. *Environmental Science and Technology*, *52*(7), 4072–4078. Available from https://doi.org/10.1021/acs.est.7b06142.

Fang, J., Shan, X., Wen, B., & Huang, R. (2013). Mobility of TX100 suspended multiwalled carbon nanotubes (MWCNTs) and the facilitated transport of phenanthrene in real soil columns. *Geoderma*, *207–208*(1), 1–7. Available from https://doi.org/10.1016/j.geoderma.2013.04.035.

Fang, J., Xu, M., Wang, D., Wen, B., & Han, J. (2013). Modeling the transport of TiO_2 nanoparticle aggregates in saturated and unsaturated granular media: Effects of ionic strength and pH. *Water Research*, *47*(3), 1399–1408. Available from https://doi.org/10.1016/j.watres.2012.12.005.

Feng, D., Wu, K., Wang, X., Li, J., Shi, J., Zhang, Y., Qi, P., & Li, X. (2019). Effects of temperature and pressure on spontaneous counter-current imbibition in unsaturated porous media. *Energy & Fuels*, *33*(9), 8544–8556. Available from https://doi.org/10.1021/acs.energyfuels.9b02310.

Gad, G. M. A., & Hegazy, M. A. (2019). Optoelectronic properties of gold nanoparticles synthesized by using wet chemical method. *Materials Research Express*, *6*(8), 1–7. Available from https://doi.org/10.1088/2053-1591/ab1bb8.

Galdames, A., Ruiz-Rubio, L., Orueta, M., Sánchez-Arzalluz, M., & Vilas-Vilela, J. L. (2020). Zero-valent iron nanoparticles for soil and groundwater remediation. *International Journal of Environmental Research and Public Health*, *17*(16), 1–23. Available from https://doi.org/10.3390/ijerph17165817.

Gao, B., Steenhuis, T. S., Zevi, Y., Morales, V. L., Nieber, J. L., Richards, B. K., McCarthy, J. F., & Parlange, J.-Y. (2008). Capillary retention of colloids in unsaturated porous media. *Water Resources Research*, *44*(4), 135–157. Available from https://doi.org/10.1029/2006WR005332.

Giese, B., Klaessig, F., Park, B., Kaegi, R., Steinfeldt, M., Wigger, H., Von Gleich, A., & Gottschalk, F. (2018). Risks, release and concentrations of engineered nanomaterial in the environment. *Scientific Reports*, *8*(1), 1565–1582. Available from https://doi.org/10.1038/s41598-018-19275-4.

Grolimund, D., & Borkovec, M. (2006). Release of colloidal particles in natural porous media by monovalent and divalent cations. *Journal of Contaminant Hydrology*, *87*(3–4), 155–175. Available from https://doi.org/10.1016/j.jconhyd.2006.05.002.

Grolimund, D., Elimelech, M., Borkovec, M., Barmettler, K., Kretzschmar, R., & Sticher, H. (1998). Transport of in situ mobilized colloidal particles in packed soil columns. *Environmental Science and Technology*, *32*(22), 3562–3569. Available from https://doi.org/10.1021/es980356z.

Han, J., Zhao, D., Li, D., Wang, X., Jin, Z., & Zhao, K. (2018). Polymer-based nanomaterials and applications for vaccines and drugs. *Polymers*, *10*(1), 31–44. Available from https://doi.org/10.3390/polym10010031.

Harvey, R. W., & Garabedlan, S. P. (1991). Use of colloid filtration theory in modeling movement of bacteria through a contaminated sandy aquifer. *Environmental Science and Technology*, *25*(1), 178–185. Available from https://doi.org/10.1021/es00013a021.

Hashem, N. M., & Gonzalez-Bulnes, A. (2020). State-of-the-art and prospective of nanotechnologies for smart reproductive management of farm animals. *Animals*, *10*(5), 840–854. Available from https://doi.org/10.3390/ani10050840.

Hauser, M., Li, G., & Nowack, B. (2019). Environmental hazard assessment for polymeric and inorganic nanobiomaterials used in drug delivery. *Journal of Nanobiotechnology*, *17*(1), 56–67. Available from https://doi.org/10.1186/s12951-019-0489-8.

He, D., Sun, Y., Xin, L., & Feng, J. (2014). Aqueous tetracycline degradation by non-thermal plasma combined with nano-TiO_2. *Chemical Engineering Journal*, *258*, 18–25. Available from https://doi.org/10.1016/j.cej.2014.07.089.

He, J., Wang, D., & Zhou, D. (2019). Transport and retention of silver nanoparticles in soil: Effects of input concentration, particle size and surface coating. *Science of the Total Environment*, *648*, 102–108. Available from https://doi.org/10.1016/j.scitotenv.2018.08.136.

He, L., Wu, D., Rong, H., Li, M., Tong, M., & Kim, H. (2018). Influence of nano- and microplastic particles on the transport and deposition behaviors of bacteria in Quartz Sand. *Environmental Science and Technology*, *52*(20), 11555–11563. Available from https://doi.org/10.1021/acs.est.8b01673.

He, F., & Zhao, D. (2007). Manipulating the size and dispersibility of zerovalent iron nanoparticles by use of carboxymethyl cellulose stabilizers. *Environmental Science & Technology*, *41* (17), 6216–6221.

Hendren, S., Englesbe, M. J., Brooks, L., Kubus, J., Yin, H., & Campbell, D. A. (2011). Prophylactic antibiotic practices for colectomy in Michigan. *American Journal of Surgery*, *201*(3), 290–294. Available from https://doi.org/10.1016/j.amjsurg.2010.08.024.

Henry, C., Minier, J.-P., Lefèvre, G., & Hurisse, O. (2011). Numerical study on the deposition rate of hematite particle on polypropylene walls: Role of surface roughness. *Langmuir: The ACS Journal of Surfaces and Colloids*, *27*(8), 4603–4612. Available from https://doi.org/10.1021/la104488a.

Hoek, E. M. V., & Agarwal, G. K. (2006). Extended DLVO interactions between spherical particles and rough surfaces. *Journal of Colloid and Interface Science*, *298*(1), 50–58. Available from https://doi.org/10.1016/j.jcis.2005.12.031.

Hofmann, T., & von der Kammer, F. (2009). Estimating the relevance of engineered carbonaceous nanoparticle facilitated transport of hydrophobic organic contaminants in porous media. *Environmental Pollution*, *157*(4), 1117–1126. Available from https://doi.org/10.1016/j.envpol.2008.10.022.

Hou, J., Zhang, M., Wang, P., Wang, C., Miao, L., Xu, Y., You, G., Lv, B., Yang, Y., & Liu, Z. (2017). Transport and long-term release behavior of polymer-coated silver nanoparticles in saturated quartz sand: The impacts of input concentration, grain size and flow rate. *Water Research, 127*, 86–95. Available from https://doi.org/10.1016/j.watres.2017.10.017.

Iftikhar, N., Rehman, A., & Sadaf, H. (2021). Theoretical investigation for convective heat transfer on Cu/water nanofluid and (SiO_2-copper)/water hybrid nanofluid with MHD and nanoparticle shape effects comprising relaxation and contraction phenomenon. *International Communications in Heat and Mass Transfer, 120*, 105012–105029. Available from https://doi.org/10.1016/j.icheatmasstransfer.2020.105012.

Jia, J., Li, F., Zhai, S., Zhou, H., Liu, S., Jiang, G., & Yan, B. (2017). Susceptibility of overweight mice to liver injury as a result of the ZnO nanoparticle-enhanced liver deposition of Pb2 + . *Environmental Science and Technology, 51*(3), 1775–1784. Available from https://doi.org/10.1021/acs.est.6b05200.

Jiang, Y., Zhang, X., Yin, X., Sun, H., & Wang, N. (2018). Graphene oxide-facilitated transport of Pb2 + and Cd2 + in saturated porous media. *Science of the Total Environment, 631–632*, 369–376. Available from https://doi.org/10.1016/j.scitotenv.2018.03.036.

John, A. C., Küpper, M., Manders-Groot, A. M. M., Debray, B., Lacome, J.-M., & Kuhlbusch, T. A. J. (2017). Emissions and possible environmental implication of engineered nanomaterials (ENMs) in the atmosphere. *Atmosphere, 8*(84), 1–29. Available from https://doi.org/10.3390/atmos8050084.

Jośko, I., Oleszczuk, P., Dobrzyńska, J., Futa, B., Joniec, J., & Dobrowolski, R. (2019). Long-term effect of ZnO and CuO nanoparticles on soil microbial community in different types of soil. *Geoderma, 352*, 204–212. Available from https://doi.org/10.1016/j.geoderma.2019.06.010.

Kamrani, S., Rezaei, M., Kord, M., & Baalousha, M. (2018). Transport and retention of carbon dots (CDs) in saturated and unsaturated porous media: Role of ionic strength, pH, and collector grain size. *Water Research, 133*, 338–347. Available from https://doi.org/10.1016/j.watres.2017.08.045.

Kanel, S. R., Misak, H., Nepal, D., Mall, S., Brittle, S. W., Sizemore, I., Kempisty, D. M., & Goltz, M. N. (2016). The use of carbon nanotube yarn as a filter medium to treat nitroaromatic-contaminated water. *Xinxing Tan Cailiao/New Carbon Materials, 31*(4), 415–423. Available from https://doi.org/10.1016/S1872-5805(16)60021-5.

Kango, S., Kalia, S., Celli, A., Njuguna, J., Habibi, Y., & Kumar, R. (2013). Surface modification of inorganic nanoparticles for development of organic-inorganic nanocomposites—A review. *Progress in Polymer Science, 38*(8), 1232–1261. Available from https://doi.org/10.1016/j.progpolymsci.2013.02.003.

Kasel, D., Bradford, S. A., Šimůnek, J., Heggen, M., Vereecken, H., & Klumpp, E. (2013). Transport and retention of multi-walled carbon nanotubes in saturated porous media: Effects of input concentration and grain size. *Water Research, 47*(2), 933–944. Available from https://doi.org/10.1016/j.watres.2012.11.019.

Keller, A. A., McFerran, S., Lazareva, A., & Suh, S. (2013). Global life cycle releases of engineered nanomaterials. *Journal of Nanoparticle Research, 15*(6), 1692–1708. Available from https://doi.org/10.1007/s11051-013-1692-4.

Khan, I., Saeed, K., & Khan, I. (2019). Nanoparticles: Properties, applications and toxicities. *Arabian Journal of Chemistry, 12*(7), 908–931. Available from https://doi.org/10.1016/j.arabjc.2017.05.011.

Kim, H.-J., Phenrat, T., Tilton, R. D., & Lowry, G. V. (2012). Effect of kaolinite, silica fines and pH on transport of polymer-modified zero valent iron nano-particles in heterogeneous porous media. *Journal of Colloid and Interface Science, 370*(1), 1–10. Available from https://doi.org/10.1016/j.jcis.2011.12.059.

Kretzschmar, R., Barmettler, K., Grolimund, D., Yan, Y., Borkovec, M., & Sticher, H. (1997). Experimental determination of colloid deposition rates and collision efficiencies in natural porous media. *Water Resources Research, 33*(5), 1129–1137. Available from https://doi.org/10.1029/97WR00298.

Levard, C., Hotze, E. M., Lowry, G. V., & Brown, G. E., Jr. (2012). Environmental transformations of silver nanoparticles: Impact on stability and toxicity. *Environmental Science & Technology, 46*(13), 6900–6914. Available from https://doi.org/10.1021/es2037405.

Liang, Y., Bradford, S. A., Simunek, J., Vereecken, H., & Klumpp, E. (2013). Sensitivity of the transport and retention of stabilized silver nanoparticles to physicochemical factors. *Water Research, 47*(7), 2572–2582. Available from https://doi.org/10.1016/j.watres.2013.02.025.

Liang, Y., Luo, Y., Lu, Z., Klumpp, E., Shen, C., & Bradford, S. A. (2021). Evidence on enhanced transport and release of silver nanoparticles by colloids in soil due to modification of grain surface morphology and co-transport. *Environmental Pollution, 276*(12), 116661–116670. Available from https://doi.org/10.1016/j.envpol.2021.116661.

Liang, Y., Zhou, J., Dong, Y., Klumpp, E., Šimůnek, J., & Bradford, S. A. (2020). Evidence for the critical role of nanoscale surface roughness on the retention and release of silver nanoparticles in porous media. *Environmental Pollution, 258*, 113803–113811. Available from https://doi.org/10.1016/j.envpol.2019.113803.

Lin, D., Tian, X., Wu, F., & Xing, B. (2010). Fate and transport of engineered nanomaterials in the environment. *Journal of Environmental Quality, 39*(6), 1896–1908. Available from https://doi.org/10.2134/jeq2009.0423.

Liu, G., Zhong, H., Ahmad, Z., Yang, X., & Huo, L. (2020). Transport of engineered nanoparticles in porous media and its enhancement for remediation of contaminated groundwater. *Critical Reviews in Environmental Science and Technology, 50*(22), 2301–2378. Available from https://doi.org/10.1080/10643389.2019.1694823.

Liu, J., Liu, A., & Zhang, W. (2016). The influence of polyelectrolyte modification on nanoscale zero-valent iron (nZVI): Aggregation, sedimentation, and reactivity with Ni(II) in water. *Chemical Engineering Journal, 303*, 268–274. Available from https://doi.org/10.1016/j.cej.2016.05.132.

Liu, L., Gao, B., Wu, L., Morales, V. L., Yang, L., Zhou, Z., & Wang, H. (2013). Deposition and transport of graphene oxide in saturated and unsaturated porous media. *Chemical Engineering Journal, 229*(4), 444–449. Available from https://doi.org/10.1016/j.cej.2013.06.030.

Liu, L., Gao, B., Wu, L., Yang, L., Zhou, Z., & Wang, H. (2013). Effects of pH and surface metal oxyhydroxides on deposition and transport of carboxyl-functionalized graphene in saturated porous media. *Journal of Nanoparticle Research, 15*(11), 2079–2086. Available from https://doi.org/10.1007/s11051-013-2079-2.

Lohwacharin, J., Takizawa, S., & Punyapalakul, P. (2015). Carbon black retention in saturated natural soils: Effects of flow conditions, soil surface roughness and soil organic matter. *Environmental Pollution, 205*, 131–138. Available from https://doi.org/10.1016/j.envpol.2015.05.036.

Lowry, G. V., Gregory, K. B., Apte, S. C., & Lead, J. R. (2012). Transformations of nanomaterials in the environment. *Environmental Science and Technology, 46*(13), 6893–6899. Available from https://doi.org/10.1021/es300839e.

Lu, H., Dong, J., Xi, B., Cai, P., Xia, T., & Zhang, M. (2021). Transport and retention of porous silicon-coated zero-valent iron in saturated porous media. *Environmental Pollution, 276*, 116700–116708. Available from https://doi.org/10.1016/j.envpol.2021.116700.

Luo, X., Yu, L., Wang, C., Yin, X., Mosa, A., Lv, J., & Sun, H. (2017). Sorption of vanadium (V) onto natural soil colloids under various solution pH and ionic strength conditions. *Chemosphere*, *169*, 609–617. Available from https://doi.org/10.1016/j.chemosphere.2016.11.105.

Lv, X., Gao, B., Sun, Y., Dong, S., Wu, J., Jiang, B., & Shi, X. (2016). Effects of grain size and structural heterogeneity on the transport and retention of nano-TiO_2 in saturated porous media. *Science of the Total Environment*, *563–564*, 987–995. Available from https://doi.org/10.1016/j.scitotenv.2015.12.128.

Lyklema, J., Van Leeuwen, H. P., & Minor, M. (1999). DLVO-theory, a dynamic re-interpretation. *Advances in Colloid and Interface Science*, *83*(1), 33–69. Available from https://doi.org/10.1016/S0001-8686(99)00011-1.

Ma, C., Huangfu, X., He, Q., Ma, J., & Huang, R. (2018). Deposition of engineered nanoparticles (ENPs) on surfaces in aquatic systems: A review of interaction forces, experimental approaches, and influencing factors. *Environmental Science and Pollution Research*, *25*(33), 33056–33081. Available from https://doi.org/10.1007/s11356-018-3225-2.

Malakar, A., Kanel, S. R., Ray, C., Snow, D. D., & Nadagouda, M. N. (2021). Nanomaterials in the environment, human exposure pathway, and health effects: A review. *Science of the Total Environment*, *759*, 143470–143488. Available from https://doi.org/10.1016/j.scitotenv.2020.143470.

Malakar, A., Snow, D. D., & Ray, C. (2019). Irrigation water quality—A contemporary perspective. *Water (Switzerland)*, *11*(7), 1482–1505. Available from https://doi.org/10.3390/w11071482.

Meseret, A., Liu, J., Tan, Z., & Deribachew, B. (2017). Transformation and bioavailability of metal oxide nanoparticles in aquatic and terrestrial environments. A review. *Environmental Pollution*, *230*, 250–267. Available from https://doi.org/10.1016/j.envpol.2017.06.064.

Misra, R. P., De Souza, J. P., Blankschtein, D., & Bazant, M. Z. (2019). Theory of surface forces in multivalent electrolytes. *Langmuir: The ACS Journal of Surfaces and Colloids*, *35*(35), 11550–11565. Available from https://doi.org/10.1021/acs.langmuir.9b01110.

Mitzel, M. R., Sand, S., Whalen, J. K., & Tufenkji, N. (2016). Hydrophobicity of biofilm coatings influences the transport dynamics of polystyrene nanoparticles in biofilm-coated sand. *Water Research*, *92*(5), 113–120. Available from https://doi.org/10.1016/j.watres.2016.01.026.

Mudunkotuwa, I. A., Pettibone, J. M., & Grassian, V. H. (2012). Environmental implications of nanoparticle aging in the processing and fate of copper-based nanomaterials. *Environmental Science and Technology*, *46*(13), 7001–7010. Available from https://doi.org/10.1021/es203851d.

Navarro, E., Baun, A., Behra, R., Hartmann, N. B., Filser, J., Miao, A.-J., Quigg, A., Santschi, P. H., & Sigg, L. (2008). Environmental behavior and ecotoxicity of engineered nanoparticles to algae, plants, and fungi. *Ecotoxicology (London, England)*, *17*(5), 372–386. Available from https://doi.org/10.1007/s10646-008-0214-0.

Nowack, B., & Bucheli, T. D. (2007). Occurrence, behavior and effects of nanoparticles in the environment. *Environmental Pollution*, *150*(1), 5–22. Available from https://doi.org/10.1016/j.envpol.2007.06.006.

Oomen, A. G., Steinhäuser, K. G., Bleeker, E. A. J., Van Broekhuizen, F., Sips, A., Dekkers, S., Wijnhoven, S. W. P., & Sayre, P. G. (2018). Risk assessment frameworks for nanomaterials: Scope, link to regulations, applicability, and outline for future directions in view of needed increase in efficiency. *NanoImpact*, *9*, 1–13. Available from https://doi.org/10.1016/j.impact.2017.09.001.

Park, C. M., Chu, K. H., Heo, J., Her, N., Jang, M., Son, A., & Yoon, Y. (2016). Environmental behavior of engineered nanomaterials in porous media: A review. *Journal of Hazardous Materials*, *309*, 133–150. Available from https://doi.org/10.1016/j.jhazmat.2016.02.006.

Patil, S. S., Shedbalkar, U. U., Truskewycz, A., Chopade, B. A., & Ball, A. S. (2016). Nanoparticles for environmental clean-up: A review of potential risks and emerging solutions. *Environmental Technology and Innovation*, *5*, 10–21. Available from https://doi.org/10.1016/j.eti.2015.11.001.

Peng, Z., Doroodchi, E., & Evans, G. (2010). DEM simulation of aggregation of suspended nanoparticles. *Powder Technology*, *204*(1), 91–102. Available from https://doi.org/10.1016/j.powtec.2010.07.023.

Phenrat, T., Kim, H. J., Fagerlund, F., Illangasekare, T., Tilton, R. D., & Lowry, G. V. (2009). Particle size distribution, concentration, and magnetic attraction affect transport of polymer-modified Fe0 nanoparticles in sand columns. *Environmental Science and Technology*, *43*(13), 5079–5085. Available from https://doi.org/10.1021/es900171v.

Phenrat, T., Navid, S., Kevin, S., Hye-Jin, K., Tilton, R. D., & Lowry, G. V. (2008). Stabilization of aqueous nanoscale zerovalent iron dispersions by anionic polyelectrolytes: adsorbed anionic polyelectrolyte layer properties and their effect on aggregation and sedimentation. *Journal of Nanoparticle Research*, *10*(5), 795–814. Available from https://doi.org/10.1007/s11051-007-9315-6.

Pokhrel, L. R., Ettore, N., Jacobs, Z. L., Zarr, A., Weir, M. H., Scheuerman, P. R., Kanel, S. R., & Dubey, B. (2017). Novel carbon nanotube (CNT)-based ultrasensitive sensors for trace mercury(II) detection in water: A review. *Science of the Total Environment*, *574*, 1379–1388. Available from https://doi.org/10.1016/j.scitotenv.2016.08.055.

Prabhavathy, S., & Arivuoli, D. (2021). Enhanced photocatalytic activity on Vanadium-doped NiO nanostructures in natural sunlight. *Journal of Materials Science: Materials in Electronics*, *32*(1), 1105–1120. Available from https://doi.org/10.1007/s10854-020-04885-4.

Prajitha, N., Athira, S. S., & Mohanan, P. V. (2019). Bio-interactions and risks of engineered nanoparticles. *Environmental Research*, *172*, 98–108. Available from https://doi.org/10.1016/j.envres.2019.02.003.

Prédélus, D., Lassabatere, L., Louis, C., Gehan, H., Brichart, T., Winiarski, T., & Angulo-Jaramillo, R. (2017). Nanoparticle transport in water-unsaturated porous media: Effects of solution ionic strength and flow rate. *Journal of Nanoparticle Research*, *19*(3), 1–17. Available from https://doi.org/10.1007/s11051-017-3755-4.

Rasmuson, A., Pazmino, E., Assemi, S., & Johnson, W. P. (2017). Contribution of nano- to microscale roughness to heterogeneity: Closing the gap between unfavorable and favorable colloid attachment conditions. *Environmental Science & Technology*, *51*(4), 2151–2160. Available from https://doi.org/10.1021/acs.est.6b05911.

Rousk, J., Ackermann, K., Curling, S. F., & Jones, D. L. (2012). Comparative toxicity of nanoparticulate CuO and ZnO to soil bacterial communities. *PLoS One*, *7*(3), e34197. Available from https://doi.org/10.1371/journal.pone.0034197. (1-8).

Sangani, M. F., Owens, G., & Fotovat, A. (2018). Transport of engineered nanoparticles in soils and aquifers. *Environmental Reviews*, *27*, 43–70. Available from https://doi.org/10.1139/er-2018-0022.

Sangani, M. F., Owens, G., Nazari, B., Astaraei, A., Fotovat, A., & Emami, H. (2019). Different modelling approaches for predicting titanium dioxide nanoparticles mobility in intact soil media. *Science of the Total Environment*, *665*, 1168–1181. Available from https://doi.org/10.1016/j.scitotenv.2019.01.345.

Schijven, J. F., Hassanizadeh, S. M., & De Bruin, R. H. A. M. (2002). Two-site kinetic modeling of bacteriophages transport through columns of saturated dune sand. *Journal of Contaminant Hydrology*, *57*(3–4), 259–279. Available from https://doi.org/10.1016/S0169-7722(01)00215-7.

Shaniv, D., Dror, I., & Berkowitz, B. (2020). Effects of particle size and surface chemistry on plastic nanoparticle transport in saturated natural porous media. *Chemosphere*, *262*, 127854–127862. Available from https://doi.org/10.1007/s11051-018-4265-8.

Shen, C., Li, B., Wang, C., Huang, Y., & Jin, Y. (2011). Surface roughness effect on deposition of nano- and micro-sized colloids in saturated columns at different solution ionic strengths. *Vadose Zone Journal*, *10*(3), 1071–1081. Available from https://doi.org/10.2136/vzj2011.0011.

Shi, Y., Wei, J., Qiu, J., Chu, H., Bai, W., & Wang, G. (2020). Numerical study of acoustic agglomeration process of droplet aerosol using a three-dimensional CFD-DEM coupled model. *Powder Technology*, *362*, 37–53. Available from https://doi.org/10.1016/j.powtec.2019.12.017.

Sun, Y., Gao, B., Bradford, S. A., Wu, L., Chen, H., Shi, X., & Wu, J. (2015). Transport, retention, and size perturbation of graphene oxide in saturated porous media: Effects of input concentration and grain size. *Water Research*, *68*, 24–33. Available from https://doi.org/10.1016/j.watres.2014.09.025.

Syngouna, V. I., Chrysikopoulos, C. V., Kokkinos, P., Tselepi, M. A., & Vantarakis, A. (2017). Cotransport of human adenoviruses with clay colloids and TiO_2 nanoparticles in saturated porous media: Effect of flow velocity. *Science of the Total Environment*, *598*, 160–167. Available from https://doi.org/10.1016/j.scitotenv.2017.04.082.

Tan, W., Peralta-Videa, J. R., & Gardea-Torresdey, J. L. (2018). Interaction of titanium dioxide nanoparticles with soil components and plants: Current knowledge and future research needs—a critical review. *Environmental Science: Nano*, *5*(2), 257–278. Available from https://doi.org/10.1039/c7en00985b.

Tian, Y., Gao, B., Wang, Y., Morales, V. L., Carpena, R. M., Huang, Q., & Yang, L. (2012). Deposition and transport of functionalized carbon nanotubes in water-saturated sand columns. *Journal of Hazardous Materials*, *213–214*, 265–272. Available from https://doi.org/10.1016/j.jhazmat.2012.01.088.

Tian, Y., Gao, B., & Ziegler, K. J. (2011). High mobility of SDBS-dispersed single-walled carbon nanotubes in saturated and unsaturated porous media. *Journal of Hazardous Materials*, *186*(2–3), 1766–1772. Available from https://doi.org/10.1016/j.jhazmat.2010.12.072.

Tufenkji, N., & Elimelech, M. (2004). Correlation equation for predicting single-collector efficiency in physicochemical filtration in saturated porous media. *Environmental Science and Technology*, *38*(2), 529–536. Available from https://doi.org/10.1021/es034049r.

Tufenkji, N., & Elimelech, M. (2005). Spatial distributions of *Cryptosporidium oocysts* in porous media: Evidence for dual mode deposition. *Environmental Science and Technology*, *39*(10), 3620–3629. Available from https://doi.org/10.1021/es048289y.

Van Koetsem, F., Verstraete, S., Van der Meeren, P., & Du Laing, G. (2015). Stability of engineered nanomaterials in complex aqueous matrices: Settling behaviour of CeO_2 nanoparticles in natural surface waters. *Environmental Research*, *142*, 207–214. Available from https://doi.org/10.1016/j.envres.2015.06.028.

Van Oss, C. J. (2008). Chapter Three—The extended DLVO theory. In C. J. van Oss (Ed.), *Interface Science and Technology* (Vol. 16, pp. 31–48). Elsevier. Available from https://doi.org/10.1016/S1573-4285(08)00203-2.

Verwey, E. J. W., & Overbeek, J. T. C. (1948). *Theory of the stability of lyophobic colloids: The interaction of sol particles having and electric double l layer*. Elsevier.

Wang, D., Park, C., Masud, A., Aich, N., & Su, N. (2017). Carboxymethylcellulose mediates the transport of carbon nanotube—Magnetite nanohybrid aggregates in water-saturated porous media. *Environmental Science and Technology*, *51*(21), 12405–12415. Available from https://doi.org/10.1021/acs.est.7b04037.

Wang, D., Shen, C., Jin, Y., Su, C., Chu, L., & Zhou, D. (2017). Role of solution chemistry in the retention and release of graphene oxide nanomaterials in uncoated and iron oxide-coated sand. *Science of the Total Environment*, *579*, 776–785. Available from https://doi.org/10.1016/j.scitotenv.2016.11.029.

Wang, D., Yan, J., & Jaisi, D. P. (2015). Effect of size-selective retention on the cotransport of hydroxyapatite and goethite nanoparticles in saturated porous media. *Environmental Science and Technology*, *49*(14), 8461–8470. Available from https://doi.org/10.1021/acs.est.5b01210.

Wang, L., Huang, Y., Kan, A. T., Tomson, M. B., & Chen, W. (2012). Enhanced transport of 2,2′,5,5′-polychlorinated biphenyl by natural organic matter (NOM) and surfactant-modified fullerene nanoparticles (n C60). *Environmental Science and Technology*, *46*(10), 5422–5429. Available from https://doi.org/10.1021/es300236w.

Wang, M., Gao, B., & Tang, D. (2016). Review of key factors controlling engineered nanoparticle transport in porous media. *Journal of Hazardous Materials*, *318*, 233–246. Available from https://doi.org/10.1016/j.jhazmat.2016.06.065.

Wang, M., Gao, B., Tang, D., Sun, H., Yin, X., & Yu, C. (2017). Effects of temperature on graphene oxide deposition and transport in saturated porous media. *Journal of Hazardous Materials*, *331*, 28–35. Available from https://doi.org/10.1016/j.jhazmat.2017.02.014.

Wang, M., Gao, B., Tang, D., & Yu, C. (2018). Concurrent aggregation and transport of graphene oxide in saturated porous media: Roles of temperature, cation type, and electrolyte concentration. *Environmental Pollution*, *235*, 350–357. Available from https://doi.org/10.1016/j.envpol.2017.12.063.

Wang, Y., Gao, B., Morales, V. L., Tian, Y., Wu, L., Gao, J., Bai, W., & Yang, L. (2012). Transport of titanium dioxide nanoparticles in saturated porous media under various solution chemistry conditions. *Journal of Nanoparticle Research*, *14*(9), 1095–1104. Available from https://doi.org/10.1007/s11051-012-1095-y.

Wang, Y., Kim, J. H., Baek, J. B., Miller, G. W., & Pennell, K. D. (2012). Transport behavior of functionalized multi-wall carbon nanotubes in water-saturated quartz sand as a function of tube length. *Water Research*, *46*(14), 4521–4531. Available from https://doi.org/10.1016/j.watres.2012.05.036.

Wen, C., Broholm, M. M., Dong, J., Uthuppu, B., Jakobsen, M. H., & Fjordbøge, A. S. (2020). Transport of citrate and polymer coated gold nanoparticles (AuNPs) in porous media: Effect of surface property and Darcy velocity. *Journal of Environmental Sciences*, *92*, 235–244. Available from https://doi.org/10.1016/j.jes.2020.02.026.

Wu, H., Fang, H., Xu, C., Ye, J., Cai, Q., & Shi, J. (2019). Transport and retention of copper oxide nanoparticles under unfavorable deposition conditions caused by repulsive van der Waals force in saturated porous media. *Environmental Pollution*, *256*, 113400–113408. Available from https://doi.org/10.1016/j.envpol.2019.113400.

Xia, T., Fortner, J. D., Zhu, D., Qi, Z., & Chen, W. (2015). Transport of sulfide-reduced graphene oxide in saturated quartz sand: Cation-dependent retention mechanisms. *Environmental Science and Technology*, *49*(19), 11468–11475. Available from https://doi.org/10.1021/acs.est.5b02349.

Xia, T., Lin, Y., Li, S., Yan, N., Xie, Y., He, M., Guo, X., & Zhu, L. (2021). Co-transport of negatively charged nanoparticles in saturated porous media: Impacts of hydrophobicity and surface O-functional groups. *Journal of Hazardous Materials*, *409*, 124477. Available from https://doi.org/10.1016/j.jhazmat.2020.124477.

Xia, T., Ma, P., Qi, Y., Zhu, L., Qi, Z., & Chen, W. (2019). Transport and retention of reduced graphene oxide materials in saturated porous media: Synergistic effects of enhanced attachment and particle aggregation. *Environmental Pollution*, *247*, 383–391. Available from https://doi.org/10.1016/j.envpol.2019.01.052.

Xiao, T., Huang, J., Wang, D., Meng, T., & Yang, X. (2020). Au and Au-based nanomaterials: Synthesis and recent progress in electrochemical sensor applications. *Talanta*, *206*, 120210–120228. Available from https://doi.org/10.1016/j.talanta.2019.120210.

Xin, J., Tang, F., Zheng, X., Shao, H., & Kolditz, O. (2016). Transport and retention of xanthan gum-stabilized microscale zero-valent iron particles in saturated porous media. *Water Research*, *88*, 199–206. Available from https://doi.org/10.1016/j.watres.2015.10.005.

Yang, H., Tong, M., & Kim, H. (2013). Effect of carbon nanotubes on the transport and retention of bacteria in saturated porous media. *Environmental Science and Technology*, *47*(20), 11537–11544. Available from https://doi.org/10.1021/es4022415.

Yecheskel, Y., Dror, I., & Berkowitz, B. (2019). Effect of phosphate, sulfate, arsenate, and pyrite on surface transformations and chemical retention of gold nanoparticles (Au-NPs) in partially saturated soil columns. *Environmental Science and Technology*, *53*(22), 13071–13080. Available from https://doi.org/10.1021/acs.est.9b03241.

Yin, X., Jiang, Y., Tan, Y., Meng, X., Sun, H., & Wang, N. (2019). Co-transport of graphene oxide and heavy metal ions in surface-modified porous media. *Chemosphere*, *218*, 1–13. Available from https://doi.org/10.1016/j.chemosphere.2018.11.089.

Yuan, C., Sadikov, K., Varfolomeev, M., Khaliullin, R., Pu, W., Al-Muntaser, A., & Saeed Mehrabi-Kalajahi, S. (2020). Low-temperature combustion behavior of crude oils in porous media under air flow condition for in-situ combustion (ISC) process. *Fuel*, *259*, 1–11. Available from https://doi.org/10.1016/j.fuel.2019.116293.

Zevi, Y., Gao, B., Zhang, W., Morales, V. L., Cakmak, M. E., Medrano, E. A., Sang, W., & Steenhuis, T. S. (2012). Colloid retention at the meniscus-wall contact line in an open microchannel. *Water Research*, *4*(2), 295–306. Available from https://doi.org/10.1016/j.watres.2011.09.046.

Zhang, C., Hu, Z., & Deng, B. (2016). Silver nanoparticles in aquatic environments: Physiochemical behavior and antimicrobial mechanisms. *Water Research*, *88*, 403–427. Available from https://doi.org/10.1016/j.watres.2015.10.025.

Zhang, L., Wang, L., Zhang, P., Kan, A. T., Chen, W., & Tomson, M. B. (2011). Facilitated transport of 2,2′,5,5′-polychlorinated biphenyl and phenanthrene by fullerene nanoparticles through sandy soil columns. *Environmental Science and Technology*, *45*(4), 1341–1348. Available from https://doi.org/10.1021/es102316m.

Zhang, M., Bradford, S. A., Šimůnek, J., Vereecken, H., & Klumpp, E. (2019). Co-transport of multi-walled carbon nanotubes and sodium dodecylbenzenesulfonate in chemically heterogeneous porous media. *Environmental Pollution*, *247*, 907–916. Available from https://doi.org/10.1016/j.envpol.2019.01.106.

Zhang, Z., Wang, P., Wu, Z., Yue, C., Wei, X., Zheng, J., Xiang, M., & Liu, B. (2020). Efficient synthesis of niobium pentoxide nanowires and application in ethanolysis of furfuryl alcohol. *RSC Advances*, *10*(10), 5690–5696. Available from https://doi.org/10.1039/d0ra00085j.

Zhang, T., Zhu, Z., Gong, W., et al. (2016). Characteristics of fine particles in an urban atmosphere—relationships with meteorological parameters and trace gases. *International Journal of Environmental Research and Public Health, 13*(8), 807–822.

Zhao, X., Wen, L., Zhengqing, C., Bing, H., Tianwei, Q., & Dongye, Z. (2016). An overview of preparation and applications of stabilized zero-valent iron nanoparticles for soil and groundwater remediation. *Water Research, 100*, 245–266. Available from https://doi.org/10.1016/j.watres.2016.05.019.

Zou, Y., Wang, X., Khan, A., Wang, P., Liu, Y., Alsaedi, A., Hayat, T., & Wang, X. (2016). Environmental remediation and application of nanoscale zero-valent iron and its composites for the removal of heavy metal ions: A review. *Environmental Science and Technology, 50*(14), 7290–7304. Available from https://doi.org/10.1021/acs.est.6b01897.

Further reading

Chen, L., Sabatini, D. A., & Kibbey, T. C. G. (2010). Retention and release of TiO_2 nanoparticles in unsaturated porous media during dynamic saturation change. *Journal of Contaminant Hydrology, 118*(3–4), 199–207. Available from https://doi.org/10.1016/j.jconhyd.2010.07.010.

Fajardo, C., Saccà, M. L., Martinez-Gomariz, M., Costa, G., Nande, M., & Martin, M. (2013). Transcriptional and proteomic stress responses of a soil bacterium *Bacillus cereus* to nanosized zero-valent iron (nZVI) particles. *Chemosphere, 93*(6), 1077–1083. Available from https://doi.org/10.1016/j.chemosphere.2013.05.082.

Fang, J., Wang, M., Shen, B., Zhang, L., & Lin, D. (2017). Distinguishable co-transport mechanisms of phenanthrene and oxytetracycline with oxidized-multiwalled carbon nanotubes through saturated soil and sediment columns: Vehicle and competition effects. *Water Research, 108*, 271–279. Available from https://doi.org/10.1016/j.watres.2016.11.004.

Hou, S., Wu, B., Peng, D., Wang, Z., Wang, Y., & Xu, H. (2019). Remediation performance and mechanism of hexavalent chromium in alkaline soil using multi-layer loaded nano-zero-valent iron. *Environmental Pollution, 252*, 553–561. Available from https://doi.org/10.1016/j.envpol.2019.05.083.

Jiang, Y., Yin, X., Xi, X., Guan, D., Sun, H., & Wang, N. (2021). Effect of surfactants on the transport of polyethylene and polypropylene microplastics in porous media. *Water Research, 196*, 117016–117026. Available from https://doi.org/10.1016/j.watres.2021.117016.

Keller, A. S., Jimenez-Martinez, J., & Mitrano, D. M. (2020). Transport of nano- and microplastic through unsaturated porous media from sewage sludge application. *Environmental Science and Technology, 54*(2), 911–920. Available from https://doi.org/10.1021/acs.est.9b06483.

Lacalle, R. G., Garbisu, C., & Becerril, J. M. (2020). Effects of the application of an organic amendment and nanoscale zero-valent iron particles on soil Cr(VI) remediation. *Environmental Science and Pollution Research, 27*(25), 31726–31736. Available from https://doi.org/10.1007/s11356-020-09449-x.

Liu, B., Ma, R., & Fan, H. (2021). Evaluation of the impact of freeze-thaw cycles on pore structure characteristics of black soil using X-ray computed tomography. *Soil and Tillage Research, 206*, 104810–104820. Available from https://doi.org/10.1016/j.still.2020.104810.

May, R., & Li, Y. (2013). The effects of particle size on the deposition of fluorescent nanoparticles in porous media: Direct observation using laser scanning cytometry. *Colloids and Surfaces A: Physicochemical and Engineering Aspects, 418*, 84–91. Available from https://doi.org/10.1016/j.colsurfa.2012.11.028.

Rastghalam, Z. S., Yan, C., Shang, J., & Cheng, T. (2020). The role of Fe oxyhydroxide coating, illite clay, and peat moss in nanoscale titanium dioxide ($nTiO_2$) retention and transport in geochemically heterogeneous media. *Environmental Pollution*, *257*, 113625–113634. Available from https://doi.org/10.1016/j.envpol.2019.113625.

Shen, C., Bradford, S. A., Li, T., Li, B., & Huang, Y. (2018). Can nanoscale surface charge heterogeneity really explain colloid detachment from primary minima upon reduction of solution ionic strength? *Journal of Nanoparticle Research*, *20*(6). Available from https://doi.org/10.1007/s11051-018-4265-8.

Wu, X., Lyu, X., Li, Z., Gao, B., Zeng, X., Wu, J., & Sun, Y. (2020). Transport of polystyrene nanoplastics in natural soils: Effect of soil properties, ionic strength and cation type. *Science of the Total Environment*, *707*, 136065–136071. Available from https://doi.org/10.1016/j.scitotenv.2019.136065.

Xu, X., Shao, L., Huang, J., Xu, X., Liu, D., Xian, Z., & Jian, W. (2021). Effect of wet-dry cycles on shear strength of residual soil. *Soils and Foundations*. Available from https://doi.org/10.1016/j.sandf.2021.03.001.

Yang, K., Chen, B., Zhu, X., & Xing, B. (2016). Aggregation, adsorption, and morphological transformation of graphene oxide in aqueous solutions containing different metal cations. *Environmental Science and Technology*, *50*(20), 11066–11075. Available from https://doi.org/10.1021/acs.est.6b04235.

Zhang, M., Bradford, S. A., Simunek, J., Vereecken, H., & Klumpp, E. (2016). Do goethite surfaces really control the transport and retention of multi-walled carbon nanotubes in chemically heterogeneous porous media? *Environmental Science and Technology*, *50*(23), 12713–12721. Available from https://doi.org/10.1021/acs.est.6b03285.

Chapter 8

Occurrences and impacts of microplastics in soils and groundwater

Jinsheng Huang[1], Panfeng Wu[2], Shunan Dong[3] and Bin Gao[1]
[1]*Department of Agricultural and Biological Engineering, University of Florida, Gainesville, FL, United States,* [2]*Jiangsu Nantong Environmental Monitoring Center, Nantong, P.R. China,* [3]*College of Agricultural Science and Engineering, Hohai University, Nanjing, P.R. China*

Chapter Outline

8.1 Overview

Due to the cost efficiency, durability and stability, plastics and their related products have been widely used in packaging, industry, agriculture, clothing, and medicine since the 1950s (Wu et al., 2019; Zhang et al., 2020). According to Plastics Europe (PlasticsEurope, 2020), the global production of plastics has reached 368 million tons in 2019 whereas the value was only 0.5 in 1960 (Wu et al., 2019). This rapid increase has resulted in a great number of plastic wastes accumulated in the environment, leading to

Emerging Contaminants in Soil and Groundwater Systems. DOI: https://doi.org/10.1016/B978-0-12-824088-5.00009-4

pollution and potential risks to the ecosystems. Concerns have been raised and laws and regulations have been proposed to regulate plastic dumping and promote waste management and recycling (Barnes et al., 2009). However, due to their good stability, plastic particles are degraded in a slow way and can be further broken down into smaller particles in the environment through fragmentation, abrasion, weathering, and biodegradation naturally (ultraviolet radiation, erosion, runoff, etc.) or anthropogenically (landfill, waste management, wastewater treatment, etc.) (Zhou et al., 2020). These small particles fragmented from larger plastics become microplastics (MPs) when their diameters or lengths are smaller than 5 mm (Hidalgo-Ruz et al., 2012; Prata et al., 2019; Thompson et al., 2004). MPs have a wide size range and include nanoplastics (NPs) with a size <100 nm.

Based on their origins, MPs can be classified into primary and secondary MPs (Wu et al., 2019) (Fig. 8.1). The former refers to MPs released directly from manufactures, industries, personal care products (cosmetic abrasives, toothpastes, facial cleansers, etc.) and medicine (Guo et al., 2020; Wu et al., 2019). They are hard to be removed by the wastewater treatment plants (WWTPs) and may enter the environment inevitably. A recent study even found that plastic tea bags can release billions of primary MPs and NPs into the tea after brewing (Hernandez et al., 2019). Secondary MPs refer to those plastics derived from fragmentation of large plastic products (e.g., ropes, agricultural mulching films, road markings, and tires) by natural weathering processes such as erosion, UV, and ocean currents (Fig. 8.1) (Guo et al., 2020). Kole et al. (2017) estimated a 0.81 kg/year/capita of MP emission from road vehicle tires' abrasion globally and a 2% of MP emission from airplane-sourced tires in the Netherlands. Based on their shapes, MPs can be divided into microbeads, fibers, foams, films, fragments, and pellets (Hidalgo-Ruz et al., 2012). Furthermore, they can be classified based on their compositions into polyethylene (PE), polypropylene (PP), polystyrene (PS), polyvinylchloride (PVC), and polyethylene terephthalate (PET) (He et al., 2018).

MPs have been demonstrated to pose potential side effects to animals, plants, microbiota, matrix function, and structure (e.g., soil), and even human (de Souza Machado, Kloas, et al., 2018; Guo et al., 2020; Prata et al., 2020; Tan et al., 2020; Zhou et al., 2020). Since the introduction of MPs in 2004 (Thompson et al., 2004), they have been found in varied matrices including water, air, and soil all over the world (Zhang et al., 2020). However, most research on MPs has mainly focused on marine (Andrady, 2011; Browne et al., 2011; Rillig, 2012; Sharma & Chatterjee, 2017; Wright et al., 2013) and freshwater (Hurley et al., 2018; Klein et al., 2015; Lechner et al., 2014); while studies on MPs in soils are limited. In fact, most marine MPs are derived from terrestrial emissions because soil can serve as a sink of MPs (Zhou et al., 2020). Meanwhile, several studies have shown vertical transport of MPs from soils to groundwater (Bläsing & Amelung, 2018; Chae & An,

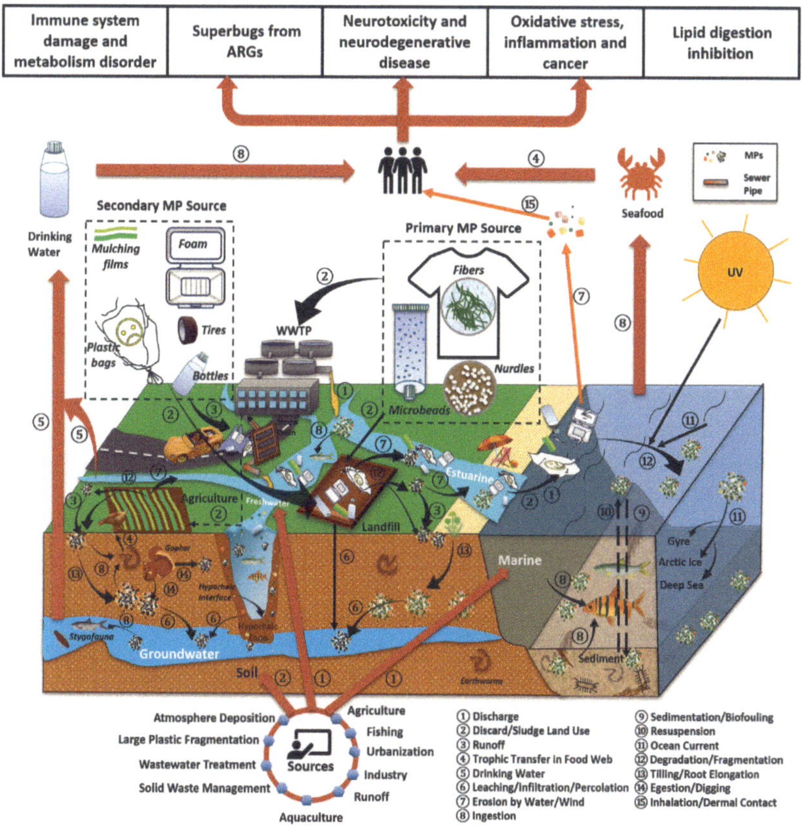

FIGURE 8.1 Human exposure to MPs in the environment through different sources and pathways. *WWTPs*, Wastewater treatment plants. *Modified from Wu, P., Huang, J., Zheng, Y., Yang, Y., Zhang, Y., He, F., Chen, H., Quan, G., Yan, J., Li, T., & Gao, B. (2019). Environmental occurrences, fate, and impacts of microplastics.* Ecotoxicology and Environmental Safety, 184. *https://doi.org/10.1016/j.ecoenv.2019.109612.*

2018; Guo et al., 2020; Hurley & Nizzetto, 2018; Zhou et al., 2020). Unfortunately, research on MPs in groundwater is even less than that in soils. Considering soil and groundwater as an inseparable system, soil MPs may enter groundwater through vertical transport and thus act as a source of groundwater MP pollution. The next chapter of this book provides insights into the vertical transport of MPs from soils into groundwater.

To date, knowledge of MPs in soils and groundwater is limited and the related research is still in the early stage. The overarching goal of this chapter is to survey current knowledge on the sources, occurrences, and impacts of MPs in the subsurface. Specifically, it focuses on the following topics: (1) occurrences, sources, and pathways of MPs in soils and groundwater, (2) analytical methods of MPs in soils and groundwater including sampling, extraction, and identification and quantification; (3) impacts of MPs on soil properties, animals, plants, and microbiota; and (4) human exposure to MPs from soil and groundwater matrices. Case studies and future perspectives are also included at the end of the chapter.

8.2 Occurrences and sources

8.2.1 Occurrences of MPs in soils and groundwater

Most MP studies have focused on water body systems especially surface water and sediments. Although soils also sever as a major sink and storage for MPs entering the terrestrial environment, MP contamination of soils is poorly understood (Rillig, 2012; Xu, Liu, et al., 2020; Xu, Zhang, et al., 2020). China, as the main producer and consumer of plastics, results in approximately 30% of global usage (PlasticsEurope, 2016). As a result, most existing studies on MPs in soils were conducted in China (Table 8.1).

The sampling types of MPs from soils in China include agricultural and nonagricultural soils. The former includes farmlands, crop lands, pastures, rangelands, greenhouse, vegetable, fruit, and grassland soils. In Shanghai (eastern China), MP abundances were found to be 78.00 ± 12.91 numbers per kg (n/kg) in shallow soil (3 cm) and 62.50 ± 12.97 n/kg in deep soil (6 cm) from farmlands, and 10.3 ± 2.2 n/kg (10 cm) from farmland paddy soils (Liu et al., 2018; Lv et al., 2019). Zhou et al. (2019) reported a MP concentration of 43,000–620,000 n/kg (vegetable) in Wuhan farmland areas. In the Dian Lake of southwestern China (Yunnan), MP abundances reached 18,760 (7100–42,960) n/kg (5–10 cm depth) in greenhouse soils (including forest buffer zone) for agricultural use due to the sludge and sewage irrigation (Zhang & Liu, 2018). Nonagricultural soils include coastal, wetland, campus lawn, industrial, and residential soils (Table 8.1). Zhou et al. (2016) first reported the abundance of MPs at 634 n/kg (2 cm depth) in coastal soils at Hebei Province, China. Duan et al. (2020) reported MP contamination in wetland soils in Shandong Province, China (136–2060 n/kg, 2 cm depth). MP content was found at 2697 (1360–4960) n/kg in sandy soils in Shaanxi Province, China. In a campus lawn soil of Tianjin, China, 95 n/kg of MPs were detected at a depth of 10 cm (Han et al., 2019).

Besides China, other countries also report MP pollution in soils (Table 8.1), especially agricultural soils. Vollertsen and Hansen (2017) investigated the effect of wastewater sludge application on MPs in agricultural soils. They found that MP abundances were 82,000 (0–165,000) and 236,000 (53,000–528,000) n/kg in the soils with and without sludge, respectively. In Chile, two studies were conducted: the first one found 600–10,400 n/kg MPs in agricultural land in Mellipilla (Corradini, Bartholomeus, et al., 2019; Corradini, Meza, et al., 2019) and the second study reported 306, 184, and 0 n/kg MPs in the crop, pasture, and rangeland

TABLE 8.1 Natural occurrences of MPs in soils and groundwater (Spiking excluded).

Location	Sample type	Sample depth	Concentration (mass or numbers)	Size	Main shape	Composition	Separation and digestion	Identification	References
Soil (N = 31)									
Hebei, China	Coastal soils	2 cm	634[a] n/kg[b] (DW)	1.56 ± 0.63[c] mm (~50% <1 mm)	Granules (76%), fragment (21%), fiber (2%), & film (1%)	–	DS (NaCl, NaI); CD (HCl)	SEM-EDS	Zhou et al. (2016)
Shandong, China	Coastal soils	2 cm	740 (1.3–14712.5) n/kg (DW)	<5 mm (~60% <1 mm)	Flakes (69%), foams (28%), fragment (1%), & fibers (1%)	PE, PP, PS & PU	DS (NaCl, NaI)	Microscope, SEM & FTIR	Zhou, (2018)
Shanghai, China	Farmland paddy soils	10 cm	10.3 ± 2.2 n/kg	0.02–1 mm	Fiber (domain), granules, fragment, & film	PE (61%), PP (35%), & PVC (4%)	DS (NaCl); CD (H_2O_2)	Microscope & μ-FTIR	Lv et al. (2019)
Franconia, Germany	Farmland	5 cm	0.34 ± 0.36 (0–1.25) n/kg (DW) or 35,000 mg/ha	1–5 mm	Film (44%), fragment (44%), & fiber (12%)	PE (62.5%), PP (25%), & PS (12.5%)	CD (H_2O_2)	Microscope & FTIR	Piehl, (2018)
Shanghai, China	Farmland	3 cm (shallow) to 6 cm (deep)	Shallow: 78.00 ± 12.91 n/kg; deep: 62.50 ± 12.97 n/kg	0.03–5 mm (93.3%); shallow: 1.91 ± 0.13 mm; deep: 1.48 ± 0.11 mm	Shallow: Fiber (53%), fragment (38%), film (7%), & pellets (2%); Deep: Fiber (38%), film (34%), fragment (28%), & pellets (<1%)	PP (51%), PE (43%), & PET (6%)	DS (NaCl); CD (H_2O_2)	Microscope & μ-FTIR	Liu et al. (2018)
Jiangsu, China	Farmland	–	420–1290 n/kg	0.02–0.25 mm (domain)	Fiber (65%, domain), bulk, & fragment	PP & PE (domain); PET & PAN	Pre-CD (H_2O_2); DS (NaCl, NaI & $ZnCl_2$); CD (H_2O_2, H_2SO_4, & HNO_3)	Microscope & μ-FTIR	Li, (2019)

(*Continued*)

TABLE 8.1 (Continued)

Location	Sample type	Sample depth	Concentration (mass or numbers)	Size	Main shape	Composition	Separation and digestion	Identification	References
Wuhan, China	Farmland (vegetable)	5 cm	320–12,560 n/kg (DW)	<0.2 mm (domain)	Microbeads (48%), fibers (37%), fragments (15%), & foams (1%)	Domain: PA (33%) & PP (29%)	DS ($ZnCl_2$)	Microscope & micro-Raman spectroscopy	Chen, (2020)
Hangzhou Bay, China	Farmland (mulch & nonmulch)	10 cm	571 n/kg (mulch) & 263 n/kg (nonmulch)	1–3 mm	Fragment (domain), fiber, & film	Domain: PP & PE (film & fragment); PET & rayon (fiber)	DS (NaCl, NaI); CD (H_2O_2)	Microscope & μ-FTIR	Zhou et al. (2020)
Xinjiang, China	Farmland (cotton field)	0–5, 5–20, & 20–40 cm	80.3–1075.6 n/kg	~2 mm	Fragment & film	PE	DS (NaI); CD (H_2O_2)	Microscope, SEM, & FTIR spectroscopy	Huang et al. (2020)
Hebei, China	Farmland	–	–	0–0.025 mm (~60% < 0.015 mm)	–	PP (34%), PA6 (28%), PET (25%), & PVC (13%)	CD (H_2O_2)	TOF-SIMS	Du, Wu, et al. (2020)
Yunnan, China	Forest buffer zone & greenhouse soils	5–10 cm	18,760 (7100–42,960) n/kg	0.05–1 mm (95%)	Fiber (92%) & film/ fragment (8%)	–	DS (NaI); CD (H_2O_2, $FeSO_4$, NaOH)	Microscope	Zhang and Liu (2018)
Loess plateau, China	Agricultural, fruit, & greenhouse soils	10–30 cm	40 ± 126 to 100 ± 141 n/kg (agriculture); 120 ± 169 to 320 ± 329 n/kg (fruit); 80 ± 193 to 100 ± 254 n/kg (greenhouse)	0.05–5 mm (>0.1 mm dominant)	–	PP & PE	DS (H_2O)	Microscope	Zhang et al. (2018)
Wuhan, China	Woodland, vegetable, & vacant land soils	5 cm	2.2×10^5 (2.2×10^4–6.9×10^5) n/kg	<0.1 mm (82%)	Fragments (52%), beads (14%), foam (10%), film (1.6%), & fiber (14%)	PE (36%), PA (17%), PS (9%), PVC (8.5%), PP (12%), & others (17%)	DS (NaCl, $ZnCl_2$); CD: (KOH:NaClO)	Microscope, SEM, & Raman spectroscopy	Zhou et al. (2019)

Yongin suburb, Korea	Rice-cultivation, mulching, & greenhouse soils	5 cm	160 ± 93 n/kg (rice); 81 ± 77 n/kg (mulching); 1880 ± 1563 n/kg (greenhouse inside); 1302 ± 2389 n/kg (greenhouse outside)	Domain: 0.2–0.3 mm (rice); 1–2 mm (mulching); 0.1–0.2 mm (greenhouse inside); 0.2–0.3 mm (greenhouse outside)	Fragment > sheet >fiber (rice); fiber > fragment > sheet (mulching); fragment > fiber > sheet (greenhouse inside); fiber > sheet > fragment (greenhouse outside)	PE > PP (rice); PP > PE > PS > PET (mulching); PE > PET > PP (greenhouse inside); PP > PE > PET (greenhouse outside)	DS (NaCl & $CaCl_2$); CD (H_2O_2)	Microscope & FTIR	Kim, (2020)
Región Metropolitana, Chile	Crop lands, pastures, rangeland, & grassland soils	–	306 ± 360 n/kg (crop); 184 ± 266 n/kg (pasture); 0 for rangeland and grassland	0.3 mm (fiber) & 0.08 mm (others); Mean area: $0.076 \pm 0.292\ mm^2$	Fiber (68%), film (23%), fragment (7%), & pellets (2%)	Dominance: Acrylates, PU, & varnish (crop); PE (pasture)	DS (NaCl, $ZnCl_2$)	Microscope & μ-FTIR	Corradini, (2021)
Mellipilla, Chile	Agricultural soils	25 cm	600–10,400 n/kg	0.02^d mm (width) × 0.97^d mm (length)	Fiber (> 97%), film, fragment, & pellets	–	DS (H_2O, NaCl, $ZnCl_2$)	Microscope	Corradini, Meza, et al. (2019)
Valencia, Spain	Agricultural soils (crop)	10–30 cm	With sludge: 2130 ± 950 n/kg (LD); 3060 ± 1680 /kg (HD); W/O sludge: 930 ± 740 n/kg (LD); 1100 ± 570 n/kg (HD)	0.15–0.25 mm (domain)	Fragments domain; fragment (sludge < w/o sludge); film & fiber (sludge > w/o sludge)	PP & PVC	DS (H_2O & NaI)	Microscope & μ-FTIR	van den Berg et al. (2020)
Shaanxi, China	Agricultural soils	10 cm	1430–3410 n/kg	0–0.49 mm (81%)	Fiber (49%), film (24%), pellets (5%), & fragment (22%)	PS, PE, PP, HDPE, PVC, & PET	DS (NaCl & $CaCl_2$); CD (H_2O_2)	Microscope, SEM, & FTIR spectroscopy	Ding, (2020)
Northern Jutland, Denmark	Agricultural soils	15 cm	With sludge: 82,000[a] (0–165,000) n/kg; W/O sludge: 236,000 (53,000–528,000) n/kg	With sludge: 0–0.13 mm; W/O sludge: 0.02–0.245 mm	–	With sludge: PP (56%), PE (39%), & Nylon (5%); W/O sludge: PE (89%), Nylon (10%), & PP (1%)	DS ($ZnCl_2$); CD (H_2O_2 + catalyst)	FTIR	Vollertsen & Hansen, 2017

(Continued)

TABLE 8.1 (Continued)

Location	Sample type	Sample depth	Concentration (mass or numbers)	Size	Main shape	Composition	Separation and digestion	Identification	References
Ontario, Canada	Agricultural soils	0–5, 5–10, & 10–15 cm	With biosolid: 18–541 n/kg; W/O biosolid: 25–298 n/kg	0.1–1 mm	Fibers (domain) & fragments	PP, PE, PET (45%, domain), & acrylic	DS (H_2O, NaI); CD (H_2O_2 & Fe (II))	Microscope & FTIR	Crossman et al. (2020)
Madrid, Spain	Soil amendments from sludge	–	302,000 ± 83,000 n/kg as heat dry pellets	Fragment: 0.029–0.533 mm; Fiber: 0.071–2.224 mm	Fiber (84%) & fragment (16%)	PET (domain), PE, PP, & others	DS ($NaCl_2$); CD (H_2O_2)	Microscope & ATR-FTIR	Edo, (2020)
New York, United States	Soil amendments from sludge	0–10 cm	0–1210 n/kg	< 2 mm	Fibers	PET, PA, & others	DS (H_2O)	Polarized light microscopy	Zubris and Richards (2005)
Feucherolles, France	Soil amendments from compost	0–27 cm	–	0.05 – 5 mm	Fragments, fibers, & films	PS (as form of styrene)	–	TEM-EDX, py-GCMS	Watteau et al. (2018)
Shandong, China	Wetland soils	2 cm	136–2060 n/kg	0.013–5 mm	Film (minor), fragment, fiber, & granule	PET (0.536–0.66 mg/kg) & PC (0.839–0.196 mg/kg)	DS ($ZnCl_2$); CD (H_2O_2)	Microscope	Duan et al. (2020)
Washing DC, United States	Wetland soils	5 cm	1270 ± 150 n/kg	0.075–5 mm	Fragment & fiber	PS (29%), PE (8%), & synthetic rubber (8%)	CD (H_2O_2 & Fe (II))	Microscope & FTIR	Helcoski, (2020)
Shaanxi, China	Sand soils	10 cm	2697 (1360–4960) n/kg	<0.5 mm (88%) & 0.5–1 mm (8%)	Fiber (39%), fragments (29%), films (20%), & pellets (11%)	PP (53%), PE (25%), PS (16%), PVC (6%), & PET (<1%)	DS ($CaCl_2$); CD (H_2O_2)	Microscope & microtransformed infrared spectroscope	Ding, (2021)

Campech, Mexico	Home garden soils	20 cm	870 ± 1900 n/kg	0.01–0.02 mm (~60%), 0.02–0.05 mm (~35%), & >0.05 mm (~5%)	–	–	DS (H_2O)	Microscope	Lwanga et al. (2017)
Tianjin, China	Campus lawn soil	–	95 n/kg (DW)	0.1–3.2 mm	Fragment	PP	DS (NaCl, NaI); CD (H_2O_2)	ATR-FTIR	Han et al. (2019)
Sydney, Australia	Industrial soils	–	4191[a] (300–67,500) mg/kg	<1 mm	–	PVC (>80%), PE, & PS	Pressurized fluid extraction	FTIR	Fuller and Gautam (2016)
Switzerland	Floodplain soils	5 cm	<593 n/kg; 5 (0–55.5) mg/kg	0.125–0.500 mm (88%)	–	PE (88%), PS, PVC, SBR, Latex, & PP	DS (NaCl & $CaCl_2$); CD (HNO_3)	FTIR & Raman spectroscopy	Scheurer and Bigalke (2018)
Hebei, China	Industrial, farmland, residential, & forest land	10 cm	–	<0.015 mm (>50%)	–	PET & PA6 (30%), PP (23%), & PVC (16%)	CD (H_2O_2)	TOF-SIMS	Du, Liang, et al. (2020)
Groundwater (N = 11)									
Krakow, Poland	Untreated potable water	–	Detected but not quantified	0–0.045 mm (microscope range)	Irregularly shaped particles (fragments)	–	–	Microscope, SEM-EDS, & Raman spectroscopy	Połeć, (2018)
Illinois, United States	Wells & springs in karst systems	<65 m (wells)	6.4[e] n/L (mean); 15.2 n/L (max)	<1.5 mm (fibers)	Fibers (100%)	PE (springs)	–	Microscope & py-GCMS	Panno et al. (2019)
Tamil Nadu, India	Local groundwater from Kovur	-	5.3 (4–7) n/L	–	Fragments & fibers	PET & PA	–	Microscope, SEM-EDS, & FTIR spectroscopy	Ganesan et al. (2019)
Holdorf, Germany	Groundwater wells	30 m	0.0007 (0–0.007) n/L (drinking water + ground water)	0.05–0.15 mm (drinking water + ground water)	Fragments	PE	DS ($ZnCl_2$); CD (H_2O_2 & HCl)	Microscope & FTIR	Mintenig et al. (2019)

(Continued)

TABLE 8.1 (Continued)

Location	Sample type	Sample depth	Concentration (mass or numbers)	Size	Main shape	Composition	Separation and digestion	Identification	References
Tamil Nadu, India	Wells & borewells	2–5 m	4.2 n/L (mean); 10.1 n/L (max)	0.6 ± 1.4 (0.12–2.5) mm; <1 mm (34%, domain)	Wells: fibers, foam, pellets, film, & fragments; Borewells: fibers	Nylon (PA, 35%), PE (55%), & PET (10%)	CD (H_2O_2 and Fe (II))	Microscope, AFM, & μ-FTIR	Selvam, (2020)
England & Wales, United Kingdom	Treated potable water	–	0–0.003 n/L (>LOQ); 0–0.011 n/L (>LOD)	>0.025 mm	–	ABS & PS (domain, >LOQ)	Enzyme digestions (cellulase & trypsin)	FTIR	Johnson et al. (2020)
Norway	Tap water	–	<1 n/L (below LOD)	>0.1 mm	–	–	DS ($ZnCl_2$); CD (H_2O_2)	Microscope & FTIR	Uhl et al. (2018)
Denmark	Tap water	–	0.3 n/L	<0.3 mm (mainly 0.02–0.1 mm)	Fibers (82%), fragments (14%), & films (4%)	PET, PP, PS, ABS, & PU	CD (acetic acid)	Microscope & FTIR	Strand, (2018)
Mexico City, Mexico	Public drinking water fountains	–	18 ± 7	0.5–5 mm (50%); <0.5 mm (50%)	Fibers (domain) and fragments	PTT & epoxy resin	–	Microscope, SEM-EDS, & micro-Raman	Shruti et al. (2020)
Skåne, Sweden	Potable water	–	0.174	<0.15 mm (32% < 0.02 mm)	Fibers (19%) & Fragments (81%)	PA, PET, & acrylates (majority)	DS (sodium polytungstate)	μ-FTIR & py-GCMS	Kirstein, (2021)
Rüsselsheim, Germany	Treated potable water	–	<1 n/L	>0.01 mm	Fragments & fibers	PET, PP, PS, & PE	CD (HCl)	Micro-Raman spectroscopy	Weber, (2021)

Notes: *W/O*, Without; *CD*, chemical digestion; *DS*, density separation; *DW*, dry weight; *LD*, light density microplastics with $\rho < 1$ g/cm^3; *HD*, heavy density microplastics with $\rho > 1$ g/cm^3; *μ-FTIR*, micro-Fourier transform infrared; *ATR-FTIR*, attenuated total reflectance-FTIR; *SEM-EDS*, scanning electron microscopy-energy dispersive X-ray spectroscopy; *TEM-EDX*, transmission electron microscopy-energy-dispersive X-ray spectroscopy; *py-GCMS*, pyrolysis gas chromatography mass spectrometry; *AFM*, atomic force microscopy; *TOF-SIMS*, time-of-flight secondary ion mass spectrometry; *PVC*, polyvinyl chloride; *PA(6)*, polyamide (6); *PE*, polyethylene; *PS*, polystyrene; *SBR*, styrene butadiene; *PP*, polypropylene; *PU*, polyurethane; *PC*, polycarbonate; *PET*, polyethylene terephthalate (a member of polyester); *ABS*, acrylonitrile butadiene styrene; *PBT*, poly butylene terephthalate; *PVCA*, polyvinyl chloride acetate; *PAN*, polyacrylonitrile; *PTT*, polytrimethylene terephthalate; *LOD*, limit of detection (1 particle detected); *LOQ*, limit of quantification (>3 particles detected).

[a] *Mean.*

[b] *n/kg means numbers/items/particles per kg.*

[c] *Mean ± standard error.*

[d] *Median.*

[e] *n/L means numbers/items/particles per liter.*

and grassland soils, respectively, in the Region Metropolitana (Corradini, 2021). Other types of soil samples have also been recorded. Fuller and Gautam (2016) utilized a pressurized fluid extraction method and found that there were 4191 (300–67,500) mg/kg MPs in industrial soils in Sydney, Australia. Scheurer and Bigalke (2018) reported a maximum MP concentration of 593 n/kg in floodplain soils in Switzerland (5 cm depth). Although some countries have not been reported, it is obvious that there is a widespread of MP pollution in soil globally.

Current research on MP occurrence in groundwater is still in its infancy. So far, only few records can be found. Panno et al. (2019) reported a MP abundance of 6.4 numbers per liter (n/L) from wells (<65 m depth) and springs in karst systems in Illinois, United States. In Tamil Nadu, India, MP contents were reported to be 5.3 (4–7) n/L and 4.2 n/L from a local groundwater resource and wells, respectively (Ganesan et al., 2019; Selvam, 2020). Johnson et al. (2020) measured MP concentrations in treated potable water in England and Wales, United Kingdom and recorded the MP abundance ranging from 0 to 0.003 n/L as a limit of quantification (LOQ; >3 particles detected), and 0–0.011 n/L as a limit of detection (1 particle detected).

8.2.2 Sources of MPs in soils

MPs can enter the soil environment via different sources and pathways (Fig. 8.1) including organic fertilizers from wastewater sludge and compost, agricultural plastic mulch films, landfill atmospheric deposition, irrigation and flooding, and other sources (Bläsing & Amelung, 2018; Zhu et al., 2019). Among them, wastewater sludge, plastic mulching, and compost (Fig. 8.2) are identified as the dominant sources of MPs entering soils (Xu, Zhang, et al., 2020; Zhou et al., 2020).

8.2.2.1 Wastewater sludge and compost

Wastewater sludge, also called sewage sludge, is one of the major sources of MPs entering the soil media. Wastewater is a liquid waste removed from residential, industrial, and commercial establishments with the surface, ground, and stormwater (Werther & Ogada, 1999). During the treatment of wastewater, a semisolid with rich nutrients is produced and named as wastewater sludge (Turovskiy & Mathai, 2006). According to the United States Environmental Protection Agency (USEPA), after the wastewater sludge is treated to meet related standards or criteria, it is renamed as biosolids. It can then be used as a soil amendment or an organic fertilizer because of its high organic matter (OM) and nutrient contents (Zhou et al., 2020). However, biosolids may also contain heavy metals and persistent organic pollutants (POPs) (Gaylor et al., 2013). In addition, up to 90% of MPs may be retained

FIGURE 8.2 MPs in the compost in Australia. *From Kumar, M., Xiong, X., He, M., Tsang, D., Gupta, J., Khan, E., Harrad, S., Hou, D., Ok, Y., & Bolan, N. (c.2020). Microplastics as pollutants in agricultural soils.* Environmental Pollution, 265. *https://doi.org/10.1016/j.envpol.2020.114980.*

and accumulated in wastewater sludge with the concentration ranges from 10^3 to 10^5 n/kg, posing potential risks (van den Berg et al., 2020).

Likewise, compost is also widely used as an organic fertilizer in agriculture. Though compost may elevate the levels of harmful trace metals and POPs in soils, it also presents a bunch of advantages including low disposal cost, nutrient element recycle, microbial activity promotion, and soil OM enrichment (García-Gil et al., 2000; Giusquiani et al., 1995). In Germany, the land application rate of compost is up to 20 or 30 tons/ha annually (Bläsing & Amelung, 2018). Due to improper disposal and separation of wastes, plastics were found in bio-waste compost in Germany by Bläsing and Amelung (2018) at an abundance of 2.38–180 mg/kg.

8.2.2.2 Agricultural plastic mulch

Another major contributor of MP pollution in soils is plastic mulch (Fig. 8.3). Over the past decades, plastic mulch is widely used all over the world in agriculture for crop yield and quality improvement, soil moisture retention and erosion reduction, and water use conservation (Gao et al., 2019; Liu, 2014; Qi et al., 2020). Over 4 million tons of plastic films have been consumed in agriculture globally and the number keeps increasing (Xu, Zhang, et al., 2020). In China, the usage of plastic mulch has been increased from 319,000 to 1,245,000 tons from 1991 to 2011 (Liu, 2014). In Europe Union (EU) and the United States, 427,000 and 300,000 tons of mulching are applied to farmland each year, respectively (Nizzetto et al., 2016a,b). Most of the plastic mulch difficult to be removed from soils due to its thin characteristics (0.01–0.03 mm), leading to further plastics accumulation and slow degradation via UV irradiation and biodegradation in the environment

FIGURE 8.3 Plastic mulch residue from agricultural soils in Gansu Province and Shanghai, China. (A, B) Recycled film residues; (C) film residues from soil; (D) plastic residue from mulch film. *From (1) Liu, M., Lu, S., Song, Y., Lei, L., Hu, J., Lv, W., Zhou, W., Cao, C., Shi, H., Yang, X., & He, D. (2018). Microplastic and mesoplastic pollution in farmland soils in suburbs of Shanghai, China.* Environmental Pollution, 242, *855–862. https://doi.org/10.1016/j.envpol.2018.07.051. (2) Qi, R., Jones, D. L., Li, Z., Liu, Q., & Yan, C. (2020). Behavior of microplastics and plastic film residues in the soil environment: A critical review.* Science of the Total Environment, 703. *https://doi.org/10.1016/j.scitotenv.2019.134722.*

(Yuanqiao et al., 2020; Zhou et al., 2020; Zhu et al., 2019). Mulch film residues can damage soil structure, aeration, and water permeability, thus affecting plant root growth and productivity (Jiang et al., 2017). In recent years, bioplastic films, which are partially/fully biodegradable, have been developed as alternatives of plastic mulch to reduce its environmental impact. However, high cost, low stability, and large carbon footprint prevent them from real-world agricultural applications (Qi et al., 2020).

8.2.2.3 Solid waste management (landfill and littering)

During the municipal solid waste (MSW) management processes especially for landfill, MPs may be released into soils and even groundwater (Alimi et al., 2018; Cole et al., 2011; He et al., 2019). Landfill may account for 21%–42% of the global plastic waste production (Nizzetto, Bussi, et al., 2016). The fragmentation of plastics in landfill may be affected by multiple factors including pH, salinity, temperature, gas (methane and carbon dioxide), and microbial activities (Hanson et al., 2005; Pramila & Ramesh, 2011). MPs in landfill can enter other environmental compartments including lakes (Imhof et al., 2013). It may take up to 100 years for MPs to be generated and released from landfill, leading to potential long-term risks (Yang et al., 2015). Landfill can deliver MPs into soils and groundwater through landfill leachates. According to the United States Geological Survey (USGS), landfill leachate is a liquid waste that consists of different kinds of pollutants when precipitation or applied water moves through the solid waste in the landfill. Leachates may contain a great number of heavy metals, OM, ammonia-N, and chlorinated salts, and toxicity analysis and thus have potentially toxic effects (Renou et al., 2008). Treatment is thus necessary before landfill leachates are released into the environment and this has attracted much attention in the EU countries (Wiszniowski et al., 2006).

For littering, taking cigarettes as an example, almost 2/3 of smoked cigarettes (trillions in unit) with nonbiodegradable filter plastics are thrown away annually in the United States. Meanwhile, the environmental impacts of dumped plastics cannot be ignored (Xu, Zhang, et al., 2020), especially for illegal dumping (Fig. 8.4) (e.g., the waste packaging materials of pesticides and fertilizers).

8.2.2.4 Atmospheric deposition

Air deposition may also an important source of MPs in surface soils (Fig. 8.4). A MP deposition rate of 2–355 numbers per m^2 per day (n/m^2/day) in the urban area of Paris was reported by Dris et al. (2016). Liu, Wang, et al. (2019) estimated an annual weight of suspended atmospheric MPs at 120.72 kg in Shanghai, China. MPs have also been found in Swiss floodplain soils in remote unsettled high mountain areas, suggesting atmospheric deposition as a potential source of soil MPs (Scheurer & Bigalke, 2018). Crossman et al. (2020) concluded that MPs in soils may derive from

FIGURE 8.4 MP contamination from littering and air deposition in Germany. (A–C) In agricultural soils; (D–F) in protected areas. *From Bläsing, M., & Amelung, W. (2018). Plastics in soil: Analytical methods and possible sources.* Science of the Total Environment, 612, *422–435. https://doi.org/10.1016/j.scitotenv.2017.08.086.*

air deposition because of the dominance of polyester (PET) fibers in samples besides biosolid.

8.2.2.5 *Irrigation and flooding*

When precipitation is in shortage, irrigation technology is used to grow crops in those water-deficient areas. The irrigation water body includes groundwater, untreated and treated wastewater (Bläsing & Amelung, 2018). Groundwater MP contamination has been observed in several studies (Table 8.1). When it is used for irrigation in developing countries, MPs (especially NPs) may exist, pass through coarse soils, and further reach groundwater, like other contaminants [e.g., silver nanoparticles (Hoppe et al., 2015)]. It is estimated that 10% of the world population is receiving food produced from contaminated wastewater (Corcoran, 2010). However, untreated sewage contains a great number of MPs (1000–627,000 n/m^3), most of which are from washing machine effluents and personal care products (Bläsing & Amelung, 2018). Majewsky et al. (2016) found 80–260 mg/m^3 of PE and PP in wastewater. Treated wastewater used in irrigation among developed countries may contain smaller MP concentration than untreated wastewater; however, it can still pose MP contamination risks to the receiving soils.

Flooding from rivers and lakes is also considered as an input of soil MPs (Bläsing & Amelung, 2018). Meanwhile, MP concentrations of freshwater systems vary due to different MP inputs (e.g., WWTPs, flow conditions, and runoffs) and variation of MP sampling and analytical methods. Consequently, the MP abundance ranged from 0 to 13751 n/m^3 and 0–220,000 n/km^2 in rivers and lakes, respectively (Bläsing & Amelung, 2018). All in all, the sequence of MP concentrations in different water bodies for irrigation and flooding is as follows: untreated wastewater > treated water > rivers > lakes (Bläsing & Amelung, 2018).

8.2.2.6 *Other sources*

Other sources including tire wear, surface runoff, and plastic fragmentation may also contribute MP pollution in soils (Bläsing & Amelung, 2018; Guo et al., 2020; Kole et al., 2017; Zhu et al., 2019). Kilponen (2016) reported MPs existence (0.17–0.26 n/L as fibers) in urban runoffs and indicated it as a potential source and pathway entering the marine systems. Kole (2015) reported that, among the 5871 tons of tire wear in the Netherlands, the highest emission is to soil (67%), much higher than that to air (12%), surface water (6%), and sewers (15%). The emission rates of tire dust are estimated as 10,000 and 110,000 tons per year in Sweden and Germany, respectively (Bläsing & Amelung, 2018). Plastic fragmentation or degradation by soil organisms (feeding, digestion, and excretion) is also considered as a source of MP in soils (Guo et al., 2020; Xu, Liu, et al., 2020).

8.2.3 Sources of MPs in groundwater

Soil migration, landfill leachates, septic effluents, sinkholes, surface runoffs, and anthropogenic activities are the potential ways of MPs entering groundwater (Panno et al., 2019; Selvam, 2020). The leaching of MPs from landfill leachates into groundwater is similar to that into soils (Section 8.2.2.3). It is thus not included in the following sections.

8.2.3.1 Soil migration

Several studies have pointed out the potential vertical transport of MPs from the vadose zone to groundwater (Bläsing & Amelung, 2018; Chae & An, 2018; Guo et al., 2020; Hurley & Nizzetto, 2018; Zhou et al., 2020). The pathways of MP migration in soils are summarized in Fig. 8.1. The most popular example can be found in MP transport from soils to groundwater through bio-pores after the ingestion of earthworms (*Lumbricus terrestris* L.): a 10 cm depth of downward movement of MPs (PE, 0.71–2.8 mm) in soil was observed within 21 days (Rillig et al., 2017). Besides, collembolans may also promote the distribution of MPs in soils (Maaß et al., 2017; Zhu et al., 2018) (Fig. 8.5). MPs may reach groundwater systems by leaching process, which is defined as the loss of mineral and organic solutes due to percolation in soil (Zhou et al., 2020). In addition, MPs may enter the groundwater with other contaminants such as pesticides, heavy metals, and nutrients (Bläsing & Amelung, 2018). Fundamental speculative reasoning

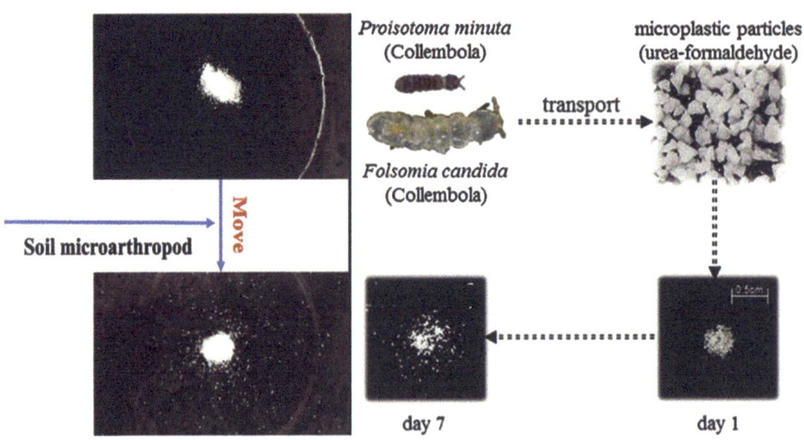

FIGURE 8.5 MP distribution by soil collembolans. *From (1) Maaß, S., Daphi, D., Lehmann, A., & Rillig, M. C. (2017). Transport of microplastics by two collembolan species.* Environmental Pollution, 225, *456–459. https://doi.org/10.1016/j.envpol.2017.03.009. (2) Zhu, D., Bi, Q. F., Xiang, Q., Chen, Q. L., Christie, P., Ke, X., Wu, L. H., & Zhu, Y. G. (2018). Trophic predator-prey relationships promote transport of microplastics compared with the single* Hypoaspis aculeifer *and* Folsomia candida. Environmental Pollution, 235, *150–154. https://doi.org/10.1016/j.envpol.2017.12.058.*

suggests that colloidal and nanosized particles (MPs and NPs) are well dispersed in pore water and thus may presumably enter at least the shallow groundwater aquifers from the soil above. In other words, soil not only is a sink but also can be a source of MPs in groundwater. If that happens, the sources of soil MPs become the indirect inputs of groundwater MP contamination.

8.2.3.2 Wastewater effluent

Another potential source of MPs in groundwater is wastewater effluent (Kim & Lee, 2020; Panno et al., 2019; Shruti et al., 2020). One of the pathways is swage leakage from WWTPs, all the way down from surface water to the hyporheic zone (a region that contains a mixture and surface water and groundwater) and finally reach shallow groundwater (Fig. 8.1). Another one is private septic tanks, which provide anaerobic digestion of OMs and reduction of suspended solids. Through septic effluent pipelines, wastewater may enter the sinkholes and immediately penetrate the HZ, reaching the underlying groundwater aquifers (Fig. 8.1).

8.2.3.3 Other sources

Surface runoffs, anthropogenic activities (littering, fishing activities, and tourism), and chemical additives (flame retardants) from the industry may also lead to MP pollution in groundwater (Ganesan et al., 2019; Panno et al., 2019; Selvam, 2020).

8.3 Analytical techniques

8.3.1 General procedures

Soils and water bodies contain minerals, OM, and nutrients. Especially for soil, which is a mixture of solids (e.g., clays) and liquids, has a great number of OM derived from soil animals and plants. OM can be further stabilized into humus, combined with other impurities in soils, making MPs analysis difficult (He et al., 2018). In addition, it is a challenge to extract MP aggregates from soils without damaging their structure and the extraction may promote the fragmentation of MPs (Möller et al., 2020). The general procedure for MP analysis in soils and groundwater includes sample collection (sites, tools, depth, approaches, and quantity), sample preparation (drying and sieving), extraction (separation, digestion, and filtration), and identification and quantification (visual sorting, chromatography, and vibration spectroscopy).

8.3.2 Sample collection

To guarantee the representativeness of sampling, suitable approaches should be considered during sample collection to generate reliable results. Soil is a heterogeneous system including many elements, from solid to liquid to air, from plants to animals, and from inorganics to organics. In general, approaches include judgmental, random, systematic, transect, unaligned grid, and stratified sampling. Table 8.2 summarizes sampling methods and related

TABLE 8.2 Sampling approaches selection based on research objectives.

Approach	Objectives	Description
Judgmental sampling	Verify a suspected contamination	A subjective choice of sampling points based on expert information (e.g., historical utilization); simple but inaccurate; may only be suitable for specific zones.
Simple random sampling	Identify the contamination of a homogeneously contaminated site	A random distribution of sampling points given that each point is equally and independently selected; allows unbiased statistical analysis (most MPs are not distributed equally).
Systematic grid sampling	Identify the contamination extent, delineate hot-spots, or spatial concentration gradients	A distribution of sampling points along the nodes of a regular pattern (e.g., triangle or square); first point is randomly selected; remaining are regularly spaced; simple and even coverage.
Transect sampling	Identify the contamination extent along linear features (e.g., drainage diches) and define concentration gradients	A one-dimensional systematic method; points are collected in regular space linearly; the linear line could be parallel but not required; define a transect is easier than a grid.
Unaligned grid sampling	Identify the contamination extent in a sampling area	A combination of random sampling and systematic grid sampling; points are randomly distributed in a grid; allows good coverage w/o losing randomness.
Stratified sampling	Identify the contamination extent in defined subareas within the total sampling area	A division of the initial sampling area into smaller strata; useful if the strata differ from each other significantly and the contamination is homogeneously distributed in each stratum.

Source: From Möller, J. N., Löder, M. G., & Laforsch, C. (2020). Finding microplastics in soils: a review of analytical methods. *Environmental Science & Technology*, *54*(4), 2078–2090.

information according to different research goals. Meanwhile, sampling sites should consider the land-use type (e.g., farmland, coastal soil, wetland soils, etc.), point source contamination (e.g., trash bins, landfill, industries, etc.), and environmental driving forces (e.g., topsoil erosion by wind and water, surface runoff, etc.) (Möller et al., 2020). Ganesan et al. (2019) found that sampling sites in canal beds of a river and solid waste dumping may contaminate nearby groundwater.

Single point or composite samples can be considered in sampling design or during sampling processes. The latter is defined as a combination and homogenization of several discrete point samples of the same size within a spatial unit to reduce the heterogeneity of MPs (Möller et al., 2020). Zhang and Liu (2018) exploited this method by combing six random sampling points into one composite for each plot (totally 5) at each site. Huang et al. (2020) collected 45 soil samples and combined them into one composite sample (three layers from 0–5, 5–20, and 20–40 cm). Likewise, Scheurer and Bigalke (2018) used a similar method to collect samples along parallel transects near the river. In addition, the number of sampling points is also important. Depending on analytical capabilities and methods, and practical conditions, the number of sampling points may be varied largely.

The most frequent sampling tool for soils is steel spade/shovel/scoop for surface samples whereas core samplers/augers (Fig. 8.6) are used for deep layers (Zhou et al., 2020). For groundwater, theoretically, it should follow the same sampling procedure of freshwater and marines. However, for most studies on MPs in groundwater, glass/plastic containers (bottle/jars) and pumping are frequently used. Johnson et al. (2020) designed a special device for treated and raw groundwater sampling (Fig. 8.7).

8.3.3 Sample preparation

Due to the moisture content, soil samples need to be dried through either air drying (25°C–35°C) or machine drying (35°C–60°C) (Carter & Gregorich, 2007). The former is often used to minimize its effect on soil properties, and the soil is placed horizontally in a metal container with a cap (Lu, 2000). After drying, samples are crushed with mortar or wood hammer to break big aggregates. However, this step may destroy MPs in the samples and cause plastic fragmentation.

Soil samples can be sorted and reduced into various sizes (usually <5 mm) through the sieving process as a preliminary step for homogeneity and even distribution. In general, the standard soil sieving (fine-earth fraction) is referring to the mesh size <2 mm (some countries <1 mm) (Bläsing & Amelung, 2018). However, it is a different story for MPs. The mesh size among the reviewed works is from 1 to 5 mm and the particles >5 mm are discarded (Zhou et al., 2020). The 5 mm mesh size is suggested for soil MP extraction (Zhou et al., 2020). It is worth noting that sieving is a method for

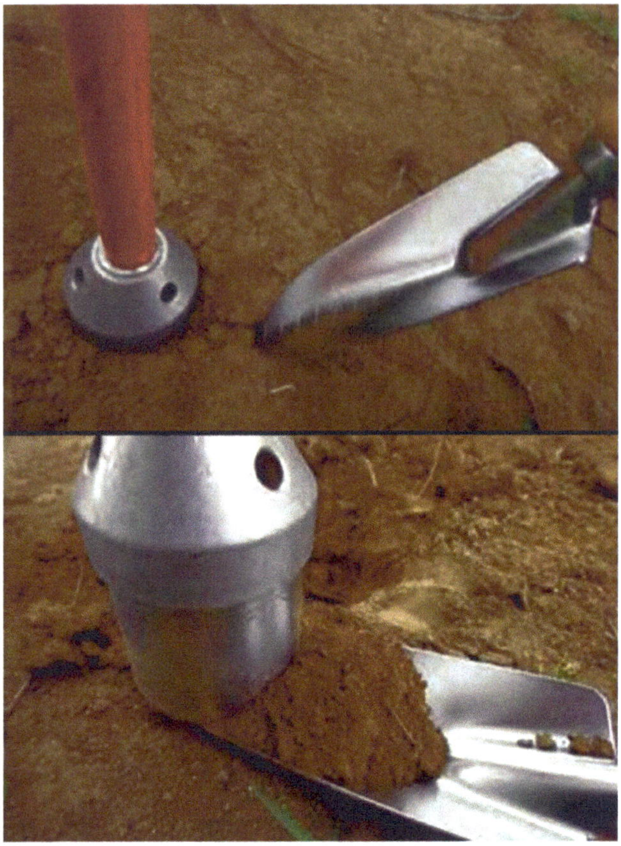

FIGURE 8.6 Scoop and core sampler applications for soil sample collection. *From Onda, Y., Kato, H., Hoshi, M., Takahashi, Y., & Nguyen, M. (2015). Soil sampling and analytical strategies for mapping fallout in nuclear emergencies based on the Fukushima Dai-ichi Nuclear Power Plant accident.* Journal of Environmental Radioactivity, 139, *300–307. https://doi.org/10.1016/j.jenvrad.2014.06.002.*

sample processing/preparation instead of separation. In water bodies, sieving is the most frequently used preliminary method. Mesh size may affect the detection of the lower size of MPs. For samples containing plastics, a metal-sieve is recommended to prevent cross-contamination. Moreover, coning and quartering or multipoint method is required for subsampling (Lu, 2000).

8.3.4 Extraction

8.3.4.1 Separation

Separation is an essential step during the extraction of MPs, separating plastic particles from water and soil aggregates. This section covers density

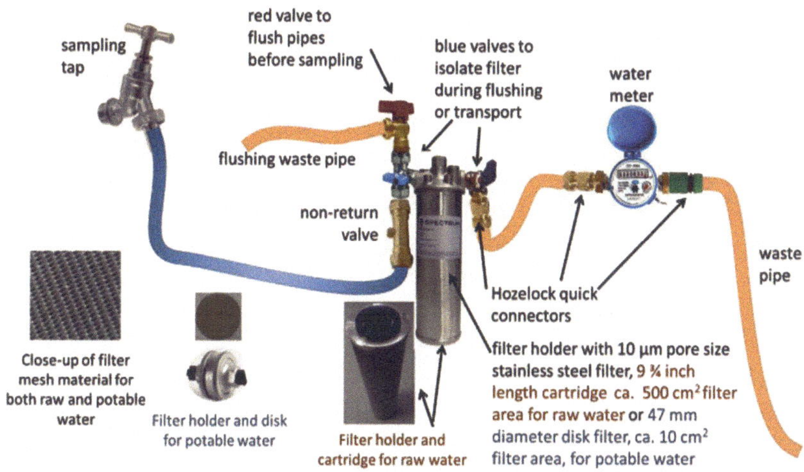

FIGURE 8.7 A special sampling tool and setup for treated potable waters. *From Johnson, A., Ball, H., Cross, R., Horton, A., Jürgens, M., Read, D., Vollertsen, J., & Svendsen, C. (2020). Identification and quantification of microplastics in potable water and their sources within water treatment works in England and Wales.* Environmental Science and Technology, 54*(19), 12326–12334. https://doi.org/10.1021/acs.est.0c03211.*

separation (dominance), oil separation, vertical density gradient separation, and electrostatic separation.

8.3.4.1.1 Density separation

Density separation is the major separation method for most studies related to MP extraction from water and sediments. This method has been modified and applied to soil and groundwater MP research. The high-density salt solution such as NaCl is often used as an extraction fluid because of its cheap price and high availability. Besides, distilled water is also used (Lwanga et al., 2017; Zhang et al., 2018) for MPs lighter than water (e.g., PP and PE). After mixing and precipitation, the plastic containing supernatant is filtered (Möller et al., 2020). However, an underestimation of the sample may occur; even worse, the saturated NaCl solution with a maximum density of 1.2 g/mL can only achieve limited extractions of low-density MP particles. For the high-density MP particles such as PET (1.5 g/mL), polybutylenes terephthalate (PBT, 1.3 g/mL), and polyvinyl chloride (PVC, 1.58 g/mL), higher density solutions are needed. Thus $ZnCl_2$ (1.6 g/mL), $CaCl_2$ (1.4–1.6 g/mL), $Na_6[H_2W_{12}O_6]$ (1.4 g/mL), NaBr (1.55 g/mL), $ZnBr_2$ (1.7 g/mL), and NaI (1.8 g/mL) are applied as alternatives (Dioses-Salinas et al., 2020; Möller et al., 2020; Prata et al., 2019). Nevertheless, some of them are costly and have side effects (toxicity of $ZnBr_2$ and $ZnCl_2$, corrosion of $ZnCl_2$ at low pH, Ca^{2+} bridges with organics, etc.) (Prata et al., 2019; Zobkov & Esiukova, 2017). Ultrasonication, stirring, and air flow are applied to prevent aggregation of soil samples. Furthermore, improved extraction methods have been developed by scientists in recent years [e.g., an automatic cycling device including separation, filtration, and circulation (Fig. 8.8)] (Liu, Song, et al., 2019).

8.3.4.1.2 Other separation methods

Other separation methods include oil separation, vertical density gradient separation, and electrostatic separation for MPs in soil and water samples. Due to the lipophilicity (the tendency of dissolving into fats) of most plastics, oil extraction has been developed. Imhof et al. (2012) used pine oil along with a froth conditioner to recover high-density plastics from sediments. Vertical density gradient separation, mainly used by the recycling industry, has also been applied to separate MPs using a liquid containing colloidal magnetic particles (Hu, 2014; Möller et al., 2020). The magnetic field near that liquid can create a vertical density gradient, moving high-density particles downward and low-density particles upward in the liquid container. Also a common method in the recycling industry, electrostatic separation has recently been used to separate MPs from water, sediment, and sand matrix (Felsing et al., 2018).

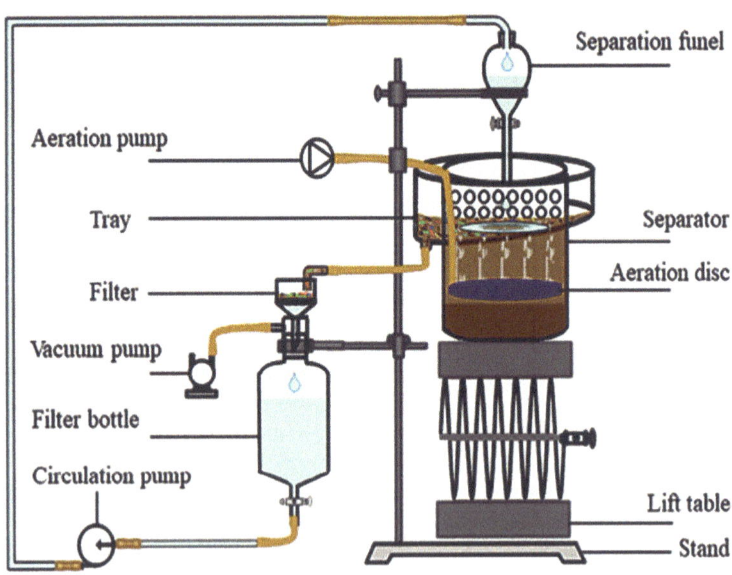

FIGURE 8.8 A developed device for circularly separation of MPs from soil. *From Liu, M., Song, Y., Lu, S., Qiu, R., Hu, J., Li, X., Bigalke, M., Shi, H., & He, D. (2019). A method for extracting soil microplastics through circulation of sodium bromide solutions.* Science of the Total Environment, 691, *341–347. https://doi.org/10.1016/j.scitotenv.2019.07.144.*

8.3.4.2 Digestion

After separation, digestion is highly recommended to further remove the biological matrix from the environmental samples, especially soils, to minimize the interference of visual inspection. The most frequently used digestion method is H_2O_2 oxidation, followed by acid/base digestion and enzymatic digestion.

8.3.4.2.1 H_2O_2 oxidation

Hydrogen peroxide is an oxidizing agent having better OM digesting efficiency than HCl and NaOH without destroying the MP structures (Prata et al., 2019). Nuelle et al. (2014) reported that PVC, PET, PP, acrylonitrile butadiene styrene (ABS), PS, PA, PE, and polyurethane can resist the oxidation of 35% H_2O_2. Hydrogen peroxide is often combined with iron (II) sulfate ($FeSO_4$) to form Fenton's reagents. Hurley et al. (2018) compared the removal efficiency among 10 M NaOH, 10% KOH, 30% H_2O_2, and Fenton's reagent for treating soil samples with MPs and found that the Fenton's reagent is the best to remove OM from soil and sewage sludge samples. They also found that temperature, pH, exposure time, and aging process can affect the digestion efficiency (Hurley et al., 2018; Tagg et al., 2017).

8.3.4.2.2 Acid/base digestion

Acidic and alkaline digestion are frequently used for organics removal in the samples for MP detection. The removal efficiency of HNO_3 is the highest among other reagents (H_2SO_4, H_2O_2, NaOH, and KClO) (Zhou et al., 2020). On the other hand, strong base is also found to break down some plastics such as PE and PA (Möller et al., 2020).

8.3.4.2.3 Enzymatic digestion

Compared to chemical digestion, enzymatic method has an advantage of not damaging the MPs during the OM removal processes (Prata et al., 2019). The general enzyme protocol includes predigestion of soils/sediments with an industrial enzyme (2.5%) at 5°C for 60 minutes followed by 30% H_2O_2 to remove debris (Crichton et al., 2017). Cole et al. (2014) used Protein K to digest plankton-rich samples from marine systems, accounting for up to 97% digestion efficiency by weight at 50°C, even better than acid (HCl) and base (NaOH) digestion without destroying the plastics. Löder et al. (2017) used the basic enzymatic purification method to realize a 98.3% removal efficiency from surface water samples. A few studies have also proposed enzyme digestion protocols for mussel tissues and fish guts with decent efficiencies and recovery rates of MPs (Catarino et al., 2017; Cole et al., 2014; Möller et al., 2020). However, due to its high cost and varied efficiencies, enzymatic digestion is mainly used in small-scale applications.

8.3.4.3 Filtration

To separate MPs from water samples or supernatants, filtration is often used as a solid–liquid separation method after digestion to facilitate the identification and quantification of MPs. A pore size between 10 and 20 μm is recommended for the effective filtration of MPs (Dioses-Salinas et al., 2020). Selection of the smallest pore size is good for the recovery test (especially for NPs) but may slow the filtration process.

8.3.4.4 Other extraction methods

Other extraction methods include manual extraction followed by microscope (Hidalgo-Ruz et al., 2012; Möller et al., 2020), pressurized fluid extraction (Fuller & Gautam, 2016), magnetic extraction (Grbic et al., 2019), froth flotation (Imhof et al., 2012), and liquid extraction (Elert et al., 2017). However, these methods are less popular than the density separation, acid/base and H_2O_2 digestion, and filtration methods. In fact, they are rarely used for MP detection in soils and groundwater.

8.3.5 Identification and quantification

When separation and digestion steps are complete, samples are ready to be identified and quantified for size, shape, color, and polymer types of MPs. The commonly used criteria for counting MPs are as follows: (1) the particle cannot be destroyed; (2) the particle should have identical and distinguishable color; (3) there is no visible cellular structure or OMs; and (4) the fibers should be unbranched and of symmetric thickness (Li et al., 2020; Zhou et al., 2020).

This section covers the main MP identification and quantification methods including visual identification [domain, usually combined with vibration spectroscopy such as Fourier transform infrared (FTIR) and Raman], vibration spectroscopy, chromatography, and other newly developed/rarely used/in situ techniques.

8.3.5.1 Visual identification

8.3.5.1.1 Light microscope

Light microscope is the earliest and to date frequently used visual identification method for MPs, accessing the surface structure and other physical characteristics (size, shape, and color) of plastics. Based on their shapes, MPs can be classified into different categories such as fragments, fibers, foams, beads, films, and the others. Combined with other technologies, scientists have made some improvements to the light microscopy identification method for MPs, including heating samples (van den Berg et al., 2020; Zhang et al., 2018), polarized light microscopy (Zubris & Richards, 2005), hot needle

(Campbell et al., 2017), and staining dyes (e.g., Solvent Red 26, Eosin B, etc.) (Prata et al., 2019).

8.3.5.1.2 Other microscopes

Scanning electron microscopy (SEM) can provide clear images of surface topography, crystalline structure, chemical composition, and electrical behavior of the top 1 μm or so of MPs (Vernon-Parry, 2000). The example of SEM images of MPs is shown in Fig. 8.9 (microscope images included). SEM is often coupled with energy dispersive X-ray spectroscopy (EDS) for surface elemental analysis of MPs (Ganesan et al., 2019; Połeć, 2018; Zhou et al., 2016). Another type of electron microscope, transmission electronic microscopy (TEM), is also used for MP identification (Watteau et al., 2018). TEM provides an inner information of the sample structures (e.g., crystal structure and morphology) due to its transmitted electrons that can pass through the sample, while SEM can only offer a surface information of the sample. TEM can also be coupled with EDX for elemental analysis. Watteau et al. (2018) used TEM-EDX to determine macro-plastics and MPs (styrene, which may come from PS) from soils amended with compost. Atomic force microscopy (AFM) has also been used in MP identification and it provides three-dimensional images under an ambient environment without coating the sample (Fig. 8.10).

8.3.5.2 Vibration spectroscopy

8.3.5.2.1 FTIR spectroscopy

Before introducing FTIR, some basic terms and fundamentals are summarized (Smith, 2011): (1) Light: an electromagnetic radiation composed of electric vector (EV) and magnetic vector waves, and it is the interaction of the EV with substances that leads to the absorbance of light; (2) Wavelength: since the EV changes over time and has a repetitive sine wave form (namely, *cycles*), a wavelength is the distance traveled by a wave during a cycle; (3) Wavenumber: the number of cycles per unit wavelength (cm) and thus is often abbreviated as cm^{-1} (Fig. 8.11); (4) Spectroscopy: the study of the interaction between light and substances; (5) Infrared (light): the electromagnetic radiation with wavelengths longer than the red end of visible light but less than that of microwaves; (6) Infrared spectroscopy: a combination of (4) and (5); (7) Spectrum/spectra: the plot of measured light intensity versus some properties of light (e.g., wavelength and wavenumber). Therefore FTIR spectroscopy is a specific type of infrared spectroscopy and its related spectra can tell you what chemical bonds (e.g., hydroxyl and carboxyl groups) are present in a sample at specific wavenumbers (Li et al., 2020). The *y*-axis of an infrared spectra can be presented as transmittance (the % of light transmitted by samples, downward peaks) or absorbance (the amount of light absorbed by samples, upward peaks) (Fig. 8.11), and most

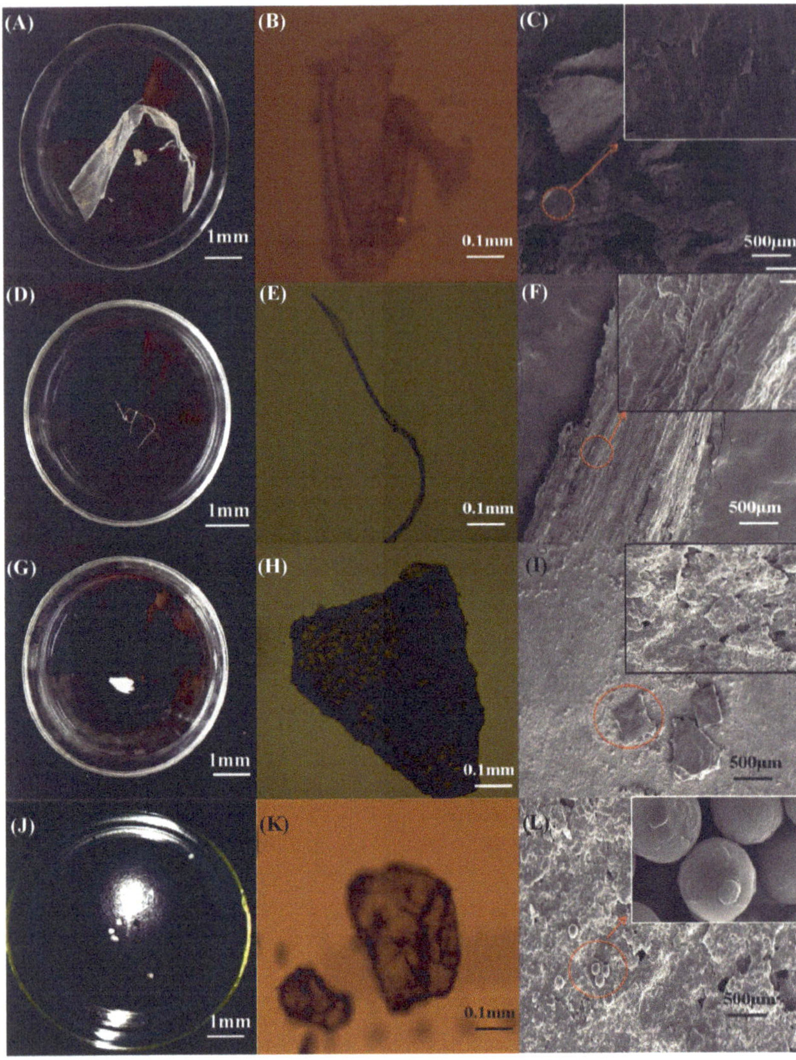

FIGURE 8.9 Identification of MPs through digital camera (A, D, G, J), microscope (B, E, H, K), and SEM (C, F, I, L). [film (A–C), fibers (D–F), fragments (G–I), and pellets (J–L)]. *From Ding, L., Zhang, S., Wang, X., Yang, X., Zhang, C., Qi, Y., & Guo, X. (2020). The occurrence and distribution characteristics of microplastics in the agricultural soils of Shaanxi Province, in north-western China.* Science of the Total Environment, 720. *https://doi.org/10.1016/j.scitotenv.2020.137525.*

infrared spectra are plotted from 4000 to 400 cm^{-1} as wavenumbers on the x-axis (mid-infrared). By comparing referenced spectrum libraries, target plastics in a sample and their polymer types can be identified.

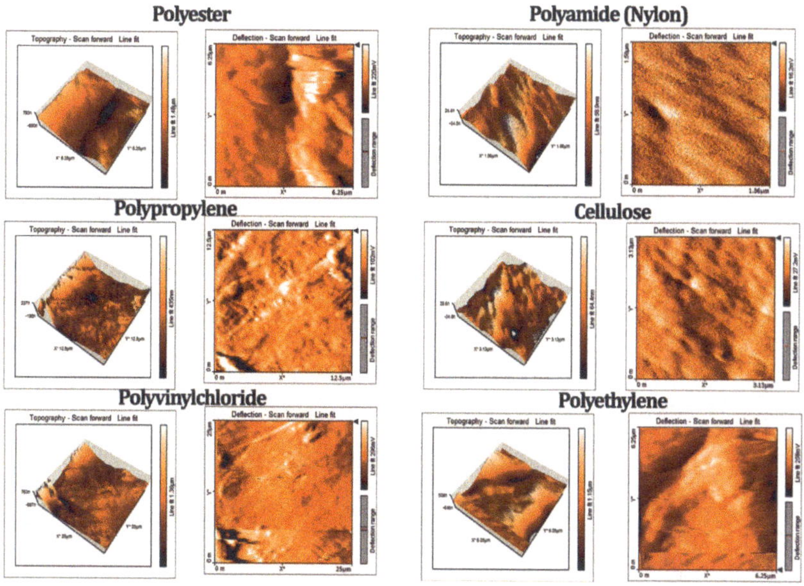

FIGURE 8.10 AFM images of MPs recovered from groundwater and surface water samples in Tamil Nadu, India. *From Selvam, S. (2020). Hazardous microplastic characteristics and its role as a vector of heavy metal in groundwater and surface water of coastal south India.* Journal of Hazardous Materials, 402.

8.3.5.2.2 Raman spectroscopy

As a nondestructive analysis, Raman spectroscopy can reveal chemical structure, phase, and polymorphism (different forms), and crystallinity based on the interaction between light and the chemical bonds within samples (Dioses-Salinas et al., 2020; Prata et al., 2019). The detection limit of size can reach as low as 1 μm (He et al., 2018) and it performs better in terms of map resolution and moisture signal interference. Using Raman spectroscopy, Zhou et al. (2019) reported an average abundance of MPs at 2.2×10^5 n/kg in woodlands, vegetable, and vacant land soils in Wuhan, China after a rarely used chemical digestion (KOH: NaClO) of samples (Fig. 8.11).

8.3.5.2.3 Near-infrared spectroscopy

Visible near-infrared (vis-NIR) spectroscopy has proposed to MP identification to avoid sample pretreatment (e.g., separation and digestion) and save time. NIR usually has a wavenumber range of 14,000–4000 cm^{-1} (Smith, 2011). Compared to FTIR (mid-infrared), vis-NIR can measure the absorbance within the wavenumber range of 28,571–4000 cm^{-1} (wavelength range of 350–2500 nm), correlating with the chemical composition of the sample (Corradini, Bartholomeus, et al., 2019; Corradini, Meza, et al., 2019). Corradini, Bartholomeus, et al. (2019) and Corradini, Meza, et al. (2019)

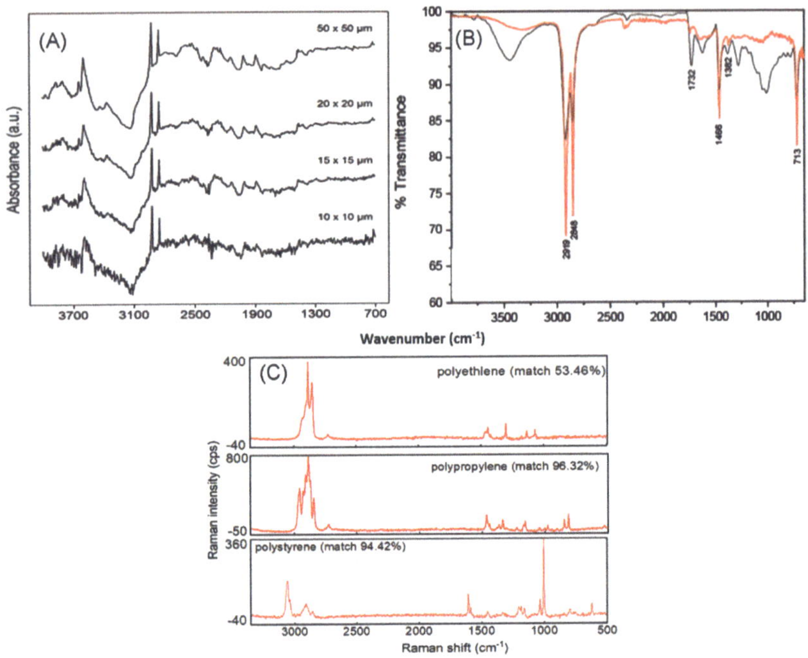

FIGURE 8.11 Examples of FTIR and Raman spectra for MP identification. (A, B) Low-density polyethylene (LDPE) plotted as absorbance (Harrison, 2012) and as percentage of transmittance (Edo et al., 2020) in FTIR spectra, respectively; and (C) PP, PS, and PE in Raman spectra (Zhou et al., 2019). *From (1) Harrison, J. P., Ojeda, J. J., & Romero-González, M. E. (2012). The applicability of reflectance micro-Fourier-transform infrared spectroscopy for the detection of synthetic microplastics in marine sediments.* Science of the Total Environment, 416, *455–463. https://doi.org/10.1016/j.scitotenv.2011.11.078. (2) Edo, C., González-Pleiter, M., Leganés, F., Fernández-Piñas, F., & Rosal, R. (2020). Fate of microplastics in wastewater treatment plants and their environmental dispersion with effluent and sludge.* Environmental Pollution, 259. *https://doi.org/10.1016/j.envpol.2019.113837. (3) Zhou, Y., Liu, X., & Wang, J. (2019). Characterization of microplastics and the association of heavy metals with microplastics in suburban soil of central China.* Science of the Total Environment, 694. *https://doi.org/10.1016/j.scitotenv.2019.133798.*

used the vis-NIR spectroscopy to detect PET, low-density PE (LDPE) and PVC particles from soils.

8.3.5.3 Chromatography

Chromatography is a physical separation method of a mixture of samples involving two phases, one of which is stationary while the other is mobile. Different components of the sample are transported by the mobile phase through the stationary phase in a definite direction at different speeds, realizing a separation by sequences (Poole & Poole, 2012). If the mobile phase is a liquid, it is called liquid chromatography (LC). While gas chromatography

(GC) uses an inert gas as the mobile phase. Here, LC and GC systems are covered for their applications in MP detection in soils and groundwater. LC or GC can be coupled with mass spectroscopy (Ms) to strengthen the MP analysis. The Ms is used to generate ions from organic/inorganic compounds and then separating these ions by their mass-to-charge ratio (m/z) so that these compounds can be quantified (Kitson et al., 1996). When they are combined, LC-Ms or GC-MC can provide not only the separation of the mixture of the components (through LC or GC), but also the identification of the chemical structure of each component (through Ms).

Wang et al. (2017) applied LC-Ms to detect polycarbonate (PC) and PET MPs in the environmental samples (landfill sludge, marine sediments, indoor dust, etc.) and obtained recovery rates between 87.2% and 97.1%. Reversed-phase liquid chromatography (Elert et al., 2017) and LC-UV (LC with ultraviolet) (Müller et al., 2020) have also been used for detecting MPs in the environment. Various GC-Ms systems including pyrolysis-GC-Ms (py-GC-Ms) and thermal extraction and desorption coupled with GC-Ms (TED-GC-Ms) are used for detecting MPs in environmental samples. Py-GC-Ms has been proposed recently for MP determination by thermally decomposing plastic particles under stationary conditions to a gas form that is cryo-captured and further separated by a chromatographic column followed by Ms identification (Prata et al., 2019). TED-GC-Ms was developed by Dümichen et al. (2017) to identify MPs in rivers and bio-gas plants. This method combines a thermogravimetric analysis (TGA, which is a method of thermal analysis of the sample mass over time vs temperature) and GC-Ms to thermally degrade polymers, extract the solid phase of the sample, and further analyze it in GC-Ms through desorption. Besides soils, TED-GC-Ms has also been applied to tire wears for MP determination (Eisentraut et al., 2018).

8.3.5.4 Other identification methods

Other methods such as TGA-Ms (David et al., 2018), proton nuclear magnetic resonance spectroscopy (^{1}H NMR) (Peez et al., 2019), hyperspectral imaging technology, and time-of-flight secondary ion mass spectrometry (TOF-SIMS) (Du, Liang, et al., 2020; Du, Wu, et al., 2020) have also been developed for detecting MPs in environmental samples. Compared to the commonly used ones (e.g., FTIR and Raman), these methods are rarely used for MP detection in soils and groundwater and their details are thus not covered here (Shan et al., 2018).

8.4 Risks and impacts

The ecological risks and impacts of MP pollution have raised serious concerns in recent years. MPs in soils and groundwater can act either as a vector of other toxicants [e.g., heavy metals, pharmaceutical and personal care products (PPCPs), hydrophobic organic contaminants (HOCs), plasticizers, and pathogens] or an environmental pollutant causing side effects to soil and groundwater quality, soil organisms, and even human beings. The impacts of MPs on animals, plants, microbiota, and biophysiochemical properties of soils and groundwater are shown in Fig. 8.12.

8.4.1 Soil properties and microbiota

It has been reported that MPs can affect the bulk density (BD), water holding capacity (WHC), water-stable aggregation (WSA), permeability, and porosity of soils (Guo et al., 2020; Zhou et al., 2020). de Souza Machado, Lau, et al. (2018) investigated the impacts of PP, PA, PE, and PET-based MPs on BD, WHC, and WSA of sandy soils and concluded that only PET affected the WSA and BD of the soil with a negative relationship whereas the others did not. Zhang et al. (2019) observed a different outcome that PET had no impact on BD but negatively affected the WHC of their soil compared to the control experiments. This may ascribe to different soil physiochemical properties

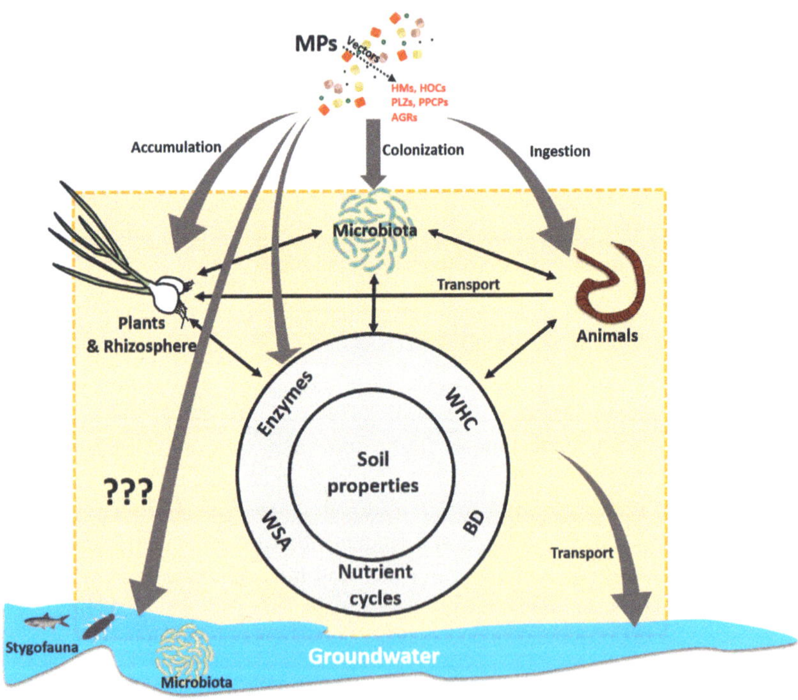

FIGURE 8.12 Interactions between animals, plants, microbiota, and biophysiochemical properties under the impact of MPs in soils and groundwater. *HMs*, Heavy metals; *HOCs*, hydrophobic organic contaminants; *PLZs*, plasticizers; *PPCPs*, pharmaceuticals and personal care products; *ARGs*, antibiotic resistance genes; *WHC*, water holding capacity; *BD*, bulk density; *WSA*, water-stable aggregation.

(pore size, minerals, humus contents, etc.) in the studies (Guo et al., 2020). MPs may also affect soil porosity through earthworms, which build denser burrow walls when exposed to MPs (Huerta Lwanga et al., 2017). In addition, soil permeability may also be changed by MPs, further affecting the evaporation (de Souza Machado, Lau, et al., 2018).

Soil bio-chemical properties [e.g., nutrient cycles of carbon, nitrogen and phosphorus (C, N & P cycles), and enzymes] can also be affected by MPs. Soil enzymes are not only regarded as fertility and nutrient cycle indicators due to their association with soil biochemical processes (Guo et al., 2020) but also can reflect microbiota. Enzymes such as fluorescein diacetate hydrolase (FDAse), urease, and catalase are shown to be stimulated by MPs in soils (Huang et al., 2019; Liu et al., 2017), further activating the beneficial C, N, and P cycles in the dissolved OM (Liu et al., 2017). Liang et al. (Liang, 2018) reported varied hydrolysis of fluorescein diacetate (FDA) affected by microfibers solely in a sandy soil. Moreover, soil water and nitrate have been reported to be affected by MP residues from plastic mulch materials (Yuanqiao et al., 2020). For soil carbon cycles, the above-mentioned BD is considered as an important index of C storage. MPs in soils not only change the status of BD, further affecting the C storage, but also make scientists misinterpret the C storage due to the high content of carbon polymers in MPs (e.g., PET contains 90% of C), most of which are different from natural soil OMs when interacted with microbiota (Rillig, 2018).

8.4.2 Plants and animals

MPs in soils can affect the growth of several types of plants including wheat (*Triticum aestivum*), perennial ryegrass (*Lolium perenne*), *Vicia faba*, cress (*Lepidium sativum*), and so on (Boots et al., 2019; Guo et al., 2020; Jiang et al., 2019; Qi et al., 2018). Qi et al. (2018) investigated the impact of plastic mulching residues on wheat growth and found that mulching residues had a higher negative influence than PE MPs. The comprehensive impact of MPs on spring onions (scallion) was investigated by de Souza Machado et al. (2019). They found that PET, PA, PP, and PS-based MPs had significant impacts on scallions, including plant biomass, element composition in tissues (water content, C/N ratio, and leaf nitrogen content), root traits (root length, mean diameter, tissue density, etc.), and soil microbial activities around root systems (root symbiosis), and these impacts varied with plastic types. Mechanisms of MPs affecting plant growth in soils are shown in Fig. 8.13 taking scallion as an example (De Souza Machado et al., 2019).

MPs can also affect soil animals including earthworms, collembolan (springtails), isopods, nematodes, mites, snails, and mice (Wang, Ge, et al., 2020; Zhou et al., 2020). Earthworms are the dominant test species to study the impacts of MPs in soils and the influences include growth inhibitions, gut damage, weight decrease, mortality increase, immune response, reproduction issues, and microbial activities alternations (Li et al., 2020; Wang, Ge, et al., 2020; Zhou et al., 2020).

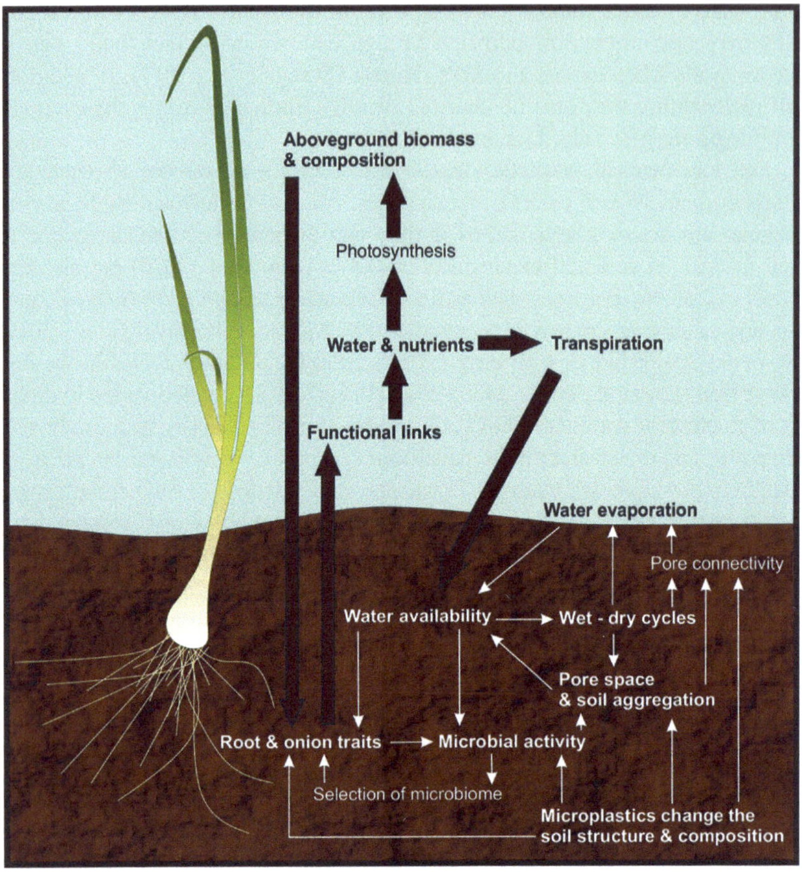

FIGURE 8.13 Mechanisms of MPs affecting the plant (scallion as an example) growth in soils. (*White arrows* indicate the MP impact on soil structure and composition; *brown arrows* indicate responses and characteristics of plants upon influence). *From De Souza Machado, A., Lau, C., Kloas, W., Bergmann, J., Bachelier, J., Faltin, E., Becker, R., Görlich, A., & Rillig, M. (2019). Microplastics can change soil properties and affect plant performance.* Environmental Science and Technology, *53(10), 6044–6052. https://doi.org/10.1021/acs.est.9b01339.*

MP impacts on other types of animals also have been recorded in publications such as snails (Song et al., 2019), mice (Jin et al., 2019; Lu et al., 2018), collembolan (Xiang et al., 2019), and isopods (Jemec Kokalj et al., 2018). Shang et al. (2020) found a negative effect on the lifespan of nematodes under the PS-MP exposure at low concentrations.

8.4.3 Human exposure

Human exposure to MPs in the environment from different sources and pathways is shown in Fig. 8.1, including inhalation, dermal contact, drinking water ingestion, and trophic transfer from food chains through air, water, and soil. A recent study found that plastic tea bags can release billions of MPs and nano-plastics at 95°C (Hernandez et al., 2019). MPs have also been detected in table salt, honey, beer, and sugar (He et al., 2018; Zhang et al., 2020), which are consumed by people directly. However, in soil and water matrices, human exposure to MPs mainly involves two potential pathways: trophic transfer and drinking water. The former pathway is through the "Earthworms-chicken-human" food chain. When earthworms are exposed to MPs in soils, they may transfer the residual MPs to their predators including chicken. As indicated in Lwanga et al. (2017), a human may ingest MPs when they eat chicken meats. They investigated MP contents in gizzards and excretion of chicken that were fed with non-MPs containing crops. However, MPs were detected, which can be attributed to the consumption of earthworms. Moreover, the research on mice shows the accumulation of MPs in the liver, kidney, and gut, which provides strong supports to the trophic transfer mechanism (Deng et al., 2017; Guo et al., 2020; Lu et al., 2018; Wang, Ge, et al., 2020). Another important pathway of MP exposure is from drinking water. Groundwater is an important source of drinking water (e.g., potable water, tap water, springs, and well). Shruti et al. (2020) reported an abundance of MPs [mainly polytrimethylene terephthalate (PTT) and epoxy resin] at 18 n/L in public drinking water fountains in Mexico using micro-Raman spectroscopy and SEM-EDS. This result should raise concerns about MP contamination to groundwater-based drinking water because 18 n/L is much higher than the LOQ at 4.1 n/L (Uhl et al., 2018).

The potential risks of MPs on human beings include oxidative stress and inflammation caused by particle toxicity that may further lead to cancer, immune systems damage, metabolism disorder, neurotoxins, and lipid digestion reduction (Guo et al., 2020; Prata et al., 2020; Tan et al., 2020; Zhou et al., 2020) (Fig. 8.1). Wang, Bai, et al. (2020) investigated the synergistic impact of MPs and bisphenol A on human Caco-2 cells and concluded that MPs increased cytotoxicity by inducing cellular oxidative stress and mitochondrial depolarization. Since Caco-2 cells' structure and function are similar to intestinal epithelial cells, which are responsible for digestion, this result suggests a potential risk of MPs on human beings through bioaccumulation. Moreover, Tan et al. (2020) lately illustrated a lipid digestion reduction under the impact of MPs in simulated human gastrointestinal systems, revealing a potential risk of MPs on digestion systems of human beings.

8.5 Case studies

8.5.1 MPs in groundwater wells in India

In India, most research of MPs has focused on marine and sediments (Selvam, 2020). In this case study, groundwater and surface resources of

Tamil Nadu state, India were evaluated and examined by Selvam et al. (Selvam, 2020), identifying and characterizing the MPs applying microscope, FTIR, and AFM methods. Influenced by industrial activities (e.g., power plants) and anthropogenic activities (e.g., fishing and tourism), groundwater samples were collected from wells and borewells (Fig. 8.14).

In the study, 20 L of groundwater (2–5 m in depth) was collected from each point by pumping as replicates and further filtered through a stainless-steel sieve (50 μm). The residues on the sieve were rinsed with water and stored in a glass jar by adding formalin solution at 4°C prior to analysis. For

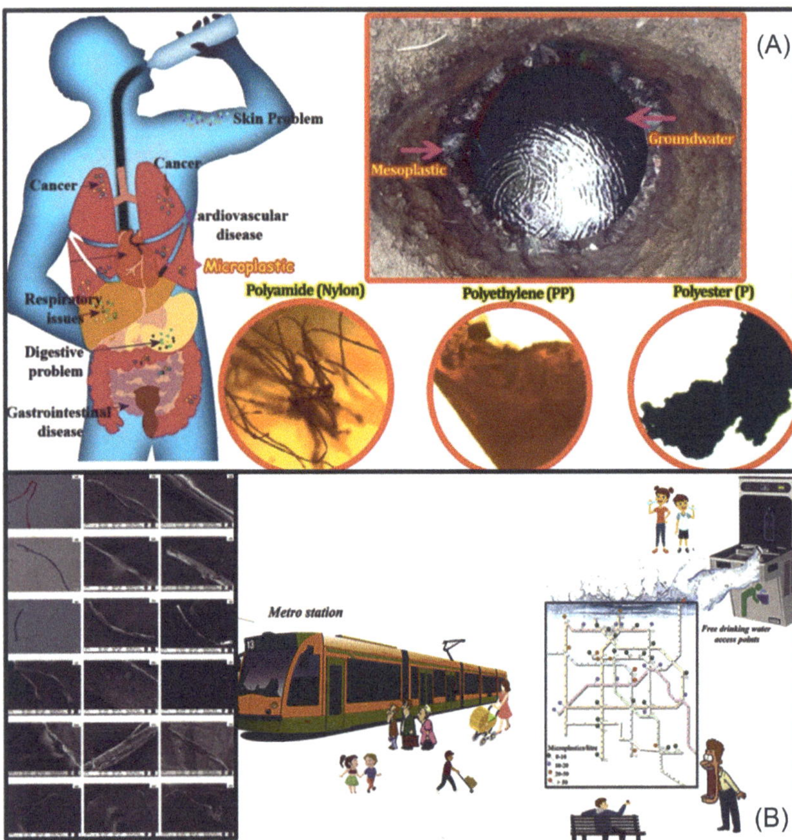

FIGURE 8.14 Case studies of groundwater MP contaminations in: (A) India (Selvam et al., 2020) and (B) Mexico (Shruti et al., 2020). *From (1) Selvam, S. (2020). Hazardous microplastic characteristics and its role as a vector of heavy metal in groundwater and surface water of coastal south India.* Journal of Hazardous Materials, 402. *(2) Shruti, V. C., Pérez-Guevara, F., & Kutralam-Muniasamy, G. (2020). Metro station free drinking water fountain—A potential "microplastics hotspot" for human consumption.* Environmental Pollution, 261. *https://doi.org/10.1016/j.envpol.2020.114227.*

chemical digestion, Fenton's reagents were used (30% H_2O_2 and Fe (II) solution) to remove OMs and the filtered paper was analyzed under a stereoscopic microscope (XRZ3, Olympus) for MPs shapes (films, foams, fragment, etc.). AFM was used to display surface images of MPs and Micro-FTIR was used to identify the chemical composition of MPs.

Eighty two types of items were tested using micro-FTIR and the major polymer types were PA (35%), PP (55%), and PET (10%) (Fig. 8.14). PA may be from fishing activities whereas PP may originate from chemical additives (e.g., dyes or flame-retardants). For AFM and microscope images, foam, plastic, fiber, film, and pellet were identified from wells whereas fibers were found in borewells (Figs. 8.10 and 8.14). Totally, PA, PE, and fibers represented the foams and they accounted for 62%–90% of MPs. Fibers may be from fragmentation of fishing nets and lines and domestic effluents. The color of MPs was also examined. Of note, 53% of transparent was found in groundwater, which might be from packaging of cloth and fishing products. In addition, the sizes of MPs ranged from 0.12 to 2.5 mm and the average abundance of MPs was 4.2 n/L (up to 10.1 n/L) in groundwater in the tested area. The higher concentration of MPs in samples compared to that of other countries (Table 8.1) may be due to more anthropogenic activities such as tourism, fishing activities, and domestic effluents.

8.5.2 MPs in public drinking water fountains in Mexico

Another case of MP occurrence as an endpoint pollution in groundwater was reported by Shruti et al. (2020), who examined public drinking water fountains in Mexico City, Mexico (Fig. 8.14). As shown in Table 8.1, only few studies have investigated MP contamination in groundwater-based drinking water. Among them, this study is the only one focusing on public fountains from 42 metro stations. The size, shape, polymer composition, surface morphology, and concentration of MPs were investigated by a microscope, SEM-EDS, and micro-Raman spectroscopy.

Among 42 metro stations (12 lines), water samples were collected as three replicates by precleaned glass bottles for 1 min prior to filling and up to the point of overflowing. In the lab, samples were filtered through 0.22 μm nitrocellulose filter paper by vacuum pumps and membrane filters and then dried at 17°C–21°C. The color and shape of the obtained samples were identified by a Nikon epifluorescence microscope (H6000L), and surface morphology and element composition were visualized by SEM-(EDS.) Moreover, micro-Raman spectroscopy was applied to identify the polymer types of MPs from the water samples.

Around 70% of MPs in the samples were transparent, dominating the identification results under the microscope. Fibers (domain) and fragments were found in all samples and the reason for fiber dominance can be attributed to the WWTPs discharge (Kim & Lee, 2020; Panno et al., 2019; Shruti et al., 2020). Meanwhile, half of the samples were exhibited a relatively small size (<0.5 mm) whereas the remaining was from 0.5 to 5 mm, which should raise concerns over potential NP pollution (Shen et al., 2020). It is worth noting that the surface roughness, cracks, and fragmentation of MPs shown in SEM results are evidence of plastic aging. Lastly, PTT and epoxy resin were the main polymer compositions among all the samples. The latter is derived from drinking water pipes for corrosion avoidance and the former may be from textiles and packaging materials. From a global point of view, the source of the textile is flowing through WWTPs (Shruti et al., 2020), entering groundwater and thus drinking water via sewage discharge, which is corresponding to the summary of groundwater MP source in Section 8.2.3.2. In addition, the abundance of MPs in this study (18 ± 7) was higher (at least several folds) than that of other groundwater MP studies in Table 8.1 (e.g., Kirstein, 2021; Mintenig et al., 2019), probably due to the differences in geographical locations, storage conditions, and analytical methods (Shruti et al., 2020).

These two case studies show examples of potential exposure of MP contamination to human beings from groundwater wells and public drinking water fountains. Special attention should be paid to promote the development of treatment technologies, monitoring of natural resources, and assessment of potential risks (Shruti et al., 2020).

8.6 Conclusion and perspectives

This chapter provides an overview of occurrences, sources, analyses, and impacts of MPs in soils and groundwater. It is obvious that the contamination of MPs is occurred globally, especially in China. The release of MPs from WWTPs to soil matrix via biosolids and composting should be concerned. The inputs from agriculture mulching films and MSW management activities may also pose MP contaminations on soils. Since groundwater is part of the soil system, any MP sources of soils may also be a direct or indirect input to the groundwater matrix. The methods for analyzing MPs in soils and groundwater, including sample collection and preparation, extraction, and identification, are systematically summarized. These analytical methods are useful for MP detection, especially the method involving density separation using salt solutions, H_2O_2 digestion, and FTIR identification, combined with appropriate sieving and filtration processes. In addition, MP impacts on soil structures and properties, animals, plants, and human beings are discussed in this chapter.

Although progresses have been made, research on MPs in soils and groundwater is still in its infancy. Therefore additional investigations are needed, especially in the following areas: (1) novel and accurate techniques to detect and characterize MPs in natural soil and groundwater samples; (2) quality assurance or quality control to minimize sample contaminations during the detection and quantification of MPs in soil and groundwater samples; (3) impacts of MPs on groundwater organisms (e.g., stygofauna and microbiota) (Fig. 8.12); (4) risks assessment of human exposure to MPs in soils and groundwater (e.g., via trophic transfer or drinking water); and (5) fragmentation and degradation mechanisms of MPs in soils and groundwater.

References

Alimi, O. S., Farner Budarz, J., Hernandez, L. M., & Tufenkji, N. (2018). Microplastics and nanoplastics in aquatic environments: Aggregation, deposition, and enhanced contaminant transport. *Environmental Science and Technology*, *52*(4), 1704–1724. Available from https://doi.org/10.1021/acs.est.7b05559.

Andrady, A. L. (2011). Microplastics in the marine environment. *Marine Pollution Bulletin*, *62* (8), 1596–1605. Available from https://doi.org/10.1016/j.marpolbul.2011.05.030.

Barnes, D. K. A., Galgani, F., Thompson, R. C., & Barlaz, M. (2009). Accumulation and fragmentation of plastic debris in global environments. *Philosophical Transactions of the Royal Society B: Biological Sciences*, *364*(1526), 1985–1998. Available from https://doi.org/10.1098/rstb.2008.0205.

Bläsing, M., & Amelung, W. (2018). Plastics in soil: Analytical methods and possible sources. *Science of the Total Environment*, *612*, 422–435. Available from https://doi.org/10.1016/j.scitotenv.2017.08.086.

Boots, B., Russell, C. W., & Green, D. S. (2019). Effects of microplastics in soil ecosystems: Above and below ground. *Environmental Science and Technology*, *53*(19), 11496–11506. Available from https://doi.org/10.1021/acs.est.9b03304.

Browne, M. A., Crump, P., Niven, S. J., Teuten, E., Tonkin, A., Galloway, T., & Thompson, R. (2011). Accumulation of microplastic on shorelines woldwide: Sources and sinks. *Environmental Science and Technology*, *45*(21), 9175–9179. Available from https://doi.org/10.1021/es201811s.

Campbell, S. H., Williamson, P. R., & Hall, B. D. (2017). Microplastics in the gastrointestinal tracts of fish and the water from an urban prairie creek. *Facets*, *2*(1), 395–409.

Carter, M. R., & Gregorich, E. G. (2007). *Soil sampling and methods of analysis*. CRC Press. Available from https://doi.org/10.1201/9781420005271.

Catarino, A. I., Thompson, R., Sanderson, W., & Henry, T. B. (2017). Development and optimization of a standard method for extraction of microplastics in mussels by enzyme digestion of soft tissues. *Environmental Toxicology and Chemistry*, *36*(4), 947–951. Available from https://doi.org/10.1002/etc.3608.

Chae, Y., & An, Y. J. (2018). Current research trends on plastic pollution and ecological impacts on the soil ecosystem: A review. *Environmental Pollution*, *240*, 387–395. Available from https://doi.org/10.1016/j.envpol.2018.05.008.

Chen, Y., et al. (2020). Microplastic pollution in vegetable farmlands of suburb Wuhan, central China. *Environmental Pollution*, *257*, 113449.

Cole, M., Lindeque, P., Halsband, C., & Galloway, T. S. (2011). Microplastics as contaminants in the marine environment: A review. *Marine Pollution Bulletin*, *62*(12), 2588–2597. Available from https://doi.org/10.1016/j.marpolbul.2011.09.025.

Cole, M., Webb, H., Lindeque, P. K., Fileman, E. S., Halsband, C., & Galloway, T. S. (2014). Isolation of microplastics in biota-rich seawater samples and marine organisms. *Scientific Reports*, *4*. Available from https://doi.org/10.1038/srep04528.

Corcoran, E. (2010). Sick water? The central role of wastewater management in sustainable development: a rapid response assessment. *UNEP/Earthprint*.

Corradini, F. (2021). Microplastics occurrence and frequency in soils under different land uses on a regional scale. *Science of the Total Environment*, *752*, 141917. Available from https://doi.org/10.1016/j.scitotenv.2020.141917.

Corradini, F., Bartholomeus, H., Lwanga, E. H., Gertsen, H., & Geissen, V. (2019). Predicting soil microplastic concentration using vis-NIR spectroscopy. *Science of the Total Environment*, *650*, 922–932.

Corradini, F., Meza, P., Eguiluz, R., Casado, F., Huerta-Lwanga, E., & Geissen, V. (2019). Evidence of microplastic accumulation in agricultural soils from sewage sludge disposal. *Science of the Total Environment*, *671*, 411–420. Available from https://doi.org/10.1016/j.scitotenv.2019.03.368.

Crichton, E. M., Noël, M., Gies, E. A., & Ross, P. S. (2017). A novel, density-independent and FTIR-compatible approach for the rapid extraction of microplastics from aquatic sediments. *Analytical Methods*, *9*(9), 1419–1428. Available from https://doi.org/10.1039/c6ay02733d.

Crossman, J., Hurley, R. R., Futter, M., & Nizzetto, L. (2020). Transfer and transport of microplastics from biosolids to agricultural soils and the wider environment. *Science of the Total Environment*, *724*. Available from https://doi.org/10.1016/j.scitotenv.2020.138334.

David, J., Steinmetz, Z., Kučerík, J., & Schaumann, G. E. (2018). Quantitative analysis of poly (ethylene terephthalate) microplastics in soil via thermogravimetry-mass spectrometry. *Analytical Chemistry*, *90*(15), 8793–8799. Available from https://doi.org/10.1021/acs.analchem.8b00355.

de Souza Machado, A. A., Kloas, W., Zarfl, C., Hempel, S., & Rillig, M. C. (2018). Microplastics as an emerging threat to terrestrial ecosystems. *Global Change Biology*, *24*(4), 1405–1416. Available from https://doi.org/10.1111/gcb.14020.

de Souza Machado, A. A., Lau, C. W., Kloas, W., Bergmann, J., Bachelier, J. B., Faltin, E., Becker, R., Görlich, A. S., & Rillig, M. C. (2019). Microplastics can change soil properties and affect plant performance. *Environmental Science and Technology*, *53*(10), 6044–6052. Available from https://doi.org/10.1021/acs.est.9b01339.

de Souza Machado, A. A., Lau, C. W., Till, J., Kloas, W., Lehmann, A., Becker, R., & Rillig, M. C. (2018). Impacts of microplastics on the soil biophysical environment. *Environmental Science and Technology*, *52*(17), 9656–9665. Available from https://doi.org/10.1021/acs.est.8b02212.

Deng, Y., Zhang, Y., Lemos, B., & Ren, H. (2017). Tissue accumulation of microplastics in mice and biomarker responses suggest widespread health risks of exposure. *Scientific Reports*, *7*(1), 1–10.

Ding, L., et al. (2020). The occurrence and distribution characteristics of microplastics in the agricultural soils of Shaanxi Province, in north-western China. *Science of The Total Environment*, *137525*.

Ding, L., et al. (2021). The occurrence of microplastic in Mu Us Sand Land soils in northwest China: Different soil types, vegetation cover and restoration years. *Journal of Hazardous Materials*, *403*, 123982.

Dioses-Salinas, D. C., Pizarro-Ortega, C. I., & De-la-Torre, G. E. (2020). A methodological approach of the current literature on microplastic contamination in terrestrial environments: Current knowledge and baseline considerations. *Science of the Total Environment*, *730*. Available from https://doi.org/10.1016/j.scitotenv.2020.139164.

Dris, R., Gasperi, J., Saad, M., Mirande, C., & Tassin, B. (2016). Synthetic fibers in atmospheric fallout: A source of microplastics in the environment? *Marine Pollution Bulletin*, *104*(1–2), 290–293. Available from https://doi.org/10.1016/j.marpolbul.2016.01.006.

Du, C., Liang, H., Li, Z., & Gong, J. (2020). Pollution characteristics of microplastics in soils in southeastern suburbs of Baoding City, China. *International Journal of Environmental Research and Public Health*, *17*(3). Available from https://doi.org/10.3390/ijerph17030845.

Du, C., Wu, J., Gong, J., Liang, H., & Li, Z. (2020). ToF-SIMS characterization of microplastics in soils. *Surface and Interface Analysis*, *52*(5), 293–300. Available from https://doi.org/10.1002/sia.6742.

Duan, Z., Zhao, S., Zhao, L., Duan, X., Xie, S., Zhang, H., Liu, Y., Peng, Y., Liu, C., & Wang, L. (2020). Microplastics in Yellow River Delta wetland: Occurrence, characteristics, human influences, and marker. *Environmental Pollution*, *258*. Available from https://doi.org/10.1016/j.envpol.2019.113232.

Dümichen, E., Eisentraut, P., Bannick, C. G., Barthel, A. K., Senz, R., & Braun, U. (2017). Fast identification of microplastics in complex environmental samples by a thermal degradation method. *Chemosphere*, *174*, 572–584. Available from https://doi.org/10.1016/j.chemosphere.2017.02.010.

Edo, C., et al. (2020). Fate of microplastics in wastewater treatment plants and their environmental dispersion with effluent and sludge. *Environmental Pollution*, *259*, 113837.

Eisentraut, P., Dümichen, E., Ruhl, A. S., Jekel, M., Albrecht, M., Gehde, M., & Braun, U. (2018). Two birds with one stone - fast and simultaneous analysis of microplastics: Microparticles derived from thermoplastics and tire wear. *Environmental Science and Technology Letters*, *5*(10), 608–613. Available from https://doi.org/10.1021/acs.estlett.8b00446.

Elert, A. M., Becker, R., Duemichen, E., Eisentraut, P., Falkenhagen, J., Sturm, H., & Braun, U. (2017). Comparison of different methods for MP detection: What can we learn from them, and why asking the right question before measurements matters? *Environmental Pollution*, *231*, 1256–1264. Available from https://doi.org/10.1016/j.envpol.2017.08.074.

Felsing, S., Kochleus, C., Buchinger, S., Brennholt, N., Stock, F., & Reifferscheid, G. (2018). A new approach in separating microplastics from environmental samples based on their electrostatic behavior. *Environmental Pollution*, *234*, 20–28. Available from https://doi.org/10.1016/j.envpol.2017.11.013.

Fuller, S., & Gautam, A. (2016). A procedure for measuring microplastics using pressurized fluid extraction. *Environmental Science and Technology*, *50*(11), 5774–5780. Available from https://doi.org/10.1021/acs.est.6b00816.

Ganesan, M., Nallathambi, G., & Srinivasalu, S. (2019). Fate and transport of microplastics from water sources. *Current Science*, *117*(11), 1879–1885. Available from https://doi.org/10.18520/cs/v117/i11/1879-1885.

Gao, H., Yan, C., Liu, Q., Ding, W., Chen, B., & Li, Z. (2019). Effects of plastic mulching and plastic residue on agricultural production: A *meta*-analysis. *Science of the Total Environment*, *651*, 484–492. Available from https://doi.org/10.1016/j.scitotenv.2018.09.105.

García-Gil, J. C., Plaza, C., Soler-Rovira, P., & Polo, A. (2000). Long-term effects of municipal solid waste compost application on soil enzyme activities and microbial biomass. *Soil Biology and Biochemistry*, *32*(13), 1907–1913. Available from https://doi.org/10.1016/S0038-0717(00)00165-6.

Gaylor, M. O., Harvey, E., & Hale, R. C. (2013). Polybrominated diphenyl ether (PBDE) accumulation by earthworms (*Eisenia fetida*) exposed to biosolids-, polyurethane foam microparticle-, and penta-BDE-amended soils. *Environmental Science and Technology*, *47*(23), 13831–13839. Available from https://doi.org/10.1021/es403750a.

Giusquiani, P. L., Pagliai, M., Gigliotti, G., Businelli, D., & Benetti, A. (1995). Urban waste compost: Effects on physical, chemical, and biochemical soil properties. *Journal of*

Environmental Quality, *24*(1), 175–182. Available from https://doi.org/10.2134/jeq1995.00472425002400010024x.

Grbic, J., Nguyen, B., Guo, E., You, J. B., Sinton, D., & Rochman, C. M. (2019). Magnetic extraction of microplastics from environmental samples. *Environmental Science and Technology Letters*, *6*(2), 68–72. Available from https://doi.org/10.1021/acs.estlett.8b00671.

Guo, J. J., Huang, X. P., Xiang, L., Wang, Y. Z., Li, Y. W., Li, H., Cai, Q. Y., Mo, C. H., & Wong, M. H. (2020). Source, migration and toxicology of microplastics in soil. *Environment International*, *137*. Available from https://doi.org/10.1016/j.envint.2019.105263.

Han, X., Lu, X., & Vogt, R. D. (2019). An optimized density-based approach for extracting microplastics from soil and sediment samples. *Environmental Pollution*, *254*. Available from https://doi.org/10.1016/j.envpol.2019.113009.

Hanson, J. L., Yesiller, N., & Kendall, L. A. (2005). Integrated temperature and gas analysis at a municipal solid waste landfill. In *Proceedings of the 16th international conference on soil mechanics and geotechnical engineering*, (pp. 2265–2268). IOS Press. Available from https://doi.org/10.3233/978-1-61499-656-9-2265.

He, D., Luo, Y., Lu, S., Liu, M., Song, Y., & Lei, L. (2018). Microplastics in soils: Analytical methods, pollution characteristics and ecological risks. *Trends in Analytical Chemistry*, *109*, 163–172. Available from https://doi.org/10.1016/j.trac.2018.10.006.

Harrison, J. P., et al. (2012). The applicability of reflectance micro-Fourier-transform infrared spectroscopy for the detection of synthetic microplastics in marine sediments. *Science of the Total Environment*, *416*, 455–463.

He, P., Chen, L., Shao, L., Zhang, H., & Lü, F. (2019). Municipal solid waste (MSW)landfill: A source of microplastics? Evidence of microplastics in landfill leachate. *Water Research*, *159*, 38–45. Available from https://doi.org/10.1016/j.watres.2019.04.060.

Helcoski, R., et al. (2020). Wetland soil microplastics are negatively related to vegetation cover and stem density. *Environmental Pollution*, *256*, 113391.

Hernandez, L. M., Xu, E. G., Larsson, H. C. E., Tahara, R., Maisuria, V. B., & Tufenkji, N. (2019). Plastic teabags release billions of microparticles and nanoparticles into tea. *Environmental Science and Technology*, *53*(21), 12300–12310. Available from https://doi.org/10.1021/acs.est.9b02540.

Hidalgo-Ruz, V., Gutow, L., Thompson, R. C., & Thiel, M. (2012). Microplastics in the marine environment: A review of the methods used for identification and quantification. *Environmental Science and Technology*, *46*(6), 3060–3075. Available from https://doi.org/10.1021/es2031505.

Hoppe, M., Mikutta, R., Utermann, J., Duijnisveld, W., Kaufhold, S., Stange, C. F., & Guggenberger, G. (2015). Remobilization of sterically stabilized silver nanoparticles from farmland soils determined by column leaching. *European Journal of Soil Science*, *66*(5), 898–909. Available from https://doi.org/10.1111/ejss.12270.

Hu, B. (2014). Magnetic density separation of polyolefin wastes. Delft: Bin HU, 172 p. https://doi.org/10.4233/uuid:0c3717fa-8000-4de0-a938-d65605bf2a96.

Huang, Y., Liu, Q., Jia, W., Yan, C., & Wang, J. (2020). Agricultural plastic mulching as a source of microplastics in the terrestrial environment. *Environmental Pollution*, *260*. Available from https://doi.org/10.1016/j.envpol.2020.114096.

Huang, Y., Zhao, Y., Wang, J., Zhang, M., Jia, W., & Qin, X. (2019). LDPE microplastic films alter microbial community composition and enzymatic activities in soil. *Environmental Pollution*, *254*. Available from https://doi.org/10.1016/j.envpol.2019.112983.

Huerta Lwanga, E., Gertsen, H., Gooren, H., Peters, P., Salánki, T., van der Ploeg, M., Besseling, E., Koelmans, A. A., & Geissen, V. (2017). Incorporation of microplastics from litter into burrows of *Lumbricus terrestris*. *Environmental Pollution, 220*, 523–531. Available from https://doi.org/10.1016/j.envpol.2016.09.096.

Hurley, R. R., Lusher, A. L., Olsen, M., & Nizzetto, L. (2018). Validation of a method for extracting microplastics from complex, organic-rich, environmental matrices. *Environmental Science and Technology, 52*(13), 7409–7417. Available from https://doi.org/10.1021/acs.est.8b01517.

Hurley, R. R., & Nizzetto, L. (2018). Fate and occurrence of micro(nano)plastics in soils: Knowledge gaps and possible risks. *Current Opinion in Environmental Science and Health, 1*, 6–11. Available from https://doi.org/10.1016/j.coesh.2017.10.006.

Imhof, H. K., Ivleva, N. P., Schmid, J., Niessner, R., & Laforsch, C. (2013). Contamination of beach sediments of a subalpine lake with microplastic particles. *Current Biology, 23*(19), R867–R868. Available from https://doi.org/10.1016/j.cub.2013.09.001.

Imhof, H. K., Schmid, J., Niessner, R., Ivleva, N. P., & Laforsch, C. (2012). A novel, highly efficient method for the separation and quantification of plastic particles in sediments of aquatic environments. *Limnology and Oceanography: Methods, 10*(July), 524–537. Available from https://doi.org/10.4319/lom.2012.10.524.

Jemec Kokalj, A., Horvat, P., Skalar, T., & Kržan, A. (2018). Plastic bag and facial cleanser derived microplastic do not affect feeding behaviour and energy reserves of terrestrial isopods. *Science of the Total Environment, 615*, 761–766. Available from https://doi.org/10.1016/j.scitotenv.2017.10.020.

Jiang, X., Chen, H., Liao, Y., Ye, Z., Li, M., & Klobučar, G. (2019). Ecotoxicity and genotoxicity of polystyrene microplastics on higher plant *Vicia faba*. *Environmental Pollution, 250*, 831–838. Available from https://doi.org/10.1016/j.envpol.2019.04.055.

Jiang, X. J., Liu, W., Wang, E., Zhou, T., & Xin, P. (2017). Residual plastic mulch fragments effects on soil physical properties and water flow behavior in the Minqin Oasis, northwestern China. *Soil and Tillage Research, 166*, 100–107. Available from https://doi.org/10.1016/j.still.2016.10.011.

Jin, Y., Lu, L., Tu, W., Luo, T., & Fu, Z. (2019). Impacts of polystyrene microplastic on the gut barrier, microbiota and metabolism of mice. *Science of the Total Environment, 649*, 308–317. Available from https://doi.org/10.1016/j.scitotenv.2018.08.353.

Johnson, A. C., Ball, H., Cross, R., Horton, A. A., Jürgens, M. D., Read, D. S., Vollertsen, J., & Svendsen, C. (2020). Identification and quantification of microplastics in potable water and their sources within water treatment works in England and Wales. *Environmental Science and Technology, 54*(19), 12326–12334. Available from https://doi.org/10.1021/acs.est.0c03211.

Kilponen, J. (2016). Microplastics and harmful substances in urban runoffs and landfill leachates: possible emission sources to marine environment. *Lahti University of Applied Sciences*.

Kim, S.-K., et al. (2020). Abundance and characteristics of microplastics in soils with different agricultural practices: Importance of sources with internal origin and environmental fate. *Journal of Hazardous Materials, 403*, 123997.

Kim, H., & Lee, J. Y. (2020). Emerging concerns about microplastic pollution on groundwater in South Korea. *Sustainability (Switzerland), 12*(13). Available from https://doi.org/10.3390/su12135275.

Kirstein, I. V. (2021). Drinking plastics?—Quantification and qualification of microplastics in drinking water distribution systems by μFTIR and Py-GCMS. *Water Research, 188*, 116519.

Kitson, F. G., Larsen, B. S., & McEwen, C. N. (1996). *Gas chromatography and mass spectrometry: A practical guide*. Academic Press.

Klein, S., Worch, E., & Knepper, T. P. (2015). Occurrence and spatial distribution of microplastics in river shore sediments of the rhine-main area in Germany. *Environmental Science and Technology, 49*(10), 6070–6076. Available from https://doi.org/10.1021/acs.est.5b00492.

Kole, P. J., et al. (2015). Autobandenslijtstof: een verwaarloosde bron van microplastics? *Milieu: opinieblad van de Vereniging van Milieuprofessionals* (5), 39–41. Available from https://www.ou.nl/documents/40554/72652/Autobandensplijtstof_een_verwaarloosde_bron_van_microplastics.pdf/dec98496-4986-4abe-8bbc-9fd2e3832bfd.

Kole, P. J., Löhr, A. J., Van Belleghem, F. G., & Ragas, A. M. (2017). Wear and tear of tyres: a stealthy source of microplastics in the environment. *International Journal of Environmental Research and Public Health, 14*(10), 1265. Available from https://doi.org/10.3390/ijerph14101265.

Lechner, A., Keckeis, H., Lumesberger-Loisl, F., Zens, B., Krusch, R., Tritthart, M., Glas, M., & Schludermann, E. (2014). The Danube so colourful: A potpourri of plastic litter outnumbers fish larvae in Europe's second largest river. *Environmental Pollution, 188*, 177–181. Available from https://doi.org/10.1016/j.envpol.2014.02.006.

Li, Q., et al. (2019). Separation and identification of microplastics from soil and sewage sludge. *Environmental Pollution, 254*, 113076.

Li, J., Song, Y., & Cai, Y. (2020). Focus topics on microplastics in soil: Analytical methods, occurrence, transport, and ecological risks. *Environmental Pollution, 257*. Available from https://doi.org/10.1016/j.envpol.2019.113570.

Liang, Y. (2018). Increasing temperature and microplastic fibers jointly influence soil aggregation by Saprobic Fungi. *Frontiers in Microbiology, 10*.

Liu, E. (2014). "White revolution" to "white pollution"—agricultural plastic film mulch in China. *Environmental Research Letters, 9*, 091001.

Liu, H., Yang, X., Liu, G., Liang, C., Xue, S., Chen, H., Ritsema, C. J., & Geissen, V. (2017). Response of soil dissolved organic matter to microplastic addition in Chinese loess soil. *Chemosphere, 185*, 907–917. Available from https://doi.org/10.1016/j.chemosphere.2017.07.064.

Liu, K., Wang, X., Fang, T., Xu, P., Zhu, L., & Li, D. (2019). Source and potential risk assessment of suspended atmospheric microplastics in Shanghai. *Science of the Total Environment, 675*, 462–471. Available from https://doi.org/10.1016/j.scitotenv.2019.04.110.

Liu, M., Lu, S., Song, Y., Lei, L., Hu, J., Lv, W., Zhou, W., Cao, C., Shi, H., Yang, X., & He, D. (2018). Microplastic and mesoplastic pollution in farmland soils in suburbs of Shanghai, China. *Environmental Pollution, 242*, 855–862. Available from https://doi.org/10.1016/j.envpol.2018.07.051.

Liu, M., Song, Y., Lu, S., Qiu, R., Hu, J., Li, X., Bigalke, M., Shi, H., & He, D. (2019). A method for extracting soil microplastics through circulation of sodium bromide solutions. *Science of the Total Environment, 691*, 341–347. Available from https://doi.org/10.1016/j.scitotenv.2019.07.144.

Löder, M. G., Imhof, H. K., Ladehoff, M., Löschel, L. A., Lorenz, C., Mintenig, S., Piehl, S., Primpke, S., Schrank, I., & Laforsch, C. (2017). Enzymatic purification of microplastics in environmental samples. *Environmental Science & Technology, 51*(24), 14283–14292.

Lu, L., Wan, Z., Luo, T., Fu, Z., & Jin, Y. (2018). Polystyrene microplastics induce gut microbiota dysbiosis and hepatic lipid metabolism disorder in mice. *Science of the Total Environment, 631–632*, 449–458. Available from https://doi.org/10.1016/j.scitotenv.2018.03.051.

Lu, R. K. (2000). *Methods for chemical analysis of soil agriculture*. Beijing, China: China Agricultural Science and Technology Press.

Lv, W., Zhou, W., Lu, S., Huang, W., Yuan, Q., Tian, M., Lv, W., & He, D. (2019). Microplastic pollution in rice-fish co-culture system: A report of three farmland stations in Shanghai, China. *Science of the Total Environment, 652*, 1209–1218. Available from https://doi.org/10.1016/j.scitotenv.2018.10.321.

Lwanga, E. H., Vega, J. M., Quej, V. K., de los Angeles Chi, J., Del Cid, L. S., Chi, C., Segura, G. E., Gertsen, H., Salánki, T., & van der Ploeg, M. (2017). Field evidence for transfer of plastic debris along a terrestrial food chain. *Scientific Reports*, *7*(1), 1–7.

Maaß, S., Daphi, D., Lehmann, A., & Rillig, M. C. (2017). Transport of microplastics by two collembolan species. *Environmental Pollution*, *225*, 456–459. Available from https://doi.org/10.1016/j.envpol.2017.03.009.

Majewsky, M., Bitter, H., Eiche, E., & Horn, H. (2016). Determination of microplastic polyethylene (PE) and polypropylene (PP) in environmental samples using thermal analysis (TGA-DSC). *Science of the Total Environment*, *568*, 507–511. Available from https://doi.org/10.1016/j.scitotenv.2016.06.017.

Mintenig, S. M., Löder, M. G. J., Primpke, S., & Gerdts, G. (2019). Low numbers of microplastics detected in drinking water from ground water sources. *Science of the Total Environment*, *648*, 631–635. Available from https://doi.org/10.1016/j.scitotenv.2018.08.178.

Möller, J. N., Löder, M. G., & Laforsch, C. (2020). Finding microplastics in soils: A review of analytical methods. *Environmental Science & Technology*, *54*(4), 2078–2090.

Müller, A., Goedecke, C., Eisentraut, P., Piechotta, C., & Braun, U. (2020). Microplastic analysis using chemical extraction followed by LC-UV analysis: a straightforward approach to determine PET content in environmental samples. *Environmental Sciences Europe*, *32*(1). Available from https://doi.org/10.1186/s12302-020-00358-x.

Nizzetto, L., Bussi, G., Futter, M. N., Butterfield, D., & Whitehead, P. G. (2016). A theoretical assessment of microplastic transport in river catchments and their retention by soils and river sediments. *Environmental Science: Processes and Impacts*, *18*(8), 1050–1059. Available from https://doi.org/10.1039/c6em00206d.

Nizzetto, L., Futter, M., & Langaas, S. (2016). Are agricultural soils dumps for microplastics of urban origin? *Environmental Science and Technology*, *50*(20), 10777–10779. Available from https://doi.org/10.1021/acs.est.6b04140.

Nizzetto, L., Langaas, S., & Futter, M. (2016). Pollution: do microplastics spill on to farm soils? *Nature*, *537*(7621), 488-488.

Nuelle, M. T., Dekiff, J. H., Remy, D., & Fries, E. (2014). A new analytical approach for monitoring microplastics in marine sediments. *Environmental Pollution*, *184*, 161–169. Available from https://doi.org/10.1016/j.envpol.2013.07.027.

Panno, S. V., Kelly, W. R., Scott, J., Zheng, W., McNeish, R. E., Holm, N., Hoellein, T. J., & Baranski, E. L. (2019). Microplastic contamination in Karst groundwater systems. *Groundwater*, *57*(2), 189–196. Available from https://doi.org/10.1111/gwat.12862.

Peez, N., Janiska, M. C., & Imhof, W. (2019). The first application of quantitative 1H NMR spectroscopy as a simple and fast method of identification and quantification of microplastic particles (PE, PET, and PS). *Analytical and Bioanalytical Chemistry*, *411*(4), 823–833. Available from https://doi.org/10.1007/s00216-018-1510-z.

Piehl, S., et al. (2018). Identification and quantification of macro-and microplastics on an agricultural farmland. *Scientific Reports*, *8*, 1–9.

PlasticsEurope, E. (2016). Plastics—the facts 2016. An analysis of European plastics production, demand and waste data. *PlasticsEurope*.

PlasticsEurope, E. (2020). Plastics – the Facts 2020. An analysis of European plastics production, demand and waste data. *PlasticsEurope*.

Połeć, M. (2018). The occurrence of microplastics in freshwater systems–preliminary results from Krakow (Poland). Geology. *Geophysics and Environment*, *44*, 391–400.

Połeć, M., et al. (2018). The occurrence of microplastics in freshwater systems–preliminary results from Krakow (Poland). *Geology, Geophysics and Environment*, *44*, 391–400.

Poole, C. F., & Poole, S. K. (2012). *Chromatography today*. Elsevier.

Pramila, R., & Ramesh, K. V. (2011). Biodegradation of low density polyethylene (LDPE) by fungi isolated from municipal landfill area. *Journal of Microbiology and Biotechnology Research*, *1*(131), 136.

Prata, J. C., da Costa, J. P., Duarte, A. C., & Rocha-Santos, T. (2019). Methods for sampling and detection of microplastics in water and sediment: A critical review. *Trends in Analytical Chemistry*, *110*, 150−159. Available from https://doi.org/10.1016/j.trac.2018.10.029.

Prata, J. C., da Costa, J. P., Lopes, I., Duarte, A. C., & Rocha-Santos, T. (2020). Environmental exposure to microplastics: An overview on possible human health effects. *Science of the Total Environment*, *702*. Available from https://doi.org/10.1016/j.scitotenv.2019.134455.

Qi, R., Jones, D. L., Li, Z., Liu, Q., & Yan, C. (2020). Behavior of microplastics and plastic film residues in the soil environment: A critical review. *Science of the Total Environment*, *703*. Available from https://doi.org/10.1016/j.scitotenv.2019.134722.

Qi, Y., Yang, X., Pelaez, A. M., Huerta Lwanga, E., Beriot, N., Gertsen, H., Garbeva, P., & Geissen, V. (2018). Macro- and micro-plastics in soil-plant system: Effects of plastic mulch film residues on wheat (*Triticum aestivum*) growth. *Science of the Total Environment*, *645*, 1048−1056. Available from https://doi.org/10.1016/j.scitotenv.2018.07.229.

Renou, S., Givaudan, J. G., Poulain, S., Dirassouyan, F., & Moulin, P. (2008). Landfill leachate treatment: Review and opportunity. *Journal of Hazardous Materials*, *150*(3), 468−493. Available from https://doi.org/10.1016/j.jhazmat.2007.09.077.

Rillig, M. C. (2012). Microplastic in terrestrial ecosystems and the soil? *Environmental Science and Technology*, *46*(12), 6453−6454. Available from https://doi.org/10.1021/es302011r.

Rillig, M. C. (2018). Microplastic disguising as soil carbon storage. *Environmental Science and Technology*, *52*(11), 6079−6080. Available from https://doi.org/10.1021/acs.est.8b02338.

Rillig, M. C., Ziersch, L., & Hempel, S. (2017). Microplastic transport in soil by earthworms. *Scientific Reports*, *7*(1). Available from https://doi.org/10.1038/s41598-017-01594-7.

Scheurer, M., & Bigalke, M. (2018). Microplastics in swiss floodplain soils. *Environmental Science and Technology*, *52*(6), 3591−3598. Available from https://doi.org/10.1021/acs.est.7b06003.

Selvam, S. (2020). Hazardous microplastic characteristics and its role as a vector of heavy metal in groundwater and surface water of coastal south India. *Journal of Hazardous Materials*, *402*, 123786.

Shan, J., Zhao, J., Liu, L., Zhang, Y., Wang, X., & Wu, F. (2018). A novel way to rapidly monitor microplastics in soil by hyperspectral imaging technology and chemometrics. *Environmental Pollution*, *238*, 121−129.

Shang, X., Lu, J., Feng, C., Ying, Y., He, Y., Fang, S., Lin, Y., Dahlgren, R., & Ju, J. (2020). Microplastic (1 and 5 μm) exposure disturbs lifespan and intestine function in the nematode *Caenorhabditis elegans*. *Science of the Total Environment*, *705*. Available from https://doi.org/10.1016/j.scitotenv.2019.135837.

Sharma, S., & Chatterjee, S. (2017). Microplastic pollution, a threat to marine ecosystem and human health: A short review. *Environmental Science and Pollution Research*, *24*(27), 21530−21547. Available from https://doi.org/10.1007/s11356-017-9910-8.

Shen, M., Song, B., Zhu, Y., Zeng, G., Zhang, Y., Yang, Y., Wen, X., Chen, M., & Yi, H. (2020). Removal of microplastics via drinking water treatment: Current knowledge and future directions. *Chemosphere*, *251*. Available from https://doi.org/10.1016/j.chemosphere.2020.126612.

Shruti, V. C., Pérez-Guevara, F., & Kutralam-Muniasamy, G. (2020). Metro station free drinking water fountain- A potential "microplastics hotspot" for human consumption. *Environmental Pollution*, *261*. Available from https://doi.org/10.1016/j.envpol.2020.114227.

Smith, B. C. (2011). *Fundamentals of Fourier transform infrared spectroscopy*. (9 (March)). CRC Press.

Song, Y., Cao, C., Qiu, R., Hu, J., Liu, M., Lu, S., Shi, H., Raley-Susman, K. M., & He, D. (2019). Uptake and adverse effects of polyethylene terephthalate microplastics fibers on terrestrial snails (*Achatina fulica*) after soil exposure. *Environmental Pollution*, *250*, 447–455. Available from https://doi.org/10.1016/j.envpol.2019.04.066.

Strand, J., et al. (2018). Analysis of microplastic particles in Danish drinking water. DCE-Danish Centre for. *Environment and Energy*.

Tagg, A. S., Harrison, J. P., Ju-Nam, Y., Sapp, M., Bradley, E. L., Sinclair, C. J., & Ojeda, J. J. (2017). Fenton's reagent for the rapid and efficient isolation of microplastics from wastewater. *Chemical Communications*, *53*(2), 372–375. Available from https://doi.org/10.1039/c6cc08798a.

Tan, H., Yue, T., Xu, Y., Zhao, J., & Xing, B. (2020). Microplastics reduce lipid digestion in simulated human gastrointestinal system. *Environmental Science and Technology*, *54*(19), 12285–12294. Available from https://doi.org/10.1021/acs.est.0c02608.

Thompson, R. C., Olson, Y., Mitchell, R. P., Davis, A., Rowland, S. J., John, A. W. G., McGonigle, D., & Russell, A. E. (2004). Lost at sea: Where is all the plastic? *Science (New York, N.Y.)*, *304*(5672), 838. Available from https://doi.org/10.1126/science.1094559.

Turovskiy, I. S., & Mathai, P. (2006). *Wastewater sludge processing*. John Wiley & Sons.

Uhl, W., Eftekhardadkhah, M., & Svendsen, C. (2018). Mapping microplastic in norwegian drinking water. *Atlantic (Boston, Mass.: 1981)*, *185*, 491–497.

van den Berg, P., Huerta-Lwanga, E., Corradini, F., & Geissen, V. (2020). Sewage sludge application as a vehicle for microplastics in eastern Spanish agricultural soils. *Environmental Pollution*, *261*, 114198.

Vernon-Parry, K. (2000). Scanning electron microscopy: An introduction. *III-Vs Review*, *13*(4), 40–44.

Vollertsen, J., & Hansen, A. A. (2017). Microplastic in Danish wastewater: Sources, occurrences and fate. *The Danish Environmental Protection Agency*, 1906.

Wang, L., Zhang, J., Hou, S., & Sun, H. (2017). A simple method for quantifying polycarbonate and polyethylene terephthalate microplastics in environmental samples by liquid chromatography-tandem mass spectrometry. *Environmental Science and Technology Letters*, *4*(12), 530–534. Available from https://doi.org/10.1021/acs.estlett.7b00454.

Wang, Q., Bai, J., Ning, B., Fan, L., Sun, T., Fang, Y., Wu, J., Li, S., Duan, C., Zhang, Y., Liang, J., & Gao, Z. (2020). Effects of bisphenol A and nanoscale and microscale polystyrene plastic exposure on particle uptake and toxicity in human Caco-2 cells. *Chemosphere*, *254*. Available from https://doi.org/10.1016/j.chemosphere.2020.126788.

Wang, W., Ge, J., Yu, X., & Li, H. (2020). Environmental fate and impacts of microplastics in soil ecosystems: Progress and perspective. *Science of the Total Environment*, *708*. Available from https://doi.org/10.1016/j.scitotenv.2019.134841.

Watteau, F., Dignac, M. F., Bouchard, A., Revallier, A., & Houot, S. (2018). Microplastic detection in soil amended with municipal solid waste composts as revealed by transmission electronic microscopy and pyrolysis/GC/MS. *Frontiers in Sustainable Food Systems*, *2*. Available from https://doi.org/10.3389/fsufs.2018.00081.

Weber, F., et al. (2021). Investigation of microplastics contamination in drinking water of a German city. *Science of The Total Environment*, *755*, 143421.

Werther, J., & Ogada, T. (1999). Sewage sludge combustion. *Progress in Energy and Combustion Science*, *25*(1), 55–116. Available from https://doi.org/10.1016/S0360-1285(98)00020-3.

Wiszniowski, J., Robert, D., Surmacz-Gorska, J., Miksch, K., & Weber, J. V. (2006). Landfill leachate treatment methods: A review. *Environmental Chemistry Letters*, *4*(1), 51–61. Available from https://doi.org/10.1007/s10311-005-0016-z.

Wright, S. L., Thompson, R. C., & Galloway, T. S. (2013). The physical impacts of microplastics on marine organisms: A review. *Environmental Pollution*, *178*, 483–492.

Wu, P., Huang, J., Zheng, Y., Yang, Y., Zhang, Y., He, F., Chen, H., Quan, G., Yan, J., Li, T., & Gao, B. (2019). Environmental occurrences, fate, and impacts of microplastics. *Ecotoxicology and Environmental Safety*, *184*. Available from https://doi.org/10.1016/j.ecoenv.2019.109612.

Xiang, Q., Zhu, D., Chen, Q. L., O'Connor, P., Yang, X. R., Qiao, M., & Zhu, Y. G. (2019). Adsorbed sulfamethoxazole exacerbates the effects of polystyrene ($\sim$2 μm) on gut microbiota and the antibiotic resistome of a Soil Collembolan. *Environmental Science and Technology*, *53*(21), 12823–12834. Available from https://doi.org/10.1021/acs.est.9b04795.

Xu, B., Liu, F., Cryder, Z., Huang, D., Lu, Z., He, Y., Wang, H., Lu, Z., Brookes, P. C., Tang, C., Gan, J., & Xu, J. (2020). Microplastics in the soil environment: Occurrence, risks, interactions and fate—A review. *Critical Reviews in Environmental Science and Technology*, *50*(21), 2175–2222. Available from https://doi.org/10.1080/10643389.2019.1694822.

Xu, C., Zhang, B., Gu, C., Shen, C., Yin, S., Aamir, M., & Li, F. (2020). Are we underestimating the sources of microplastic pollution in terrestrial environment? *Journal of Hazardous Materials*, *400*. Available from https://doi.org/10.1016/j.jhazmat.2020.123228.

Yang, N., Damgaard, A., Kjeldsen, P., Shao, L. M., & He, P. J. (2015). Quantification of regional leachate variance from municipal solid waste landfills in China. *Waste Management*, *46*, 362–372. Available from https://doi.org/10.1016/j.wasman.2015.09.016.

Yuanqiao, L., Caixia, Z., Changrong, Y., Lili, M., Qi, L., Zhen, L., & Wenqing, H. (2020). Effects of agricultural plastic film residues on transportation and distribution of water and nitrate in soil. *Chemosphere*, *242*. Available from https://doi.org/10.1016/j.chemosphere.2019.125131.

Zhang, G. S., & Liu, Y. F. (2018). The distribution of microplastics in soil aggregate fractions in southwestern China. *Science of the Total Environment*, *642*, 12–20. Available from https://doi.org/10.1016/j.scitotenv.2018.06.004.

Zhang, G. S., Zhang, F. X., & Li, X. T. (2019). Effects of polyester microfibers on soil physical properties: Perception from a field and a pot experiment. *Science of the Total Environment*, *670*, 1–7. Available from https://doi.org/10.1016/j.scitotenv.2019.03.149.

Zhang, Q., Xu, E. G., Li, J., Chen, Q., Ma, L., Zeng, E. Y., & Shi, H. (2020). A review of microplastics in table salt, drinking water, and air: Direct human exposure. *Environmental Science and Technology*, *54*(7), 3740–3751. Available from https://doi.org/10.1021/acs.est.9b04535.

Zhang, S., Yang, X., Gertsen, H., Peters, P., Salánki, T., & Geissen, V. (2018). A simple method for the extraction and identification of light density microplastics from soil. *Science of the Total Environment*, *616–617*, 1056–1065. Available from https://doi.org/10.1016/j.scitotenv.2017.10.213.

Zhou, Q., Zhang, H., Zhou, Y., Li, Y., Xue, Y., Fu, C., Tu, C., & Luo, Y. (2016). Separation of microplastics from a coastal soil and their surface microscopic features. *Chinese Science Bulletin*, *61*(14), 1604–1611.

Zhou, Q., et al. (2018). The distribution and morphology of microplastics in coastal soils adjacent to the Bohai Sea and the Yellow Sea. *Geoderma*, *322*, 201–208.

Zhou, Y., Liu, X., & Wang, J. (2019). Characterization of microplastics and the association of heavy metals with microplastics in suburban soil of central China. *Science of the Total Environment*, *694*. Available from https://doi.org/10.1016/j.scitotenv.2019.133798.

Zhou, Y., Wang, J., Zou, M., Jia, Z., Zhou, S., & Li, Y. (2020). Microplastics in soils: A review of methods, occurrence, fate, transport, ecological and environmental risks. *Science of the Total Environment*, *748*. Available from https://doi.org/10.1016/j.scitotenv.2020.141368.

Zhu, D., Bi, Q. F., Xiang, Q., Chen, Q. L., Christie, P., Ke, X., Wu, L. H., & Zhu, Y. G. (2018). Trophic predator-prey relationships promote transport of microplastics compared with the single *Hypoaspis aculeifer* and *Folsomia candida*. *Environmental Pollution*, *235*, 150–154. Available from https://doi.org/10.1016/j.envpol.2017.12.058.

Zhu, F., Zhu, C., Wang, C., & Gu, C. (2019). Occurrence and ecological impacts of microplastics in soil systems: A review. *Bulletin of Environmental Contamination and Toxicology*, *102*(6), 741–749. Available from https://doi.org/10.1007/s00128-019-02623-z.

Zobkov, M. B., & Esiukova, E. E. (2017). Evaluation of the Munich Plastic Sediment Separator efficiency in extraction of microplastics from natural marine bottom sediments. *Limnology and Oceanography: Methods*, *15*(11), 967–978. Available from https://doi.org/10.1002/lom3.10217.

Zubris, K. A. V., & Richards, B. K. (2005). Synthetic fibers as an indicator of land application of sludge. *Environmental Pollution*, *138*(2), 201–211. Available from https://doi.org/10.1016/j.envpol.2005.04.013.

Chapter 9

Fate and transport of microplastics in soils and groundwater

Shunan Dong[1], Zebin Yu[2], Jun Huang[3,4] and Bin Gao[5]

[1]*College of Agricultural Science and Engineering, Hohai University, Nanjing, P.R. China,* [2]*School of Resources, Environment and Materials, Guangxi University, Nanning, P.R. China,* [3]*Hualan Design & Consulting Group Co. Ltd., Nanning, P.R. China,* [4]*College of Civil Engineering and Architecture, Guangxi University, Nanning, P.R. China,* [5]*Department of Agricultural and Biological Engineering, University of Florida, Gainesville, FL, United States*

Chapter Outline

9.1 Overview

9.1.1 Importance of microplastic transport in soils and groundwater

Because of their recalcitrant nature, microplastics can remain for a long time in soils and groundwater, which are the most important natural resources. For instance, synthetic fibers from sludge manure can be found in soil after

Emerging Contaminants in Soil and Groundwater Systems. DOI: https://doi.org/10.1016/B978-0-12-824088-5.00001-X

being applied for 15 years (Zubris & Richards, 2005). Polystyrene, polyvinyl chloride, and urea formaldehyde resins can stay in soil for more than 32 years without any signs of biodegradation (Waldschlaeger et al., 2020). Microplastics contain several kinds of additives such as plasticizer, retardant, antioxidant, and photo-stabilizer, which have strong toxic effects on the ecosystems (Koelmans et al., 2019). Several studies have shown that microplastics can be ingested and accumulated in aquatic animals, which will then be transmitted through the food chain to threaten human health (Ha & Yeo, 2018; Han et al., 2019; Jiang et al., 2019). Previous studies also have indicated that microplastics are toxic to organisms and may result in a persistent effect (Heindler et al., 2017; Piccardo et al., 2020; Smiroldo et al., 2019). In addition, microplastics can adsorb other contaminants, leading to their combined risks to soils and groundwater.

Microplastics are mobile in soils and groundwater. Several natural and human activities, such as irrigation (Li et al., 2020b), rainfall infiltration (Waldschlaeger et al., 2020), bioturbation (Huerta Lwanga et al., 2016), and agricultural plowing (Rillig et al., 2017) and drainage (Zubris & Richards, 2005) (Fig. 9.1), can cause the transport of microplastics into the soil environment. Microplastics also can transport or deposit in groundwater with water flow (Alimi et al., 2018). Given their high mobility, the distribution and environmental behaviors of microplastics cannot be evaluated and predicted without considering their fate and transport in soils and groundwater. In addition, microplastics can be an important vector for the transport and accumulation of heavy metals and organic contaminants in soils and groundwater. Older microplastics transported from distant regions can accumulate other contaminants and thus have greater risks than those locally originated

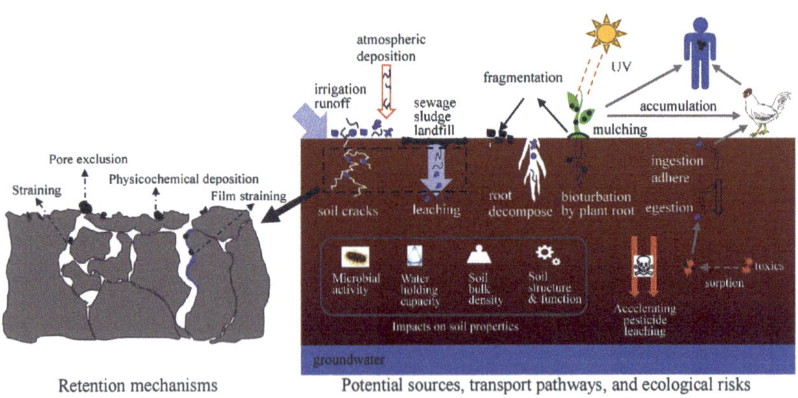

FIGURE 9.1 Sources and transport of microplastics in soils (Li et al., 2020a). *From Li, J., Song, Y., & Cai, Y. (2020). Focus topics on microplastics in soil: Analytical methods, occurrence, transport, and ecological risks.* Environmental Pollution, 257. *https://doi.org/10.1016/j.envpol.2019.113570.*

microplastics (Gorman et al., 2019). As a result, the fate and transport of microplastics in soils and groundwater should be an important issue that requires further research to advance the knowledge and understanding. Soils and groundwater are now confronted with the serious microplastic pollution (Alimi et al., 2018; O'Connor et al., 2019), understanding the fate and transport of microplastics in soils and groundwater is critical for evaluating the environmental distribution and the potential risks of these newly emerged contaminants (Re, 2019; Wu et al., 2020).

9.1.2 Chapter goals

In this chapter, fate and transport as well as the relevant mechanisms of microplastics in soil and groundwater systems are illustrated. The main objectives are as follows: (1) illustrate the stability and aggregation of microplastics in pore water; (2) assess the chemical, physical, and biological factors affecting the fate and transport of microplastics in soil and groundwater systems; (3) summarize the cotransport behaviors of microplastics with other soluble or micro/nanoparticle contaminants in soil and groundwater systems; and (4) discuss possible bottlenecks and knowledge gaps precluding current researches, meanwhile suggest potential issues necessary to be considered in future studies.

9.2 Stability and aggregation of microplastics in pore water

Aggregation means the collision and attachment of different microplastic particles. Microplastic aggregation can be divided into homoaggregation and heteroaggregation. Among the two kinds of aggregation, heteroaggregation has been regarded as the key process in controlling the environmental identity of microplastics (Oriekhova & Stoll, 2018). Aggregation of micro/nanoparticles in aqueous environment usually aggravates the filtration process, resulting in the ripening phenomena to reduce the mobility of micro/nanoparticles in soil and groundwater systems. Similarly, aggregation also plays an important role in controlling the fate and transport of microplastics in soils and groundwater, which may depend on the solution physicochemical conditions and properties of microplastics.

9.2.1 Aggregation kinetic

Interaction between microplastic particles dominants the aggregation kinetics in aqueous environment. The aggregation kinetics of microplastics are usually determined from the initial growth rate of intensity-weighted averaged hydrodynamic radius (D_h) with time (t). The initial aggregation rate constant (k_a) is proportional to the initial growth of D_h as a function of t, but inversely proportional to the initial primary particle concentration (N_0):

$$k_a \propto \frac{1}{N_0}\left(\frac{dD_h(t)}{dt}\right)_{t\to 0} \tag{9.1}$$

The $(dD_h(t)/dt)_{t\to 0}$ can be determined with the linear least-squares regression analysis on the increasing rate of D_h. The analysis usually performed over the time period that D_h increased to 1.2 or 1.5 times. Under some unfavorable conditions (fast), the aggregation of microplastics can be negligible and $(dD_h(t)/dt)_{t\to 0}$ is determined with the achieved maximum D_h.

The attachment efficiency (α) reflects the ratio of attachment collisions to total collisions of microplastics, which can be calculated by normalizing the aggregation rate constant obtained under the chemical conditions of interest to the rate constant obtained under the favorable aggregation conditions ($k_{a,\text{fast}}$):

$$\alpha = \frac{k_a}{k_{a,fast}} = \frac{\frac{1}{N_0}\left(\frac{dD_h(t)}{dt}\right)_{t\to 0}}{\frac{1}{(N_0)_{fast}}\left(\frac{dD_h(t)}{dt}\right)_{t\to 0,fast}} \tag{9.2}$$

In actual practice, α is usually determined directly by normalizing the initial slope of the aggregation profile under the chemical conditions of interest by the initial slope under the favorable aggregation condition.

9.2.2 Critical coagulation concentration

Critical coagulation concentration (CCC) values are basic parameters for evaluating the stability of microplastics, which can be determined from the intersection of extrapolated lines through the diffusion- and reaction-limited regimes mentioned above. CCC values indicate the threshold of fast aggregation. Microplastic particles will become gradually destabilized when electrolyte concentration is below CCC, indicating the process of reaction-limited regime. However, the aggregation rate of microplastics will reach the maximum when electrolyte concentration is above CCC, indicating the completely destabilize situation of microplastics and the process of diffusion-limited regime.

9.2.3 Key factors of microplastic aggregation

The aggregation behaviors of microplastics in aqueous environment strongly depend on aqueous chemistry (e.g., electrolyte, ionic strength, dissolved organic matter, and other colloids/nanoparticles) (Fig. 9.2), microorganism and microplastic properties (e.g., density, chemical composition, size, shape, functional groups, and surface coating).

Microplastics can remain stable in solution at a large range of ionic strength with the CCC higher than 100 mM NaCl and 15 mM $CaCl_2$ (Cai et al., 2018; Li et al., 2018). The aggregation of microplastics can increase with the ionic strength, resulting the gradually decreasing stability. According to DLVO

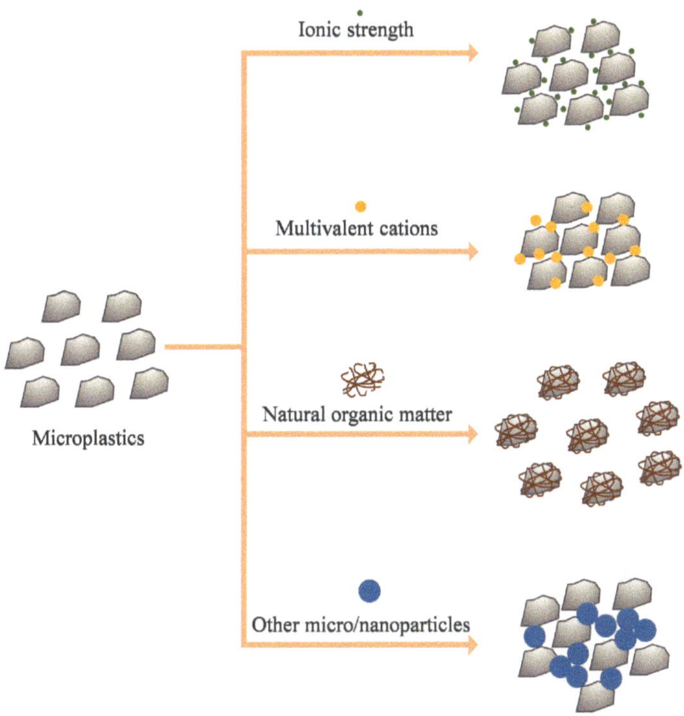

FIGURE 9.2 Effect of typical aqueous conditions on the aggregation of microplastics.

theory, the electrical double layers of microplastics are compressed with increasing ionic strength, leading to the decreasing repulsive forces and increasing aggregation of microplastics. Compared to monovalent cations (e.g., K^+, Na^+, and Li^+), multivalent cations (e.g., Ca^{2+}, Mg^{2+}, and Fe^{3+}) usually show more significant effect in aggregating microplastics, following the Schulze-Hardy rule and Hofmeister series (Oncsik et al., 2014). Other than the increasing ionic strength due to the high ion valence, multivalent cations can also build cation bridges between function groups on the surface of different microplastics to enhance the aggregation consequently.

The aggregation of microplastics increases with the decrease in pH values (Dong et al., 2021a; Jódar-Reyes et al., 2006). The protonation of functional groups on the surface of microplastics can be improved with decreasing pH, which may result in less negative zeta potentials and reduced repulsive forces between different microplastics. Other micro/nanoparticles such as diesel soot and graphene nanoparticles also show the enhanced aggregation and decreased stability with the decreasing pH values (Chen & Huang, 2017; Wu et al., 2013).

Nature organic matter (NOM) in aqueous environment may inhibit the aggregation of microplastics and thus enhance their stability (Cai et al.,

2018; Horton & Dixon, 2018; Lu et al., 2018). Humic substances are commonly negatively charged in the environment. The dispersed microplastics in aqueous environment may be coated by NOM molecules, which not only increases the negative charges and electrical repulsion between different microplastics, but also generates steric hindrance to restrain the aggregation of microplastics. In addition, NOM coated on microplastics may reduce the irregularity of shape, which also contributes to inhibit the aggregation of microplastics (Qi et al., 2014). However, the aggregation of microplastics can be enhanced when multivalent cations and NOM exist simultaneously, which can be attributed to the cross-link effect of the multivalent cations (Cai et al., 2018; Li et al., 2018; Lu et al., 2018) (Fig. 9.3). On the other hand, the variation of NOM concentrations barely affects the stability of negatively charged microplastics, but significantly enhances the stability of positively charged microplastics (Deshiikan et al., 1998).

Microorganisms should be an interesting factor in microplastics aggregation. Microplastics in aqueous environment can interact with single phytoplankton cells (e.g., alga) and residual organic matter, which potentially influence their distribution and aggregation in water environment (Long et al., 2017). Effect of microorganisms on the aggregation and stability of microplastics dependents on the species and the physiological state of the microorganisms (Long et al., 2017; Rummel et al., 2017). On the other hand, inorganic colloids (e.g., Fe_2O_3), alginate and phytoplankton in aqueous environment may influence the aggregation behaviors of microplastics (Long et al., 2017; Oriekhova & Stoll, 2018).

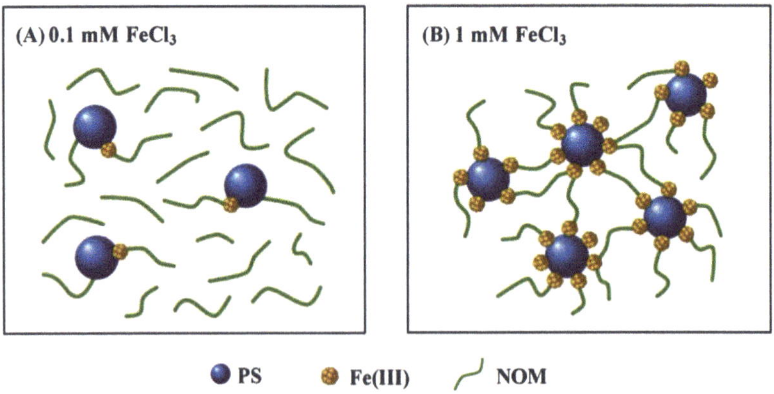

FIGURE 9.3 Schematic diagram illustrating the interactions of polystyrene microplastics and NOM in 0.1 mM (A) and 1 mM (B) $FeCl_3$ solutions (Cai et al., 2018). *From Cai, L., Hu, L., Shi, H., Ye, J., Zhang, Y., & Kim, H. (2018). Effects of inorganic ions and natural organic matter on the aggregation of nanoplastics.* Chemosphere, 197, *142–151. https://doi.org/10.1016/j.chemosphere.2018.01.052.*

9.3 Retention and transport of microplastics in soil and groundwater systems

In addition to the effect of aggregation, the fate and transport of microplastics in soils and groundwater are also strongly influenced by multiple physical, chemical, and biological factors. Aqueous chemistry, hydrodynamic conditions, biological activities, sediment, and microplastic properties all directly affect the retention and transport of microplastics. The impact factors and main mechanisms on microplastic transport in soil and groundwater systems are summarized in Table 9.1.

9.3.1 Chemical factors

9.3.1.1 Chemical properties of microplastics

Some of the basic chemical properties of microplastics, such as surface functional groups and surface coating, play important roles in influencing microplastic transport in soils and groundwater.

9.3.1.1.1 Surface functional groups

Functional groups on the surface are one of the main basic chemical factors of microplastics. The hydrophilicity/hydrophobicity of microplastics are all influenced by their surface functional groups, which will then affect the deposition and transport of microplastics in soils and groundwater. Some hydrophilic functional groups may improve the hydrophilicity of microplastics in aqueous environment, which then will inhibit the deposition of microplastics onto porous media and improve the mobility of microplastics in soils and groundwater (Wan & Wilson, 1994).

As a sensitivity factor to environmental pH, functional groups show remarkable effect on the surface charge of microplastics (Tufenkji & Elimelech, 2005). For instance, hydroxyl and carboxyl existing on the surface may strengthen the negative charges of microplastics, which can be influenced by pH conditions. The electrostatic interaction of surface charges between microplastics and porous media dominates the deposition process and then affects the mobility of microplastics.

In addition, reaction (e.g., complexation and chelation) between microplastic functional groups and media may also influence the transport behaviors of microplastics in soils and groundwater.

9.3.1.1.2 Surface coating

The modification according to manufactured purpose or the transformation in natural environment may lead to the coating of microplastics with some macromolecules. Surface coating materials, such as polyacryclic acid and polystyrene sulfonate, are generally observed to stabilize microplastics and

TABLE 9.1 Summary of impact factors and main mechanisms on microplastic transport in soils and groundwater.

Type	Subtype	Impact factors	Impact form	Main mechanisms
Chemical factors	Microplastic chemical properties	Functional groups	Mobility may increase with the amount of hydrophilic functional groups	(1) Increase hydrophilicity; (2) increase charges and electrostatic repulsion
		Surface coating	Mobility may increase when coated with some macromolecules	(1) Inhibit aggregation; (2) increase charges and electrostatic repulsion; (3) increase steric repulsion
	Aqueous chemistry	Cation type	Mobility decreases in the presence of multivalent cations	(1) Increase aggregation; (2) decrease negative charges and electrostatic repulsion; (3) build cation bridges
		Cation concentration	Mobility decreases with the increasing cation concentrations	(1) Increase aggregation; (2) compress electrical double layers and thus increase the electrostatic repulsion
		pH	Mobility increases with pH	(1) Inhibit aggregation; (2) functional groups deprotonation, increase negative charges and electrostatic repulsion
		Natural organic matter	Mobility increases with the presence of some kind of natural organic matters	(1) Inhibit aggregation; (2) increase negative charges and electrostatic repulsion; (3) increase steric repulsion; (4) cover deposition sites on media surface
Physical factors	Media chemistry	Fe/Al mineral	Mobility decreases with the increasing Fe/Al mineral contents	(1) Increase deposition sites on media surface; (2) decrease negative charges and electrostatic repulsion
		Cations release	Mobility may decrease with the releasing of cations from media surface	(1) Increase cations concentrations; (2) increase multivalent cations content

	Microplastics physical properties	Particle size	Mobility decreases with the increasing particle size	(1) Increase deposition at secondary energy well; (2) increase physical straining
		Particle morphology	Microplastics with more irregular shapes usually show lower mobility	(1) Increase physical straining
		Density	Microplastics with higher density usually show lower mobility	(1) Gravitational sedimentation increase deposition
	Flow physics	Concentration	Mobility increases with their concentrations	(1) Occupy deposition sites rapidly; (2) decrease mass transfer
		Flow rate	Mobility increases with flow rate	(1) Influence the force balance and torque; (2) reduce contact time
	Media physics	Grain size	Mobility decreases with decreasing media grain size	(1) Increase mass transfer rate; (2) increase deposition sites via larger specific surface area; (3) increase physical straining via smaller pore throat
		Saturation	Mobility decreases with decreasing saturation	(1) Air–water interface attraction; (2) increase physical straining
		Surface roughness	Mobility decreases with increasing media surface roughness	(1) Increase deposition sites; (2) increase physical straining
Biological factors	Plant	Leaves	Plant leaves may enhance the amount of microplastics in soil	(1) Intercept atmospheric microplastics
		Body	Plant body may enhance the amount of microplastics in soil	(1) Intercept microplastics from surface runoff

(*Continued*)

TABLE 9.1 (Continued)

Type	Subtype	Impact factors	Impact form	Main mechanisms
		Roots	Plant roots may increase microplastic transport in soil	(1) Provide the preferential channels
	Animal	Digger mammals	Digger mammals may increase microplastic transport in soil via dig activities	(1) Provide the preferential channels; (2) turn microplastics from topsoil to deep soil
		Soil worms	Soil worms may increase microplastic transport in soil via migration activities	(1) Attach microplastics; (2) ingest and further crush microplastics; (3) provide the preferential channels
	Microorganisms	Biofilms	Biofilms may decrease microplastic transport	(1) Trap microplastics in; (2) phagocytose microplastics.
		Growth medium	Growth medium may increase microplastic transport	(1) Cover deposition sites.
		Extracellular polymeric substances	Extracellular polymeric substances may decrease microplastic transport	(1) Decrease negative charges and electrostatic repulsion

promote their mobility in porous media via enhancing steric or electrostatic repulsion (Hierrezuelo et al., 2010).

9.3.1.2 Aqueous chemistry

Multiple aqueous chemical factors, including ionic strength, electrolyte type, pH, dissolved organic matter, and other micro/nanoparticles, significantly impact the transport and retention behaviors of microplastics in soils and groundwater.

9.3.1.2.1 Cation type and concentration

Similar to the tendency of aggregation, the mobility of microplastics usually decreases with the presence of multivalent cations (Dong et al., 2021b; Franchi & O'Melia, 2003; Quevedo & Tufenkji, 2012). With more positive charges, multivalent cations will also decrease the electrostatic repulsion between microplastics and media surface, leading to the decreased mobility of microplastics in porous media. Additionally, several investigations have suggested that multivalent cations may form the cation bridging between microplastics and media surface, which will remarkably increase the deposition and decrease the mobility of microplastics.

The mobility of microplastics usually decreases with increasing cation concentrations, which is mainly due to the increased ionic strength (Chu et al., 2019; Wu et al., 2020; Xu et al., 2016). As mentioned above, the increased aggregation with increasing ionic strength can enhance the retention of microplastics in porous media. According to the DLVO theory, electrical double layers of microplastics and media should be compressed with the increasing ionic strength, which will lead to decreased electrostatic repulsion and increased deposition of microplastics onto media surface. Cation concentration sometimes shows combined effects with the properties on microplastic transport. May and Li (2013) found that smaller microplastics are much more sensitive to the changes of cation concentrations, and their mobility is thus more susceptible (May & Li, 2013).

9.3.1.2.2 pH

The transport of microplastics is usually inhibited with decreasing pH values. Similar to engineered nanoparticles, microplastics also have the peculiar isoelectric point (IEP), which is dominated by pH conditions. The surface potential of both microplastics and porous media will become less negative with decreasing pH, which can be attributed to the deprotonation of functional groups (Sadri et al., 2017). Given the decreased absolute value of surface potential, electrostatic repulsion between microplastics and porous media subsequently decreases and provides more favorable conditions for microplastic deposition. In addition, the hydrophilicity of microplastics can be enhanced with deprotonation of functional groups, which may increase the amount of "free- microplastics" and then improve microplastic transport consequently.

On the other hand, deprotonation of functional groups also can provide more reactive sites for multivalent cations to cross-link microplastics with porous media, which then may influence the transport of microplastics.

9.3.1.2.3 Natural organic matter

NOMs in aqueous environment, such as fulvic and humic acids, commonly play an important role in improving the transport of microplastics in soil and groundwater environments (Alimi et al., 2018; Franchi & O'Melia, 2003). Microplastics coated with NOMs in aqueous environment usually show lower attachment efficiencies than those uncoated. Except for microplastics, NOMs also may adsorb to the surface of porous media via $\pi-\pi$ and hydrogen-bond interactions (Dong et al., 2017). NOMs can enhance the magnitude of negative surface charge and enhance the electrical repulsion between microplastics and porous media. Moreover, generated steric hindrance between microplastics and media surface also can be enhanced to restrain the deposition of microplastics onto porous media. Even though with the presence of NOMs, multivalent cations also may decrease microplastics mobility via building the cation bridges between microplastics and porous media.

The types of NOMs may influence the transport of microplastics in porous media. Amirbahman and Olson (1995) found that microplastics coated with the lower molecular-sized Georgetown fulvic acid were more likely to deposit onto quartz sand, which was attributed to the reduced steric-stabilizing effect of the smaller organic molecule (Amirbahman & Olson, 1995).

9.3.1.3 Media chemistry

Due to the complicated composition, the type of soil and granular media will play a remarkable role in the transport of microplastics. In general, microplastics show higher mobility in realistic media than that in some "clean" media, such as quartz sand and glass beads. For instance, Quevedo and Tufenkji (2012) found that microplastics showed a lower mobility in agricultural loamy sand than high purity quartz sand (Quevedo & Tufenkji, 2012). Bouchard et al. (2013) found that the retention of microplastics was higher in sediments from a creek in Georgia, United States than in pure Iota quartz sand (Bouchard et al., 2013). The mineralogy of the porous media and the aqueous chemistry (e.g., pH) play a combined role in affecting microplastic retention. For instance, under normal pH condition, allophane (IEP is about 7.8) usually shows positive charge and thus can provide favorable sites for microplastic deposition (negatively charge) (Quevedo & Tufenkji, 2012). However, given the relatively low IEP of albite (about 5.9) and orthoclase (about 1.0), these two minerals may offer unfavorable conditions for microplastic deposition and then improve the mobility (Zaman et al., 2002). Fe/Al oxides (IEP is about 7.5–9.0) are usually positively charged under acid condition (e.g., pH 4.97), become less positively charged under the near neutral condition (e.g., pH 6.57), but change to negatively charged under the alkaline condition (e.g., pH 9.75). The mobility of microplastics in three soils follows the order of Desert soil > Black soil > Red soil under the same conditions (Wu et al., 2020). The electrical property of clay minerals with amphoteric Al-OH and Si-OH functional group sites are pH-dependent, which may also influence the transport of microplastics (Kim et al., 2012).

On the other hand, some cations may be released from grain surface to aqueous phase due to the pH variation or water renewal, which increases the concentration of cations and the content of multivalent cations. As mentioned above, the transport of microplastics in soil and groundwater systems thus will be inhibited (Slack et al., 2005).

9.3.2 Physical factors

9.3.2.1 Physical factors of microplastics

The physical properties of microplastics, such as particle size, morphology and density, will influence the filtration process and stress condition, which then influence the transport of microplastics in soil and groundwater systems.

9.3.2.1.1 Particle size

The retention of microplastics usually increases with its particle size, leading to the decreased mobility in porous media. Particle size shows a significant impact on the deposition rate of microplastics onto the media surface. DLVO results indicate that the deposition of microplastics onto media surface at secondary energy well can be more significant with increasing particle size (Pelley & Tufenkji, 2008) (Fig. 9.4). Several studies have demonstrated that physical straining is an important retention mechanism for large microplastics (Bradford et al., 2002; Foppen et al., 2007). In addition, May and Li (2013) found that smaller microplastics seem to be more sensitive to the variation of aqueous chemistry under favorable deposition conditions, which shows more obvious mobilization and reattachment following with suddenly reduced ionic strength (May & Li, 2013) (Fig. 9.5).

9.3.2.1.2 Particle morphology

Morphology also shows a considerable impact on the retention and transport of microplastics in soil and groundwater systems. O'Connor et al. (2019) indicated that microplastics with granular morphotype show greater mobility than fragments, filaments, films, or foam morphotypes (O'Connor et al.,

2019). The irregular shape may increase the deposition of microplastics in porous media due to the surface heterogeneity and enhanced hydrophobicity during the stretching process (Seymour et al., 2013). The transport of microplastics with rod shape may be lower than that with spherical shape in porous media (Liu et al., 2010; Salerno et al., 2006). Peanut-shaped colloids show a lower mobility than spherical colloids due to the more sensitive straining rate, which resulted from the preferred orientation when peanut-shaped colloids approaching the pore-scale constriction (Xu et al., 2008). Straining of rod shape microplastics is less than spherical microplastics, indicating the minor axis may be the critical dimension in controlling the transport process (Liu et al., 2010). The straining of microplastics with irregular shape may be more significant in soil pores, and thus inhibits transport (Xu et al., 2008).

9.3.2.1.3 Density

Microplastics with high density may have lower mobility in porous media due to the high gravitational sedimentation. Chen et al. (2010) found that gravity has a remarkable impact on the deposition of microsized microplastics on media surfaces under both unfavorable and favorable conditions (Chen et al., 2010). Micro/nanoparticles with higher density are more easily affected by gravity due to the contributed hydrodynamic force balance (Cai

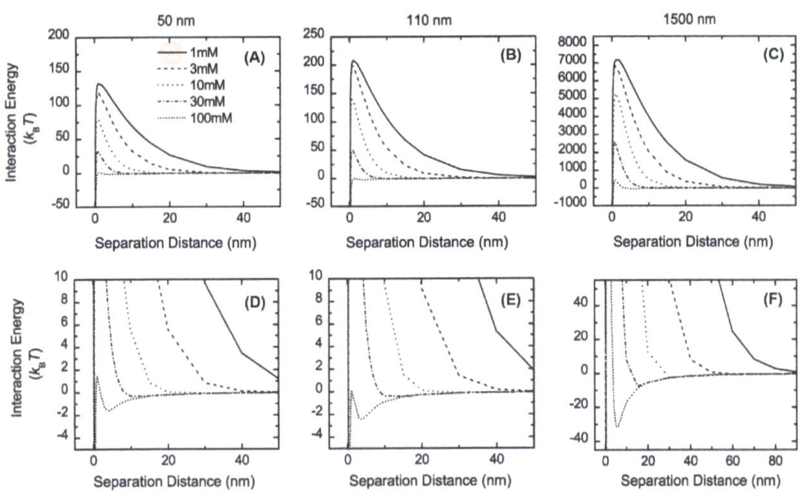

FIGURE 9.4 Calculated DLVO interaction energy for microplastics with different particle sizes at a range of ionic strengths in the absence of NOM (pH 5.7 ± 0.2): (A, D) for 50 nm; (B, E) for 110 nm; (C, F) for 1500 nm (Pelley & Tufenkji, 2008). *From Pelley, A., & Tufenkji, N. (2008). Effect of particle size and natural organic matter on the migration of nano- and micro-scale latex particles in saturated porous media.* Journal of Colloid and Interface Science, 321*(1), 74–83. https://doi.org/10.1016/j.jcis.2008.01.046.*

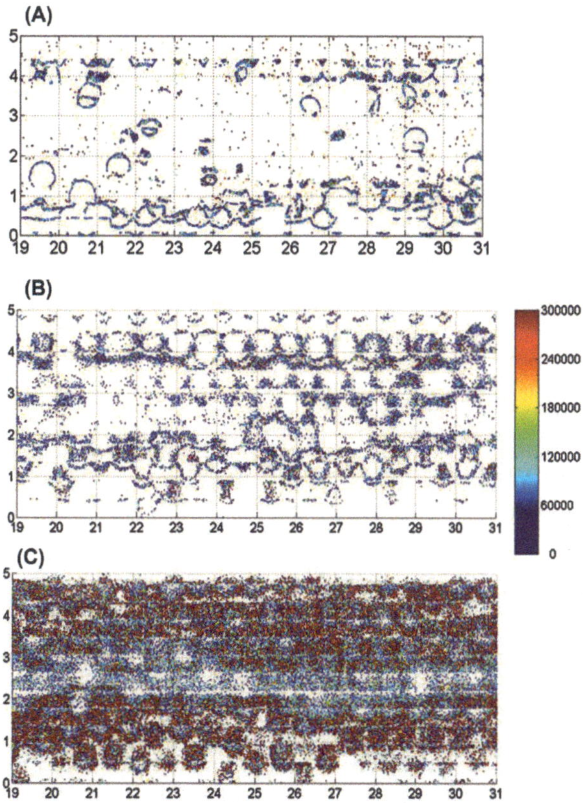

FIGURE 9.5 The distributions of retained microplastics with different particle sizes in porous media: (A) 57 nm, (B) 210 nm, and (C) 510 nm (May & Li, 2013). *From May, R., & Li, Y. (2013). The effects of particle size on the deposition of fluorescent nanoparticles in porous media: Direct observation using laser scanning cytometry.* Colloids and Surfaces A: Physicochemical and Engineering Aspects, 418, *84–91. https://doi.org/10.1016/j.colsurfa.2012.11.028.*

et al., 2015). Gravity may assist fluid and improve the approaching of micro/nanoparticles toward the media surfaces (Ma et al., 2011).

9.3.2.2 *Flow physics*

Flow provides direct motion energy for microplastic transport, whose properties (e.g., concentration and flow rate) remarkably influence the retention and transport of microplastics in soils and groundwater.

9.3.2.2.1 Concentration

The retention of microplastics in porous media is inhibited with their increasing concentration, leading to improved transport. This concentration-dependent phenomenon can be attributed to the blocking effect. The deposition sites on grain surface for microplastics can be more rapidly occupied when microplastic concentration is relatively high. In addition, higher concentrations may result in lower mass transfer to retention locations, which can be attributed to the increased numbers of collisions that knock weakly associated microplastics off the solid phase. As a result, the retention rate of microplastics in porous media will be reduced subsequently and thus increase transport. On the other hand, concentration-dependent transport also can be influenced by aqueous chemistry such as ionic strength. Bradford et al. (2009) found that the transport of microplastics in porous media increased when microplastic concentrations were at intermediate ionic strength (31 or 56 mM) (Bradford et al., 2009). However, concentration showed negligible effect on microplastic transport at extreme low (6 mM) or high (106 mM) ionic strength (Bradford et al., 2009).

9.3.2.2.2 Flow rate

Similar to other micro/nanoparticles, increasing flow rate generally promotes the transport of microplastics in porous media. The kinetic energy of microplastic collision is directly proportional with the square of particle trajectory velocity. Flow rate can influence the force balance and torque of microplastics on the media surface. The deposition of microplastics onto grain surface decreases with the increasing flow rate, leading to a higher mobility. Moreover, the contact time between microplastics and media surface becomes shorter with the increasing flow rate, which also inhibits the deposition and improves the mobility of microplastics in porous media. On the other hand, the flow rate may show different effects on microplastic transport under various hydrochemical conditions (Tong & Johnson, 2006).

9.3.2.2.3 Temperature

In recent years, temperature transformation has been found to play an important role in microplastic transport. Alimi et al. (2021) found that microplastics exposed to freeze-thaw cycles (−10°C to 10°C) showed significant aggregation and reduced mobility compared to that held at constant 10°C (Alimi et al., 2021). Even though NOM increased the mobility of microplastics in porous media under freeze-thaw cycles, the aggregation of microplastics was not inhibited. This research indicated that microplastics may aggregate and associate with soils rather than mobile in colder climates following freezing temperatures.

9.3.2.3 Medium physics

Porous media are the main occurrence environment of microplastics in soil and groundwater systems, whose physical properties (e.g., grain size, saturation and surface morphology) strongly influence microplastic transport.

9.3.2.3.1 Grain size

Grain size has a strong influence on the transport and retention of microplastics in soil and groundwater systems. The retention of microplastics in porous media usually increases with the decreasing grain size, indicating that finer sand should be more efficient for microplastics removal. The mass transfer rate of microplastics to grain surface can be increased with decreasing grain size. Smaller grain has larger specific surface areas and can provide more deposition sites for microplastics during the transport process. In addition, porous media constituting with finer grain usually has smaller pore throat, which can enhance the filtration and then inhibit the transport of microplastics in porous media.

9.3.2.3.2 Saturation

Vadose zone is a transitional area between soil and groundwater, the transport and retention of microplastics in unsaturated porous media are worthy of attention. Compared to saturated porous media, microplastics usually show lower mobility in unsaturated porous media due to the effect of air--water interface and film straining. Microplastics accumulate at the air–water interface via collision or attractive forces, which is almost irreversible unless the air–water interface eliminates. The capillary force is regarded as the major force to maintain the microplastics at air–water interface, which is much larger than the energy barrier. Small air bubbles in flow can act as the carrier and influence the transport of the attached microplastics. The air--water interface will disappear with the increasing water content, whereafter the pre-attached microplastics may be released into aqueous phase and remobilized. In addition, straining also plays an important role in microplastic retention in unsaturated porous media. On the other hand, Chen et al. (2009) found that microplastic retention decreased with the increasing particle size at high water saturation, but increased with particle size at low water saturation (Chen et al., 2009). Hydrophobic microplastics may have lower mobility in unsaturated porous media (Alimi et al., 2018).

9.3.2.3.3 Surface roughness

The rougher surface of grain can provide more deposition sites for microplastics, which will reduce their mobility in porous media (Bouchard et al., 2013). The irregular shape of grain contributes to the formation of smaller pore throats, which will then increase the filtration and inhibit the transport of microplastics in soil and groundwater systems.

9.3.3 Biological factors

9.3.3.1 Plants

Terrestrial plants can be a potential temporary sink of atmospheric microplastics during transport (Liu et al., 2010; Salerno et al., 2006) (Fig. 9.6).

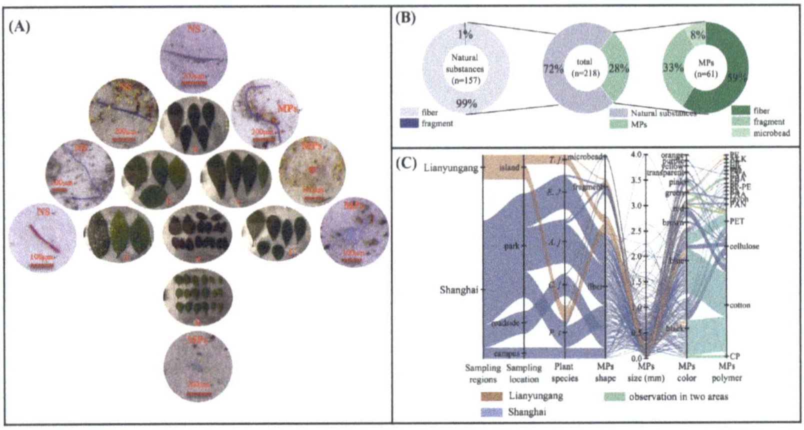

FIGURE 9.6 Deposition of atmospheric microplastics on plant leaves: (A) for sampled leaves; (B) for microplastics proportion; (C) for mapping interrelationship (Liu et al., 2020). *From Liu, K., Wang, X., Song, Z., Wei, N., & Li, D. (2020). Terrestrial plants as a potential temporary sink of atmospheric microplastics during transport.* Science of the Total Environment, 742. *https://doi.org/10.1016/j.scitotenv.2020.140523.*

Liu et al. (2020) found that leaves can indiscriminately retain atmospheric microplastics regardless of plant species, and nearly 28% of attached substances deposited on the plant leaves are microplastics (Liu et al., 2020). Low-growing plants, such as grass and shrubs, can also capture and retain some microplastics from surface runoff. With the death and decomposition of plants, the attached microplastics therewith may enter into the soil environment and even groundwater system. In addition, the growth of plant roots can introduce holes and cracks in soil structure, which may provide the preferential channels for fast transport of microplastics in the soil. On the other hand, microplastics accumulated in the soil environment may be ingested by plants and thus taken away from soil (Li et al., 2020a).

9.3.3.2 Animals

9.3.3.2.1 Digger mammals

Excavation of digger mammals (e.g., gopher or mole) may influence the transport and distribution of microplastics in soil environment. Similar to plant roots, digger mammals can dig holes in soils, which may act as preferential channels for microplastic transport. In addition, microplastics in the topsoil may be flipped into the deeper layers by digger mammals, accelerating the transport and distribution of microplastics in the soil environment. In addition to digger mammals, Beckwith and Fuentes (2018) found microplastics in turtle nest, which may negatively affect the incubating environment for marine turtles (Beckwith & Fuentes, 2018).

9.3.3.2.2 Soil worms

Life activities of soil worms (e.g., earthworm and collembolan) are regarded as one of the important factors in influencing microplastic transport in soil. Microplastics may attach onto the skin or be ingested by soil worms (Selonen et al., 2020) (Fig. 9.7). These captive microplastics will be transported to farther areas accompanied by soil worms, and then be spread in the soil environment via detaching, excreting or death of their carrier (Huerta Lwanga et al., 2017) (Fig. 9.8). The effect of soil worms on microplastic transport depends on their species. Maaß et al. (2017) found that the *Folsomia candida* can transport larger microplastics further and faster than *Proisotoma minuta* (Maaß et al., 2017). Zhu et al. (2018) demonstrated that oribatid mite can spread more microplastics than collembolan and predatory mite (Zhu et al., 2018).

On the other hand, microplastics can be further crushed into smaller pieces by the ingesting, digesting, and excreting processes of soil worms. Geophagous soil fauna (e.g., earthworm) can ingest brittle microplastics and then grind them up in gizzard. Other soil mesofauna (such as collembola or mites) can also crush microplastics by scraping or chewing activities (Rillig, 2012). As mentioned above, microplastics with smaller sizes usually have higher mobility in soil and groundwater systems. Additionally, vertical burrows created by soil worms provide preferential channels to accelerate the transport of microplastics in soil. This enhanced transport of microplastics by soil worms is nonnegligible and should be considered as one of the potential sources of microplastics in deep soil layer and groundwater.

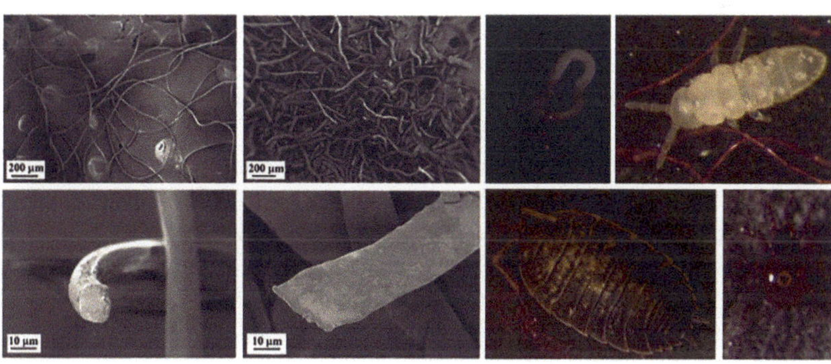

FIGURE 9.7 Microplastics captured by soil worms (Selonen et al., 2020). *From Selonen, S., Dolar, A., Jemec Kokalj, A., Skalar, T., Parramon Dolcet, L., Hurley, R., & van Gestel, C. (2020). Exploring the impacts of plastics in soil – The effects of polyester textile fibers on soil invertebrates.* Science of the Total Environment, 700. *https://doi.org/10.1016/j.scitotenv.2019.134451.*

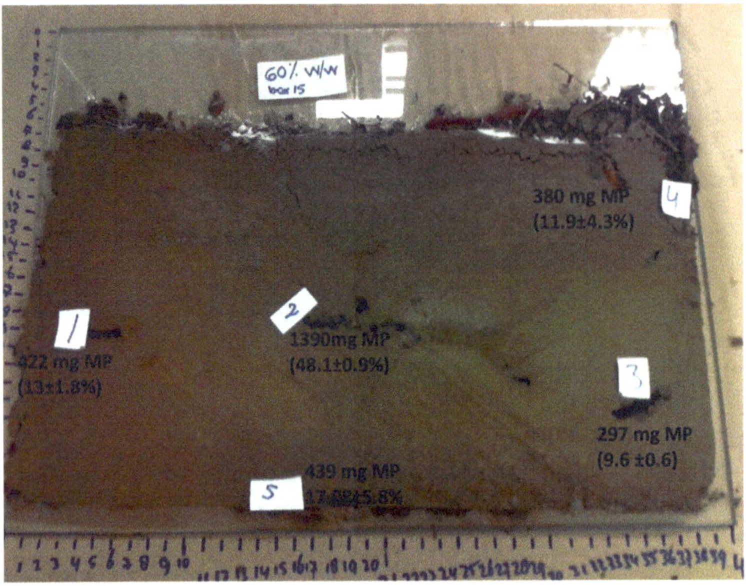

FIGURE 9.8 Microplastic vertical distribution along the gallery walls of earthworm (Huerta Lwanga et al., 2017). *From Huerta Lwanga, E., Gertsen, H., Gooren, H., Peters, P., Salánki, T., van der Ploeg, M., Besseling, E., Koelmans, A., & Geissen, V. (2017). Incorporation of microplastics from litter into burrows of* Lumbricus terrestris. Environmental Pollution, 220, *523–531. https://doi.org/10.1016/j.envpol.2016.09.096.*

9.3.3.3 Microorganisms

Microorganisms are ubiquitous in the soil and groundwater environment, which may affect the retention and transport of microplastics. Microplastics may be trapped onto the biofilms or phagocytosed by microorganisms during their transport, resulting in increased retention and reduced mobility in porous media. Medium surface coated with growth medium or extracellular polymeric substances associated with the microorganisms further influence the retention behaviors of microplastics in porous media. Growth medium coated on the medium surface may cover some deposition sites for microplastics, decreasing retention and increasing mobility. However, extracellular polymeric substances may reduce the electronegativity of medium surface, which will reduce the electrostatic repulsion between microplastics and porous media and thus enhance microplastic retention. The types of microorganisms should also be considered as an important factor. For instance, due to the different hydrophobicity, microplastic transport in *P. aeruginosa* PAO1 coated porous media is generally unchanged with time, whereby their transport in *P. aeruginosa* 9027 coated porous media fluctuates (Mitzel et al., 2016). In addition, aqueous chemistry usually shows combined effects with microorganisms on microplastic transport, which is necessary to be further investigated.

9.4 Cotransport with other contaminants

Cotransport of microplastics with other contaminants such as heavy metals, organic pollutants, and other micro/nanoparticles also needs attention. The cotransport behaviors of microplastics and other contaminants are more complicated than their individual transport in soil and groundwater systems.

9.4.1 Soluble contaminants

Soluble contaminants (e.g., heavy metals and organic pollutants) may be adsorbed on microplastics via multiple physicochemical mechanisms such as $\pi-\pi$, complexation, and electrostatic interactions, which can be strengthened by the high surface area and aboundant functional groups of microplastics. Other factors, such as aqueous conditions (e.g., electrolyte type, ionic strength, pH, and NOM) and contaminant status also influence the adsorption. Due to the colloid properties, the mobility of microplastics may be higher than that of the soluble contaminants in soils and groundwater. As a result, microplastics can act as a vector and thus influence transport behaviors of these contaminants via cotransport process, facilitating the transport and wider distribution of these contaminants in soil and groundwater systems. The cotransport of microplastics and soluble contaminants can be influenced by several factors. For example, Liu et al. (2019) found that aging by UV or O-3 exposure enhanced the mobility and contaminant-mobilizing ability (pyrene and 4-nonylphenol) of microplastics (Liu et al., 2019). On the other hand, the adsorbed contaminants also may affect the transport behaviors of microplastics in porous media in return. Zhao et al. (2020) reported that tetracycline slightly inhibited microplastics mobility in K^+ solutions but facilitated microplastic transport in Ca^{2+} solutions, moreover, tetracycline adsorption on microplastics was regarded as the critical factor (Zhao et al., 2020).

9.4.2 Micro/nanoparticles

Microplastics may aggregate with other micro/nanoparticles in aqueous environment, leading to the interactional transport behaviors in soil and groundwater systems. For instance, microplastics can enhance fullerene (C_{60}) dispersion and facilitate C_{60} transport with the mass concentration ratio of one, which effectiveness will reduce with the decreasing microplastic/C_{60} ratio (Dong et al., 2019). Microplastics with smaller size show a stronger effect in facilitating biochar transport, while biochar can significantly decrease the transport of microplastics in porous media (Tong et al., 2020). Microplastic transport in porous media can be remarkably inhibited by $nTiO_2$ due to the microplastic-$nTiO_2$ hetero-aggregates, additional deposition sites provided by previously deposited nTiO, and increased surface roughness of porous media (Cai et al., 2019). Graphene oxide may adsorb microplastics, and then influence the retention and transport of microplastics in porous media (Peng et al., 2017) (Fig. 9.9).

9.4.3 Released contaminants by microplastics

In addition to the contaminants in external environment, microplastics may self-release some contaminants during the transport process. The additives in microplastics, such as phthalic acid esters and polycyclic aromatic hydrocarbons, are physically dispersed and can be easily released from microplastics (Kida & Koszelnik, 2021; Ye et al., 2020). Till now, the cotransport behaviors of microplastics with these released contaminants are unclear, even though they have strong influence on soil and groundwater quality. Presumably, four processes may occur: (1) the released contaminants may re-adsorb onto microplastics and change their surface properties and thus influence microplastic transport; (2) the released contaminants may leave microplastics and be "free," showing negligible effect on microplastic transport; (3) the released contaminants may compete or cover the microplastic deposition sites on media surface, and thus inhibit the retention of microplastics; and (4) the released contaminants may act as the bridges to link microplastics and media, enhancing the retention of microplastics and thus reducing their transport in porous media.

9.5 Future prospects

In general, the fate and transport of microplastics in the soil and groundwater environment are controlled by multiple physicochemical factors. Even

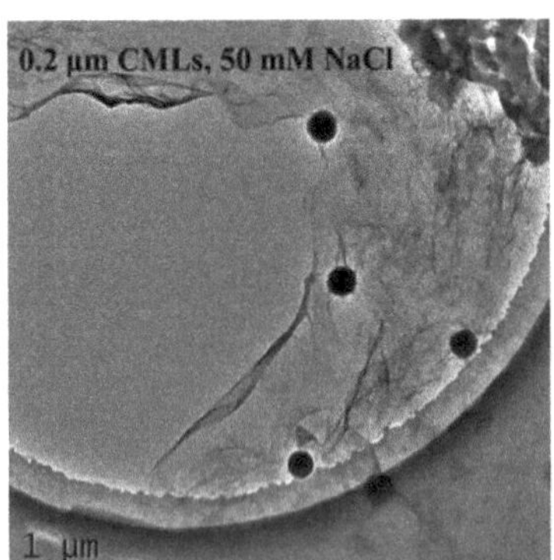

FIGURE 9.9 TEM images of microplastics on graphene oxide (Peng et al., 2017). *From Peng, S., Wu, D., Ge, Z., Tong, M., & Kim, H. (2017). Influence of graphene oxide on the transport and deposition behaviors of colloids in saturated porous media.* Environmental Pollution, 225, *141–149. https://doi.org/10.1016/j.envpol.2017.03.064.*

though several studies have explored some aspects of the fate and transport of microplastics in soils and groundwater, most of the current investigations are still at an initial stage. Further studies thus are necessary needed. Based on current knowledge, four major research questions are suggested to be important to future research:

1. Almost all of the current studies are limited on the laboratory scale; thus, field-scale investigations are urgently needed. Even though laboratory experiments have successfully illustrated some transport mechanisms of microplastics in porous media, the dimensional and complexities of actual soil and groundwater systems still cannot be well represented. Upscaling from laboratory scale to field scale is an essential step toward understanding the fate and transport of microplastics in natural soil and groundwater systems.
2. Transformation of microplastics during transport needs more research attention. The properties of microplastics may change over time due to degradation, fragmentation, and biofouling during their transport in soils and groundwater; however, the relevant process and effect have rarely been studied. Microplastics may release plasticizer, antioxidants and dye due to their transformation. The secondary transport behaviors of these toxic contaminants and their influences on microplastic transport in soils and groundwater are still unclear.
3. Diversified microplastics should be widely considered in future studies. Till now, most of the previous studies have chosen model microplastics (i.e., polystyrene spheres) in research. However, traditional model microplastics with uniform shape and property cannot well represent the real microplastics in the environment. This may limit the understanding of the fate and impact of microplastics in soils and groundwater.
4. Models of the retention and transport of microplastics in soil and groundwater systems should be advanced. More accurate and multidimensional models are needed to accurately simulate the transport and cotransport of microplastics and other contaminants in soil and groundwater systems.

Acknowledgment

This work was supported by the National Natural Science Foundation of China (41907160), the National Key R&D Program of China (2018YFD0900805), and the Fundamental Research Funds for the Central Universities (B200202098).

References

Alimi, O. S., Farner, J. M., & Tufenkji, N. (2021). Exposure of nanoplastics to freeze-thaw leads to aggregation and reduced transport in model groundwater environments. *Water Research*, 189. Available from https://doi.org/10.1016/j.watres.2020.116533.

Alimi, O. S., Farner Budarz, J., Hernandez, L. M., & Tufenkji, N. (2018). Microplastics and nanoplastics in aquatic environments: Aggregation, deposition, and enhanced contaminant transport. *Environmental Science and Technology*, *52*(4), 1704–1724. Available from https://doi.org/10.1021/acs.est.7b05559.

Amirbahman, A., & Olson, T. M. (1995). The role of surface conformations in the deposition kinetics of humic matter-coated colloids in porous media. *Colloids and Surfaces A: Physicochemical and Engineering Aspects*, *95*(2–3), 249–259. Available from https://doi.org/10.1016/0927-7757(94)03015-R.

Beckwith, V. K., & Fuentes, M. M. P. B. (2018). Microplastic at nesting grounds used by the northern Gulf of Mexico loggerhead recovery unit. *Marine Pollution Bulletin*, *131*, 32–37. Available from https://doi.org/10.1016/j.marpolbul.2018.04.001.

Bouchard, D., Zhang, W., & Chang, X. (2013). A rapid screening technique for estimating nanoparticle transport in porous media. *Water Research*, *47*(12), 4086–4094. Available from https://doi.org/10.1016/j.watres.2012.10.026.

Bradford, S. A., Kim, H. N., Haznedaroglu, B. Z., Torkzaban, S., & Walker, S. L. (2009). Coupled factors influencing concentration-dependent colloid transport and retention in saturated porous media. *Environmental Science and Technology*, *43*(18), 6996–7002. Available from https://doi.org/10.1021/es900840d.

Bradford, S. A., Yates, S. R., Bettahar, M., & Simunek, J. (2002). Physical factors affecting the transport and fate of colloids in saturated porous media. *Water Resources Research*. Available from https://doi.org/10.1029/2002wr001340, 63-1-63-12.

Cai, L., He, L., Peng, S., Li, M., & Tong, M. (2019). Influence of titanium dioxide nanoparticles on the transport and deposition of microplastics in quartz sand. *Environmental Pollution*, *253*, 351–357. Available from https://doi.org/10.1016/j.envpol.2019.07.006.

Cai, L., Hu, L., Shi, H., Ye, J., Zhang, Y., & Kim, H. (2018). Effects of inorganic ions and natural organic matter on the aggregation of nanoplastics. *Chemosphere*, *197*, 142–151. Available from https://doi.org/10.1016/j.chemosphere.2018.01.052.

Chen, C., & Huang, W. (2017). Aggregation kinetics of diesel soot nanoparticles in wet environments. *Environmental Science and Technology*, *51*(4), 2077–2086. Available from https://doi.org/10.1021/acs.est.6b04575.

Cai, L., Zhu, J., Hou, Y., Tong, M., & Kim, H. (2015). Influence of gravity on transport and retention of representative engineered nanoparticles in quartz sand. *J Contam Hydrol*, *181*, 153–160. Available from https://doi.org/10.1016/j.jconhyd.2015.02.005.

Chen, G., Hong, Y., & Walker, S. L. (2010). Colloidal and bacterial deposition: Role of gravity. *Langmuir: The ACS Journal of Surfaces and Colloids*, *26*(1), 314–319. Available from https://doi.org/10.1021/la903089x.

Chen, G., Liu, J., Tawfiq, K., Yang, K., & Banks, C. (2009). Colloid retention in unsaturated porous media as impacted by colloid size. *Particulate Science and Technology*, *27*(1), 35–49. Available from https://doi.org/10.1080/02726350802611754.

Chu, X., Li, T., Li, Z., Yan, A., & Shen, C. (2019). Transport of microplastic particles in saturated porous media. *Water (Switzerland)*, *11*(12). Available from https://doi.org/10.3390/w11122474.

Deshiikan, S. R., Eschenazi, E., & Papadopoulos, K. D. (1998). Transport of colloids through porous beds in the presence of natural organic matter. *Colloids and Surfaces A: Physicochemical and Engineering Aspects*, *145*(1–3), 93–100. Available from https://doi.org/10.1016/S0927-7757(98)00666-9.

Dong, S., Cai, W., Xia, J., Sheng, L., Wang, W., & Liu, H. (2021a). Aggregation kinetics of fragmental PET nanoplastics in aqueous environment: Complex roles of electrolytes, pH and humic acid. *Environ Pollut*, *268*, 115828. Available from https://doi.org/10.1016/j.envpol.2020.115828.

Dong, S., Sun, Y., Gao, B., Shi, X., Xu, H., Wu, J., & Wu, J. (2017). Retention and transport of graphene oxide in water-saturated limestone media. *Chemosphere*, *180*, 506–512. Available from https://doi.org/10.1016/j.chemosphere.2017.04.052.

Dong, S., Xia, J., Sheng, L., Wang, W., Liu, H., & Gao, B. (2021b). Transport characteristics of fragmental polyethylene glycol terephthalate (PET) microplastics in porous media under various chemical conditions. *Chemosphere*, *276*, 130214. Available from https://doi.org/10.1016/j.chemosphere.2021.130214.

Dong, Z., Zhang, W., Qiu, Y., Yang, Z., Wang, J., & Zhang, Y. (2019). Cotransport of nanoplastics (NPs) with fullerene (C-60) in saturated sand: Effect of NPs/C-60 ratio and seawater salinity. *Water Res*, *148*, 469–478. Available from https://doi.org/10.1016/j.watres.2018.10.071.

Foppen, J. W., van Herwerden, M., & Schijven, J. (2007). Measuring and modelling straining of *Escherichia coli* in saturated porous media. *Journal of Contaminant Hydrology*, *93*(1–4), 236–254. Available from https://doi.org/10.1016/j.jconhyd.2007.03.001.

Franchi, A., & O'Melia, C. R. (2003). Effects of natural organic matter and solution chemistry on the deposition and reentrainment of colloids in porous media. *Environmental Science and Technology*, *37*(6), 1122–1129. Available from https://doi.org/10.1021/es015566h.

Gorman, D., Moreira, F. T., Turra, A., Fontenelle, F. R., Combi, T., Bícego, M. C., & de Castro Martins, C. (2019). Organic contamination of beached plastic pellets in the South Atlantic: Risk assessments can benefit by considering spatial gradients. *Chemosphere*, *223*, 608–615. Available from https://doi.org/10.1016/j.chemosphere.2019.02.094.

Ha, J., & Yeo, M. K. (2018). The environmental effects of microplastics on aquatic ecosystems. *Molecular and Cellular Toxicology*, *14*(4), 353–359. Available from https://doi.org/10.1007/s13273-018-0039-8.

Han, X., Lu, X., & Vogt, R. D. (2019). An optimized density-based approach for extracting microplastics from soil and sediment samples. *Environmental Pollution*, 254. Available from https://doi.org/10.1016/j.envpol.2019.113009.

Heindler, F. M., Alajmi, F., Huerlimann, R., Zeng, C., Newman, S. J., Vamvounis, G., & van Herwerden, L. (2017). Toxic effects of polyethylene terephthalate microparticles and Di(2-ethylhexyl)phthalate on the calanoid copepod, *Parvocalanus crassirostris*. *Ecotoxicology and Environmental Safety*, *141*, 298–305. Available from https://doi.org/10.1016/j.ecoenv.2017.03.029.

Hierrezuelo, J., Sadeghpour, A., Szilagyi, I., Vaccaro, A., & Borkovec, M. (2010). Electrostatic stabilization of charged colloidal particles with adsorbed polyelectrolytes of opposite charge. *Langmuir: The ACS Journal of Surfaces and Colloids*, *26*(19), 15109–15111. Available from https://doi.org/10.1021/la102912u.

Horton, A. A., & Dixon, S. J. (2018). Microplastics: An introduction to environmental transport processes. *Wiley Interdisciplinary Reviews: Water*, e1268. Available from https://doi.org/10.1002/wat2.1268.

Huerta Lwanga, E., Gertsen, H., Gooren, H., Peters, P., Salánki, T., Van Der Ploeg, M., Besseling, E., Koelmans, A. A., & Geissen, V. (2016). Microplastics in the terrestrial ecosystem: Implications for *Lumbricus terrestris* (Oligochaeta, Lumbricidae). *Environmental Science and Technology*, *50*(5), 2685–2691. Available from https://doi.org/10.1021/acs.est.5b05478.

Huerta Lwanga, E., Gertsen, H., Gooren, H., Peters, P., Salánki, T., Van der Ploeg, M., Besseling, E., Koelmans, A. A., & Geissen, V. (2017). Incorporation of microplastics from litter into burrows of *Lumbricus terrestris*. *Environmental Pollution*, *220*, 523–531. Available from https://doi.org/10.1016/j.envpol.2016.09.096.

Jiang, C., Yin, L., Li, Z., Wen, X., Luo, X., Hu, S., Yang, H., Long, Y., Deng, B., Huang, L., & Liu, Y. (2019). Microplastic pollution in the rivers of the Tibet Plateau. *Environmental Pollution*, *249*, 91–98. Available from https://doi.org/10.1016/j.envpol.2019.03.022.

Jódar-Reyes, A. B., Ortega-Vinuesa, J. L., & Martín-Rodríguez, A. (2006). Electrokinetic behavior and colloidal stability of polystyrene latex coated with ionic surfactants. *Journal of Colloid and Interface Science*, 170–181. Available from https://doi.org/10.1016/j.jcis.2005.10.033.

Kida, M., & Koszelnik, P. (2021). Investigation of the presence and possible migration from microplastics of phthalic acid esters and polycyclic aromatic hydrocarbons. *Journal of Polymers and the Environment*, *29*(2), 599–611. Available from https://doi.org/10.1007/s10924-020-01899-1.

Kim, H. J., Phenrat, T., Tilton, R. D., & Lowry, G. V. (2012). Effect of kaolinite, silica fines and pH on transport of polymer-modified zero valent iron nano-particles in heterogeneous porous media. *Journal of Colloid and Interface Science*, *370*(1), 1–10. Available from https://doi.org/10.1016/j.jcis.2011.12.059.

Koelmans, A. A., Mohamed Nor, N. H., Hermsen, E., Kooi, M., Mintenig, S. M., & De France, J. (2019). Microplastics in freshwaters and drinking water: Critical review and assessment of data quality. *Water Research*, *155*, 410–422. Available from https://doi.org/10.1016/j.watres.2019.02.054.

Li, S., Liu, H., Gao, R., Abdurahman, A., Dai, J., & Zeng, F. (2018). Aggregation kinetics of microplastics in aquatic environment: Complex roles of electrolytes, pH, and natural organic matter. *Environmental Pollution*, *237*, 126–132. Available from https://doi.org/10.1016/j.envpol.2018.02.042.

Liu, J., Zhang, T., Tian, L., Liu, X., Qi, Z., Ma, Y., Ji, R., & Chen, W. (2019). Aging significantly affects mobility and contaminant-mobilizing ability of nanoplastics in Saturated Loamy Sand. *Environmental Science and Technology*, *53*(10), 5805–5815. Available from https://doi.org/10.1021/acs.est.9b00787.

Liu, K., Wang, X., Song, Z., Wei, N., & Li, D. (2020). Terrestrial plants as a potential temporary sink of atmospheric microplastics during transport. *Science of the Total Environment*, 742. Available from https://doi.org/10.1016/j.scitotenv.2020.140523.

Li, J., Song, Y., & Cai, Y. (2020a). Focus topics on microplastics in soil: Analytical methods, occurrence, transport, and ecological risks. *Environmental Pollution*, 257. Available from https://doi.org/10.1016/j.envpol.2019.113570.

Li, L., Luo, Y., Li, R., Zhou, Q., Peijnenburg, W. J. G. M., Yin, N., Yang, J., Tu, C., & Zhang, Y. (2020b). Effective uptake of submicrometre plastics by crop plants via a crack-entry mode. *Nature Sustainability*, *3*(11), 929–937. Available from https://doi.org/10.1038/s41893-020-0567-9.

Liu, Q., Lazouskaya, V., He, Q., & Jin, Y. (2010). Effect of particle shape on colloid retention and release in saturated porous media. *Journal of Environmental Quality*, *39*(2), 500–508. Available from https://doi.org/10.2134/jeq2009.0100.

Long, M., Paul-Pont, I., Hégaret, H., Moriceau, B., Lambert, C., Huvet, A., & Soudant, P. (2017). Interactions between polystyrene microplastics and marine phytoplankton lead to species-specific hetero-aggregation. *Environmental Pollution*, *228*, 454–463. Available from https://doi.org/10.1016/j.envpol.2017.05.047.

Lu, S., Zhu, K., Song, W., Song, G., Chen, D., Hayat, T., Alharbi, N. S., Chen, C., & Sun, Y. (2018). Impact of water chemistry on surface charge and aggregation of polystyrene microspheres suspensions. *Science of the Total Environment*, *630*, 951–959. Available from https://doi.org/10.1016/j.scitotenv.2018.02.296.

Ma, H., Pazmino, E. F., & Johnson, W. P. (2011). Gravitational settling effects on unit cell predictions of colloidal retention in porous media in the absence of energy barriers. *Environmental Science and Technology*, *45*(19), 8306–8312. Available from https://doi.org/10.1021/es200696x.

Maaß, S., Daphi, D., Lehmann, A., & Rillig, M. C. (2017). Transport of microplastics by two collembolan species. *Environmental Pollution*, *225*, 456–459. Available from https://doi.org/10.1016/j.envpol.2017.03.009.

May, R., & Li, Y. (2013). The effects of particle size on the deposition of fluorescent nanoparticles in porous media: Direct observation using laser scanning cytometry. *Colloids and Surfaces A: Physicochemical and Engineering Aspects*, *418*, 84–91. Available from https://doi.org/10.1016/j.colsurfa.2012.11.028.

Mitzel, M. R., Sand, S., Whalen, J. K., & Tufenkji, N. (2016). Hydrophobicity of biofilm coatings influences the transport dynamics of polystyrene nanoparticles in biofilm-coated sand. *Water Research*, *92*, 113–120. Available from https://doi.org/10.1016/j.watres.2016.01.026.

O'Connor, D., Pan, S., Shen, Z., Song, Y., Jin, Y., Wu, W. M., & Hou, D. (2019). Microplastics undergo accelerated vertical migration in sand soil due to small size and wet-dry cycles. *Environmental Pollution*, *249*, 527–534. Available from https://doi.org/10.1016/j.envpol.2019.03.092.

Oncsik, T., Trefalt, G., Csendes, Z., Szilagyi, I., & Borkovec, M. (2014). Aggregation of negatively charged colloidal particles in the presence of multivalent cations. *Langmuir: The ACS Journal of Surfaces and Colloids*, *30*(3), 733–741. Available from https://doi.org/10.1021/la4046644.

Oriekhova, O., & Stoll, S. (2018). Heteroaggregation of nanoplastic particles in the presence of inorganic colloids and natural organic matter. *Environmental Science: Nano*, *5*(3), 792–799. Available from https://doi.org/10.1039/c7en01119a.

Pelley, A. J., & Tufenkji, N. (2008). Effect of particle size and natural organic matter on the migration of nano- and microscale latex particles in saturated porous media. *Journal of Colloid and Interface Science*, *321*(1), 74–83. Available from https://doi.org/10.1016/j.jcis.2008.01.046.

Peng, S., Wu, D., Ge, Z., Tong, M., & Kim, H. (2017). Influence of graphene oxide on the transport and deposition behaviors of colloids in saturated porous media. *Environmental Pollution*, *225*, 141–149. Available from https://doi.org/10.1016/j.envpol.2017.03.064.

Piccardo, M., Provenza, F., Grazioli, E., Cavallo, A., Terlizzi, A., & Renzi, M. (2020). PET microplastics toxicity on marine key species is influenced by pH, particle size and food variations. *Science of the Total Environment*, 715. Available from https://doi.org/10.1016/j.scitotenv.2020.136947.

Qi, Z., Zhang, L., Wang, F., Hou, L., & Chen, W. (2014). Factors controlling transport of graphene oxide nanoparticles in saturated sand columns. *Environmental Toxicology and Chemistry*, *33*(5), 998–1004. Available from https://doi.org/10.1002/etc.2525.

Quevedo, I. R., & Tufenkji, N. (2012). Mobility of functionalized quantum dots and a model polystyrene nanoparticle in saturated quartz sand and loamy sand. *Environmental Science and Technology*, *46*(8), 4449–4457. Available from https://doi.org/10.1021/es2045458.

Re, V. (2019). Shedding light on the invisible: Addressing the potential for groundwater contamination by plastic microfibers. *Hydrogeology Journal*, 2719–2727. Available from https://doi.org/10.1007/s10040-019-01998-x.

Rillig, M. C. (2012). Microplastic in terrestrial ecosystems and the soil. *Environ Sci Technol, 46* (12), 6453–6454. Available from https://doi.org/10.1021/es302011r.

Rillig, M. C., Ziersch, L., & Hempel, S. (2017). Microplastic transport in soil by earthworms. *Scientific Reports*. Available from https://doi.org/10.1038/s41598-017-01594-7.

Rummel, C. D., Jahnke, A., Gorokhova, E., Kühnel, D., & Schmitt-Jansen, M. (2017). Impacts of biofilm formation on the fate and potential effects of microplastic in the aquatic environment. *Environmental Science and Technology Letters, 4*(7), 258–267. Available from https://doi.org/10.1021/acs.estlett.7b00164.

Sadri, B., Pernitsky, D., & Sadrzadeh, M. (2017). Aggregation and deposition of colloidal particles: Effect of surface properties of collector beads. *Colloids and Surfaces A: Physicochemical and Engineering Aspects, 530*, 46–52. Available from https://doi.org/10.1016/j.colsurfa.2017.07.041.

Salerno, M. B., Flamm, M., Logan, B. E., & Velegol, D. (2006). Transport of rodlike colloids through packed beds. *Environmental Science and Technology, 40*(20), 6336–6340. Available from https://doi.org/10.1021/es0614565.

Selonen, S., Dolar, A., Jemec Kokalj, A., Skalar, T., Parramon Dolcet, L., Hurley, R., & van Gestel, C. A. M. (2020). Exploring the impacts of plastics in soil – The effects of polyester textile fibers on soil invertebrates. *Science of the Total Environment*, 700. Available from https://doi.org/10.1016/j.scitotenv.2019.134451.

Seymour, M. B., Chen, G., Su, C., & Li, Y. (2013). Transport and retention of colloids in porous media: Does shape really matter? *Environmental Science and Technology, 47*(15), 8391–8398. Available from https://doi.org/10.1021/es4016124.

Slack, R. J., Gronow, J. R., & Voulvoulis, N. (2005). Household hazardous waste in municipal landfills: Contaminants in leachate. *Science of the Total Environment, 337*(1–3), 119–137. Available from https://doi.org/10.1016/j.scitotenv.2004.07.002.

Smiroldo, G., Balestrieri, A., Pini, E., & Tremolada, P. (2019). Anthropogenically altered trophic webs: Alien catfish and microplastics in the diet of *Eurasian otters*. *Mammal Research, 64* (2), 165–174. Available from https://doi.org/10.1007/s13364-018-00412-3.

Tong, M., & Johnson, W. P. (2006). Excess colloid retention in porous media as a function of colloid size, fluid velocity, and grain angularity. *Environmental Science and Technology, 40* (24), 7725–7731. Available from https://doi.org/10.1021/es061201r.

Tong, M., Li, T., Li, M., He, L., & Ma, Z. (2020). Cotransport and deposition of biochar with different sized-plastic particles in saturated porous media. *Science of the Total Environment*, 713. Available from https://doi.org/10.1016/j.scitotenv.2019.136387.

Tufenkji, N., & Elimelech, M. (2005). Breakdown of colloid filtration theory: Role of the secondary energy minimum and surface charge heterogeneities. *Langmuir: The ACS Journal of Surfaces and Colloids, 21*(3), 841–852. Available from https://doi.org/10.1021/la048102g.

Waldschlaeger, K., Lechthaler, S., Stauch, G., & Schuettrumpf, H. (2020). The way of microplastic through the environment: Application of the source-pathway-receptor model (review). *Science of the Total Environment*, 713.

Wan, J., & Wilson, J. L. (1994). Colloid transport in unsaturated porous media. *Water Resources Research, 30*(4), 857–864. Available from https://doi.org/10.1029/93WR03017.

Wu, L., Liu, L., Gao, B., Muñoz-Carpena, R., Zhang, M., Chen, H., Zhou, Z., & Wang, H. (2013). Aggregation kinetics of graphene oxides in aqueous solutions: Experiments, mechanisms, and modeling. *Langmuir: The ACS Journal of Surfaces and Colloids, 29*(49), 15174–15181. Available from https://doi.org/10.1021/la404134x.

Wu, X., Lyu, X., Li, Z., Gao, B., Zeng, X., Wu, J., & Sun, Y. (2020). Transport of polystyrene nanoplastics in natural soils: Effect of soil properties, ionic strength and cation type. *Science of the Total Environment*, 707. Available from https://doi.org/10.1016/j.scitotenv.2019.136065.

Xu, S., Liao, Q., & Saiers, J. E. (2008). Straining of nonspherical colloids in saturated porous media. *Environmental Science and Technology*, *42*(3), 771–778. Available from https://doi.org/10.1021/es071328w.

Xu, S., Qi, J., Chen, X., Lazouskaya, V., Zhuang, J., & Jin, Y. (2016). Coupled effect of extended DLVO and capillary interactions on the retention and transport of colloids through unsaturated porous media. *Science of the Total Environment*, *573*, 564–572. Available from https://doi.org/10.1016/j.scitotenv.2016.08.112.

Ye, X., Wang, P., Wu, Y., Zhou, Y., Sheng, Y., & Lao, K. (2020). Microplastic acts as a vector for contaminants: the release behavior of dibutyl phthalate from polyvinyl chloride pipe fragments in water phase. *Environmental Science and Pollution Research*, *27*(33), 42082–42091. Available from https://doi.org/10.1007/s11356-020-10136-0.

Zaman, A. A., Tsuchiya, R., & Moudgil, B. M. (2002). Adsorption of a low-molecular-weight polyacrylic acid on silica, alumina, and kaolin. *Journal of Colloid and Interface Science*, *256*(1), 73–78. Available from https://doi.org/10.1006/jcis.2001.7941.

Zhao, P., Cui, L., Zhao, W., Tian, Y., Li, M., Wang, Y., & Chen, Z. (2020). Cotransport and deposition of colloidal polystyrene microplastic particles and tetracycline in porous media: The impact of ionic strength and cationic types. *The Science of the Total Environment*, *753*, 142064, 142064.

Zhu, D., Bi, Q. F., Xiang, Q., Chen, Q. L., Christie, P., Ke, X., Wu, L. H., & Zhu, Y. G. (2018). Trophic predator-prey relationships promote transport of microplastics compared with the single Hypoaspis aculeifer and Folsomia candida. *Environmental Pollution*, *235*, 150–154. Available from https://doi.org/10.1016/j.envpol.2017.12.058.

Zubris, K. A. V., & Richards, B. K. (2005). Synthetic fibers as an indicator of land application of sludge. *Environmental Pollution*, *138*(2), 201–211. Available from https://doi.org/10.1016/j.envpol.2005.04.013.

Index

Note: Page numbers followed by "*f*" and "*t*" refer to figures and tables, respectively.

I

J

K

L

M

N

X

Y

Z

Kolibri 360 GmbH
Ferdinand-Jühlke-Straße 7
99095 Erfurt, Deutschland
produktsicherheit@kolibri360.de